Handbook of
Viscosity

Volume 1

Organic
Compounds
C_1 to C_4

LIBRARY OF PHYSICO-CHEMICAL PROPERTY DATA

Handbook of Vapor Pressure

Volume 1: C_1 to C_4 Compounds (Product #5189)
Volume 2: C_5 to C_7 Compounds (Product #5190)
Volume 3: C_8 to C_{28} Compounds (Product #5191)
Carl L. Yaws

Handbook of Viscosity

Volume 1: C_1 to C_4 Compounds (Product #5362)
Volume 2: C_5 to C_7 Compounds (Product #5364)
Volume 3: C_8 to C_{28} Compounds (Product #5368)
Carl L. Yaws

Handbook of Thermal Conductivity

Volume 1: C_1 to C_4 Compounds (Product #5382)
Volume 2: C_5 to C_7 Compounds (Product #5383)
Volume 3: C_8 to C_{28} Compounds (Product #5384)
Carl L. Yaws

Each of the above three-volume sets contains data for more than 1,000 organic compounds, including hydrocarbons, oxygenates, halogenates, nitrogenates, sulfur compounds, and silicon compounds. The data are presented in graphs for *vapor pressure, viscosity,* or *thermal conductivity* as a function of temperature and are arranged by carbon number and chemical formula to enable the engineer to quickly determine values at the desired temperatures.

Handbook of Transport Property Data (Product #5392)

Carl L. Yaws

Comprehensive data on viscosity, thermal conductivity, and diffusion coefficients of gases and liquids are presented in convenient tabular format.

Physical Properties of Hydrocarbons

Volume 1, Second Edition (Product #5067)
Volume 2, Third Edition (Product #5175)
Volume 3 (Product #5176)
R. W. Gallant and Carl L. Yaws

The three-volume series provides chemical, environmental, and safety engineers with quick and easy access to vital physical property data needed for production and process design calculations.

Thermodynamic and Physical Property Data (Product #5031)

Carl L. Yaws

Property data for 700 major hydrocarbons and organic chemicals, including oxygen, nitrogen, fluorine, chlorine, bromine, iodine, and sulfur compounds, are provided.

LIBRARY OF PHYSICO-CHEMICAL PROPERTY DATA

Handbook of Viscosity

Volume 1

Organic Compounds C₁ to C₄

Carl L. Yaws

Gulf Publishing Company
Houston, London, Paris, Zurich, Tokyo

Handbook of Viscosity, Volume 1

ISBN 0-88415-362-2

Gulf Publishing Company
Book Division
P.O. Box 2608, Houston, Texas 77252-2608

Library of Congress Cataloging-in-Publication Data

Yaws, Carl L.
   Handbook of viscosity / Carl L. Yaws.
       p.   cm. — (The library of physico-chemical property
   data)
   Includes bibliographical references and index.
   Contents : v. 1. Organic compounds C_1 to C_4 — v. 2. Organic
compounds C_5 to C_7 — v. 3. Organic compounds C_8 to C_{28}.
   ISBN 0-88415-362-2 (v. 1 : acid-free). — ISBN 0-88415-
364-9 (v. 2 : acid-free). — ISBN 0-88415-368-1 (v. 3 : acid-
free)
   1. Organic compounds—Handbooks, manuals, etc. 2. Viscosity—Handbooks, manuals, etc.   I. Title.   II. Series.
QD257.7.Y394   1994
547—dc20   94-33309
                                                      CIP

CONTENTS

CONTRIBUTORS

Li Bu

Graduate student, Chemical Engineering Department, Lamar University, Beaumont, Texas 77710, U.S.A.

Xiaoyan Lin

Graduate student, Chemical Engineering Department, Lamar University, Beaumont, Texas 77710, U.S.A.

Carl L. Yaws

Professor, Chemical Engineering Department, Lamar University, Beaumont, Texas 77710, U.S.A.

ACKNOWLEDGMENTS

Many colleagues and students have made contributions and helpful comments over the years. The author is grateful to each: Jack R. Hopper, Joe W. Miller, Jr., C. S. Fang, K. Y. Li, Keith C. Hansen, Daniel H. Chen, P. Y. Chiang, H. C. Yang, Xiang Pan, Xiaoyan Lin, and Li Bu.

The author wishes to acknowledge special appreciation to his wife (Annette) and family (Kent, Michele, and Chelsea; Lindsay, Rebecca, Matthew, and Sarah).

The author wishes to acknowledge that the Gulf Coast Hazardous Substance Research Center provided partial support to this work.

DISCLAIMER

PREFACE

Viscosity data are important in many engineering applications in the chemical processing and petroleum refining industries. The objective of this book is to provide the engineer with such viscosity data. The data are presented in graphs covering a wide temperature range to enable the engineer to quickly determine values at the desired temperatures of interest. The contents of the book are arranged in the following order: graphs, reference, and appendixes.

The graphs for viscosity as a function of temperature are arranged by carbon number and chemical formula to provide ease of use. Most of the graphs for the liquid cover the full range from melting point to boiling point to critical point. The graphs for viscosity of gas cover a wide temperature range and are applicable at low pressure. Common units are used for the viscosity (centipoise for liquid and micropoise for gas). For those involved in English usage, each graph displays a conversion factor to provide English units.

The coverage encompasses a wide range of organic compounds, including hydrocarbons such as alkanes, olefins, acetylenes, and cycloalkanes; oxygenates such as alcohols, aldehydes, ketones, acids, ethers, glycols, and anhydrides; halogenates such as chlorinated, brominated, fluorinated, and iodinated compounds; nitrogenates such as nitriles, amines, cyanates, and amides; sulfur compounds such as mercaptans, sulfides, and sulfates; silicon compounds such as silanes and chlorosilanes; and many other chemical types.

The graphs for liquids are based on both experimental data and estimated values. Liquid viscosities at low temperatures were primarily estimated using the Van Velzen method (26, group, and structural contributions). For liquid viscosities at high temperatures, both experimental data and estimates were extended using a modified Letsou and Stiel equation (26, corresponding states) for saturated liquids. Experimental data and estimates were then regressed to provide the same equation for all compounds.

Very limited experimental data for liquid viscosities are available at temperatures in the regions of the melting and critical point temperatures. Thus, the values in the regions of melting and critical point temperatures should be considered rough approximations. The values in the intermediate region (above melting and below critical point) are more accurate.

The graphs for gases are also based on both experimental data and estimates. In the absence of experimental data, estimates were primarily based on modified Chapman-Enskog method (26, intermolecular forces, collision diameter . . .) and Reichenberg equation (26, corresponding state, group contribution). Experimental data and estimates were then regressed to provide the same equation for all compounds. The graphs are applicable for low pressure gas viscosity. The presented values may be adjusted to provide values at higher pressure using the methods in Reid, Prausnitz and Poling (26).

The literature has been carefully searched in construction of the graphs. The references, given in the section following the graphs near the end of the book, provide full documentation for the original sources used in preparing the work.

The equations for liquid and gas viscosities are:

$$\log_{10} \eta_{liq} = A + B/T + C\,T + D\,T^2 \qquad \text{(liquid)}$$
$$\eta_{gas} = A + B\,T + C\,T^2 \qquad \text{(gas)}$$

The coefficients for the viscosity equations are provided in the appendixes near the end of the book. The tabulated values are especially arranged for quick usage with hand calculator or computer. A list of compounds is given near the end of the book to aid the user in quickly locating the compound of interest by the chemical formula or name.

Computer programs, containing coefficients for all compounds, are available for liquid and gas. Each program is in ASCII which can be accessed by other software. For information, contact Carl L. Yaws, Ph.D., P.O. Box 10053, Beaumont, Texas 77710, phone/fax (409) 880-8787.

Handbook of
Viscosity

Volume 1

Organic
Compounds
C_1 to C_4

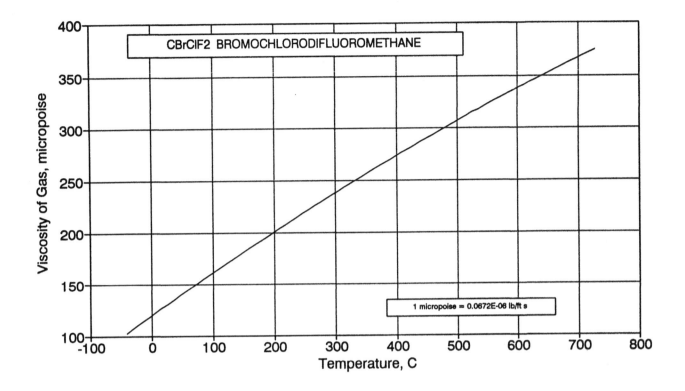

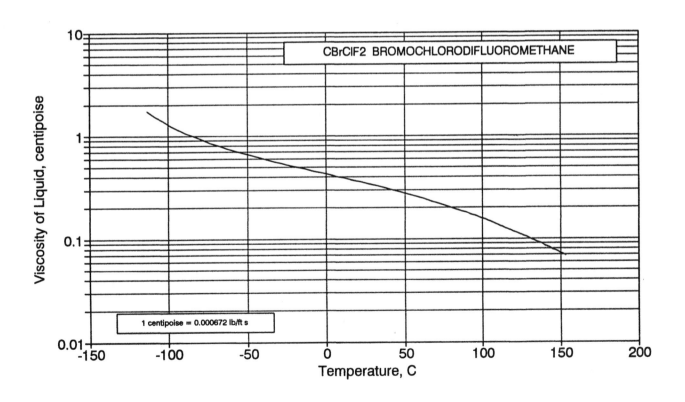

1

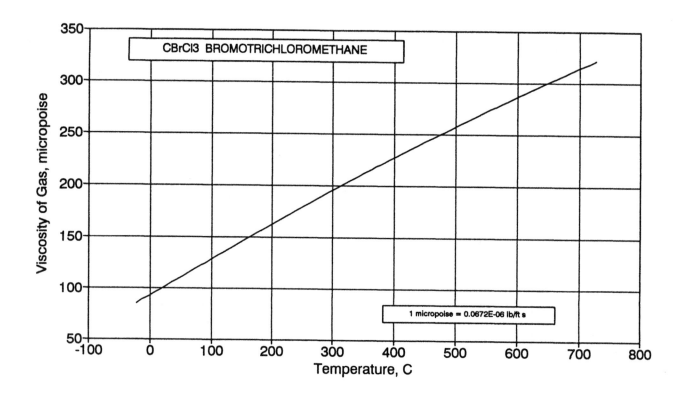

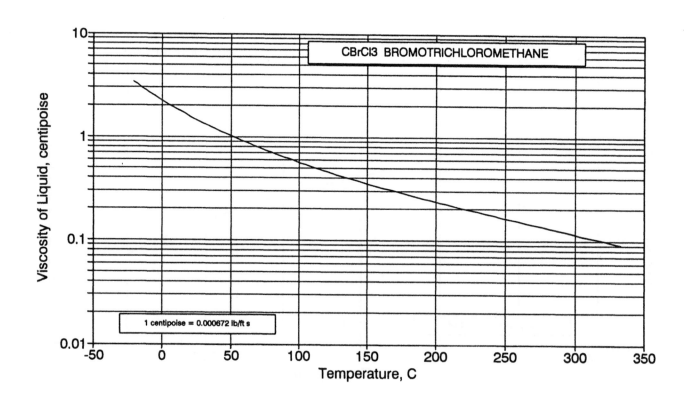

2

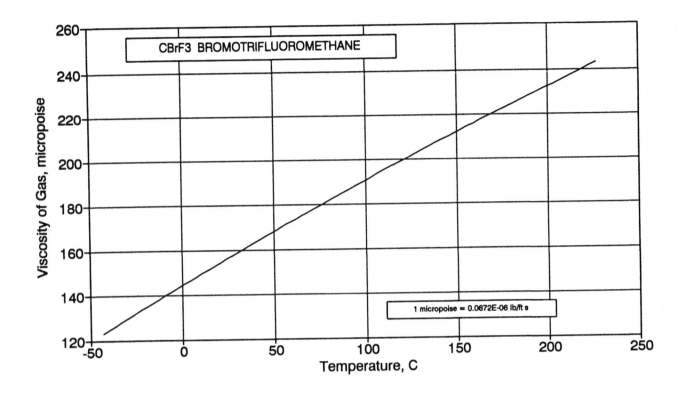

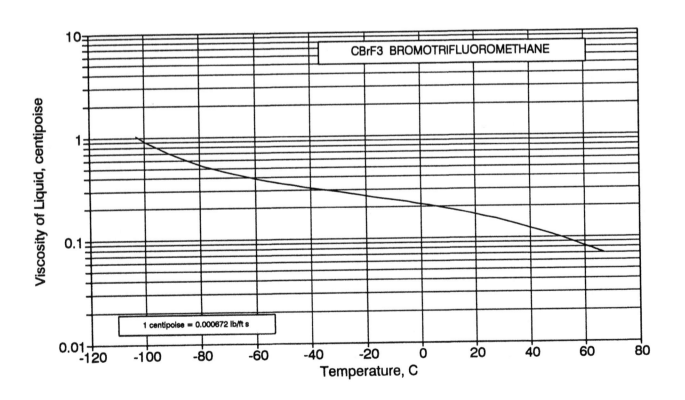

3

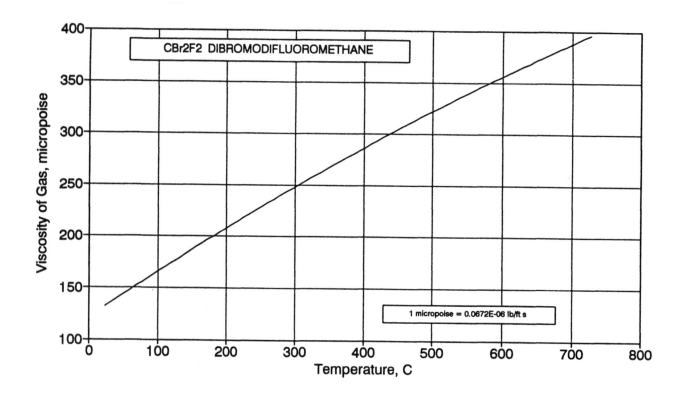

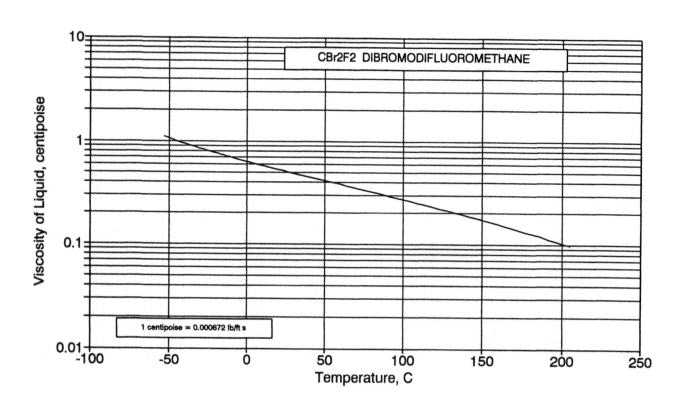

4

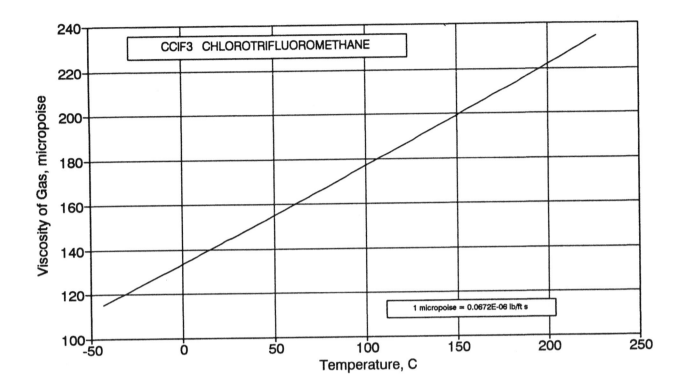

CCIF3 CHLOROTRIFLUOROMETHANE

Viscosity of Gas, micropoise vs Temperature, C

1 micropoise = 0.0672E-06 lb/ft s

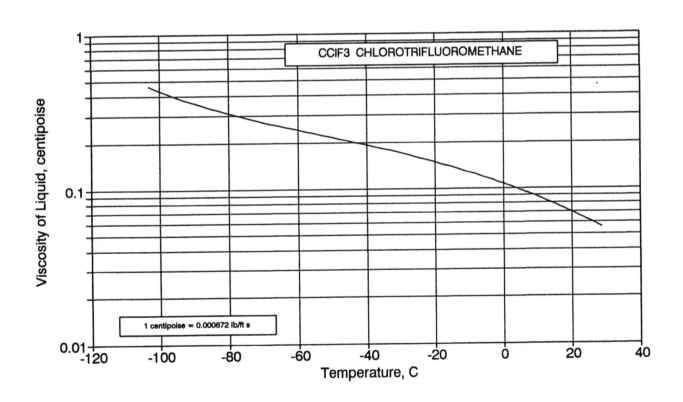

CCIF3 CHLOROTRIFLUOROMETHANE

Viscosity of Liquid, centipoise vs Temperature, C

1 centipoise = 0.000672 lb/ft s

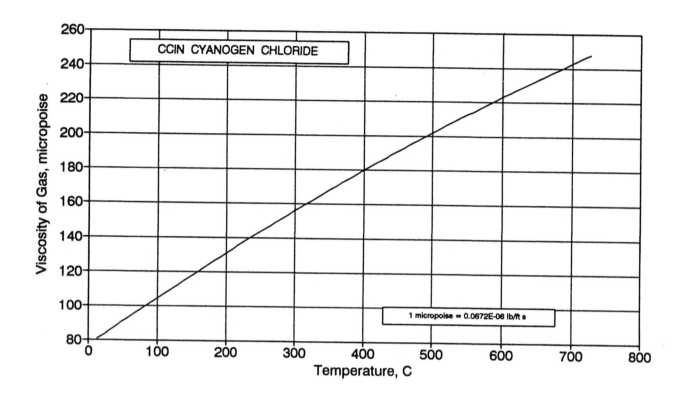

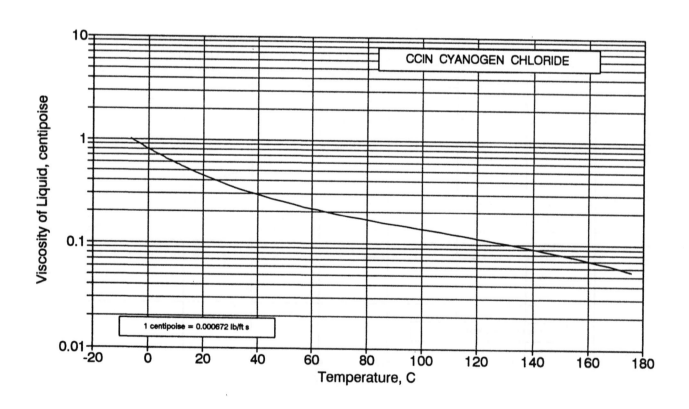

6

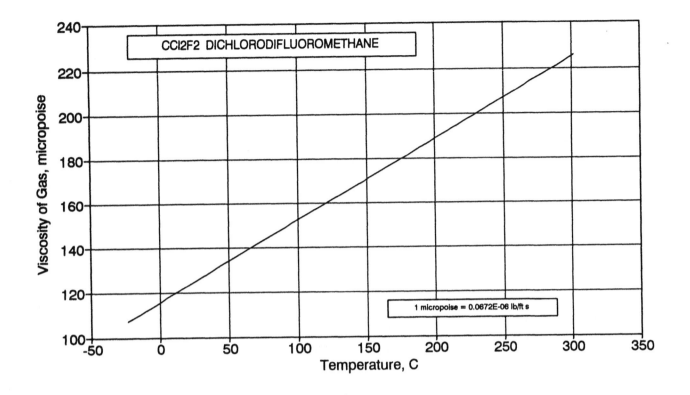

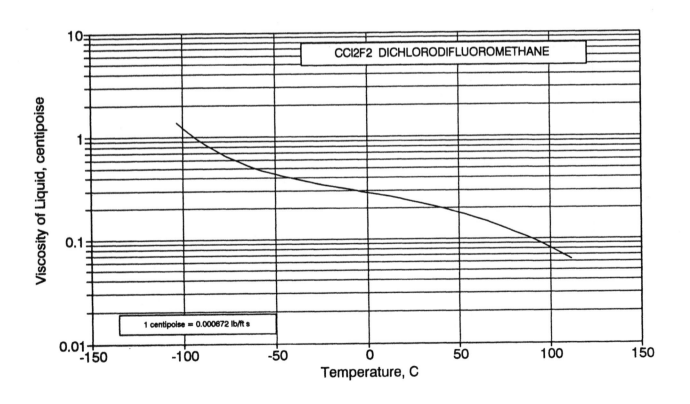

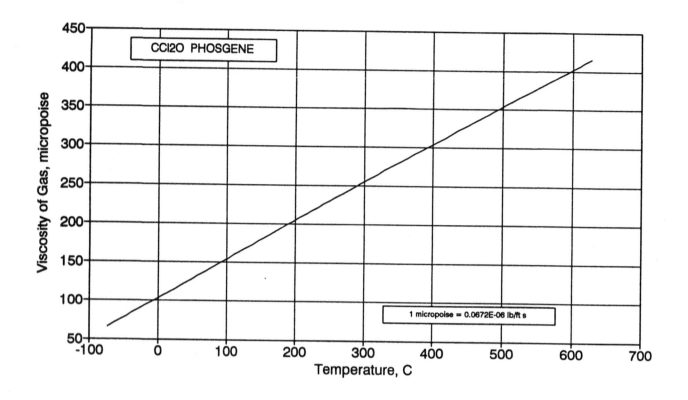

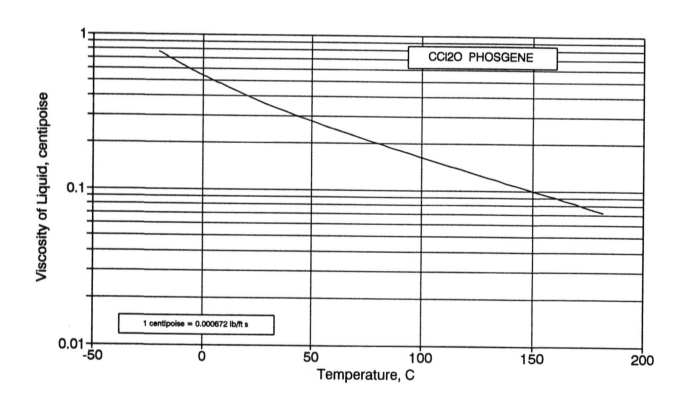

8

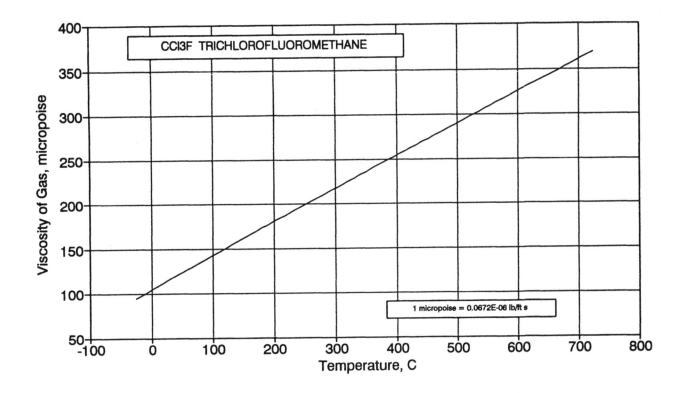

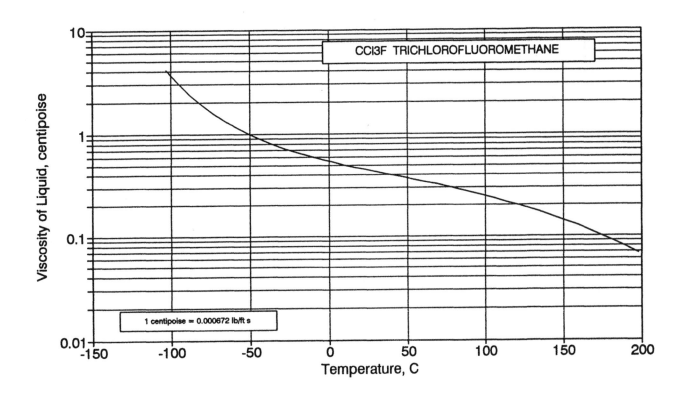

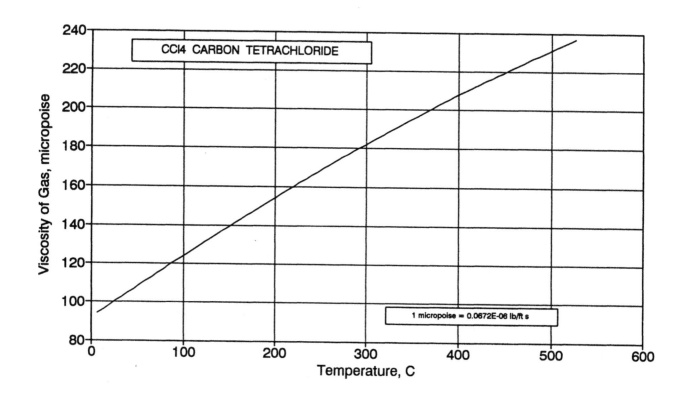

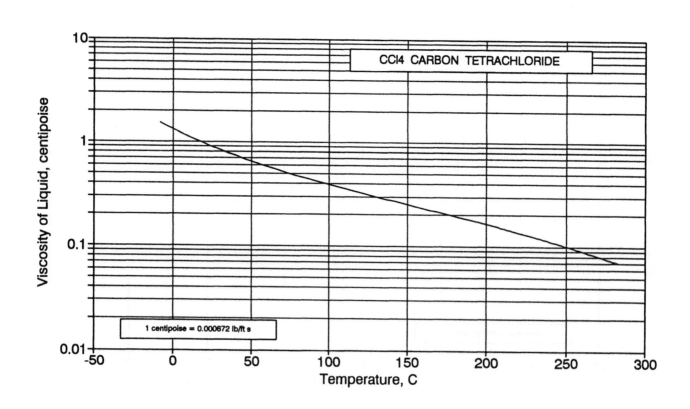

10

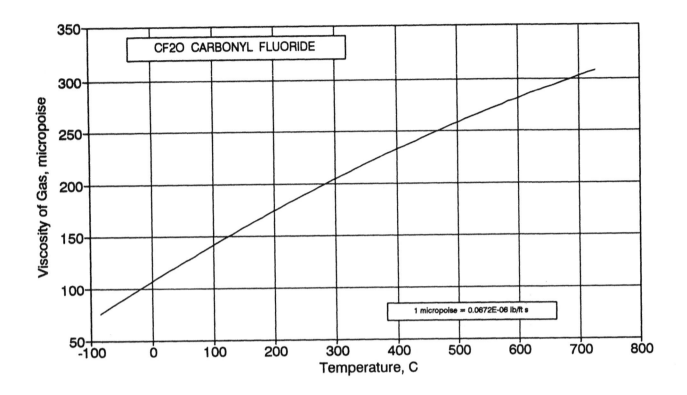

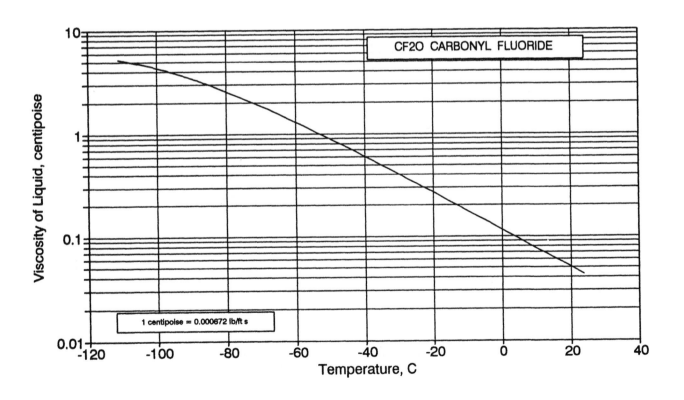

11

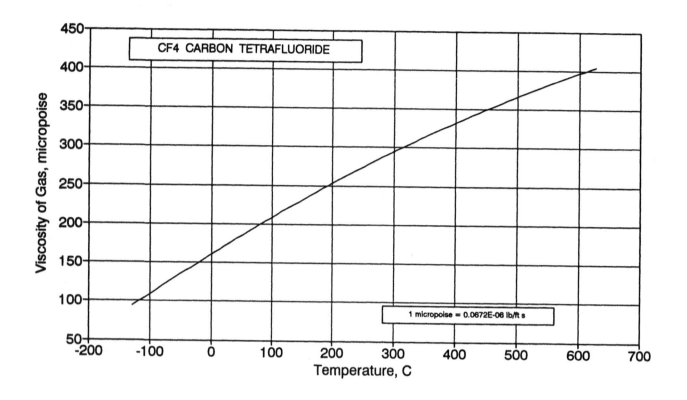

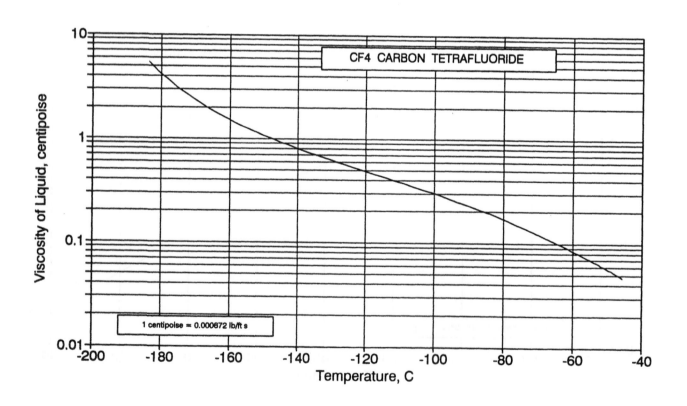

12

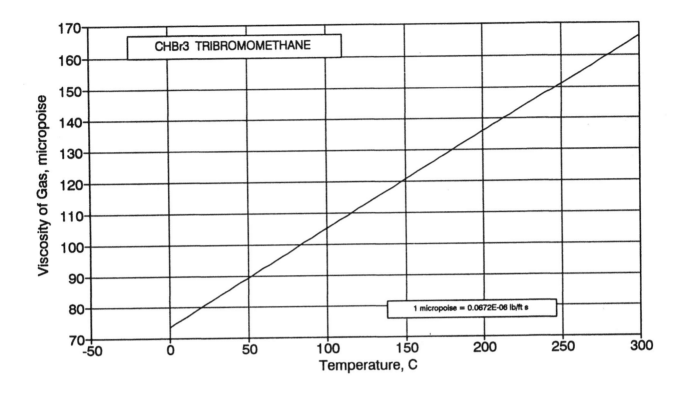

Viscosity of Gas, micropoise vs Temperature, C

CHBr3 TRIBROMOMETHANE

1 micropoise = 0.0672E-06 lb/ft s

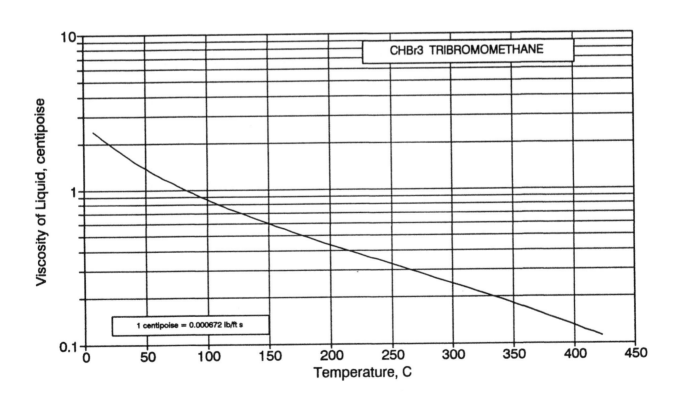

Viscosity of Liquid, centipoise vs Temperature, C

CHBr3 TRIBROMOMETHANE

1 centipoise = 0.000672 lb/ft s

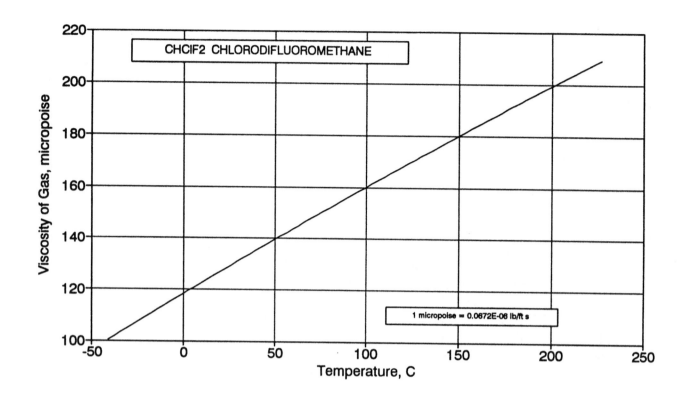

CHCIF2 CHLORODIFLUOROMETHANE

Viscosity of Gas, micropoise

Temperature, C

1 micropoise = 0.0672E-06 lb/ft s

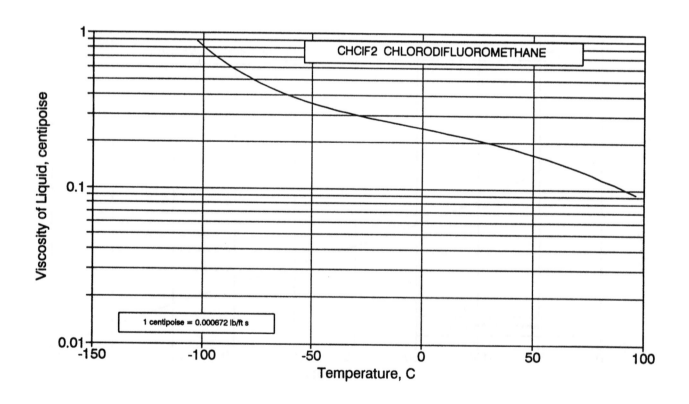

CHCIF2 CHLORODIFLUOROMETHANE

Viscosity of Liquid, centipoise

Temperature, C

1 centipoise = 0.000672 lb/ft s

14

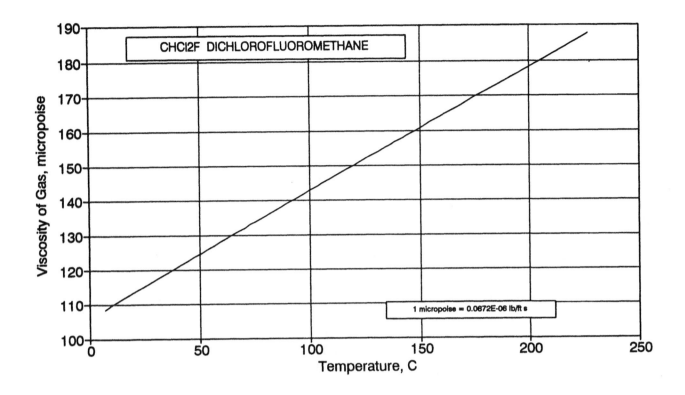

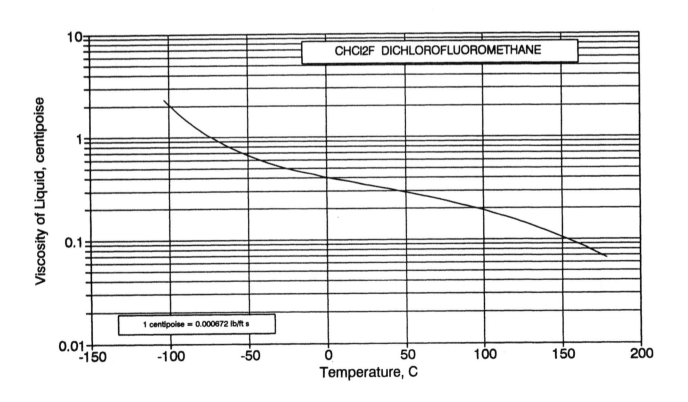

15

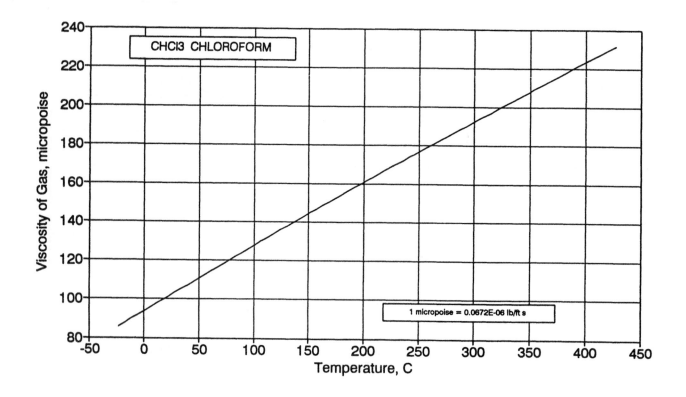

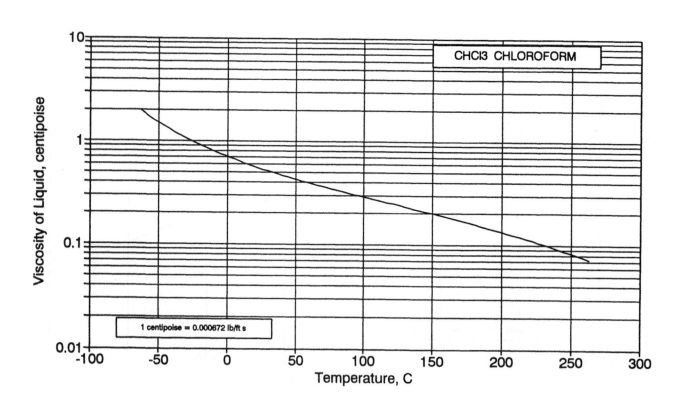

16

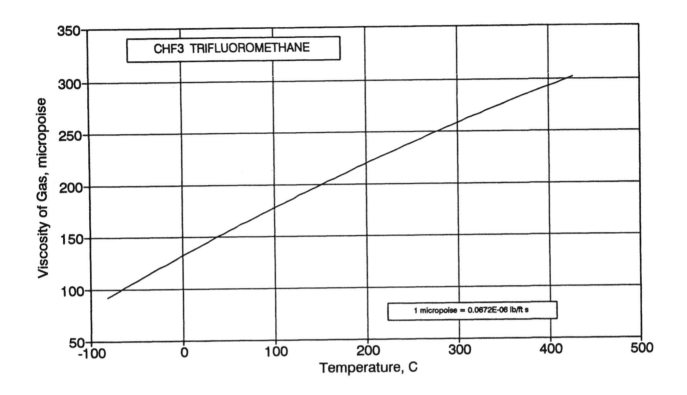

CHF3 TRIFLUOROMETHANE

1 micropoise = 0.0672E-06 lb/ft s

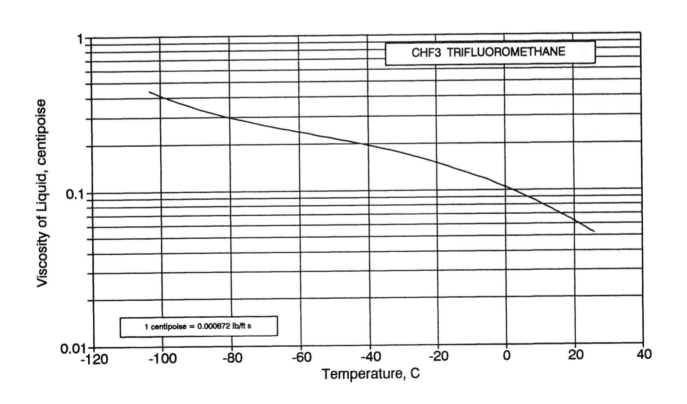

CHF3 TRIFLUOROMETHANE

1 centipoise = 0.000672 lb/ft s

17

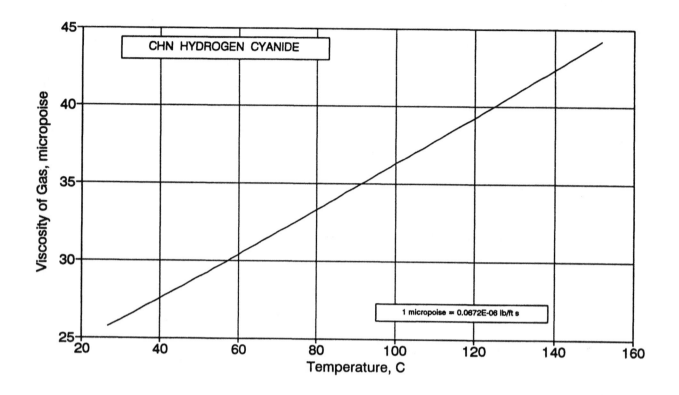

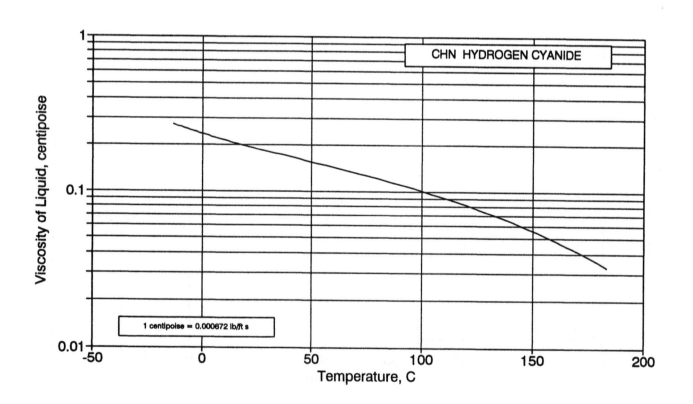

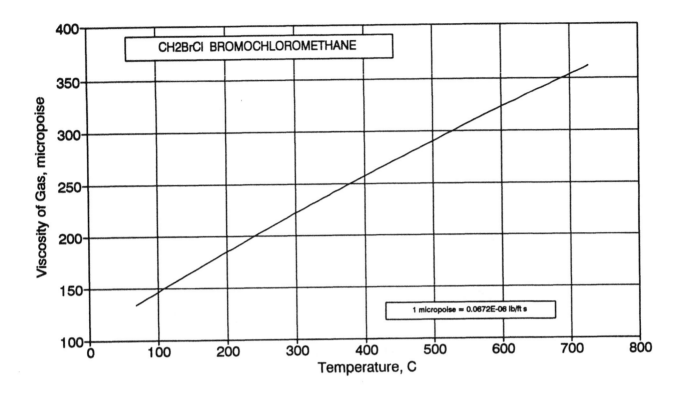

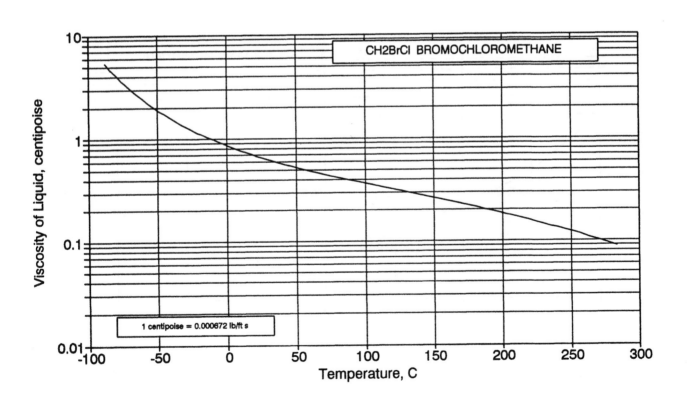

19

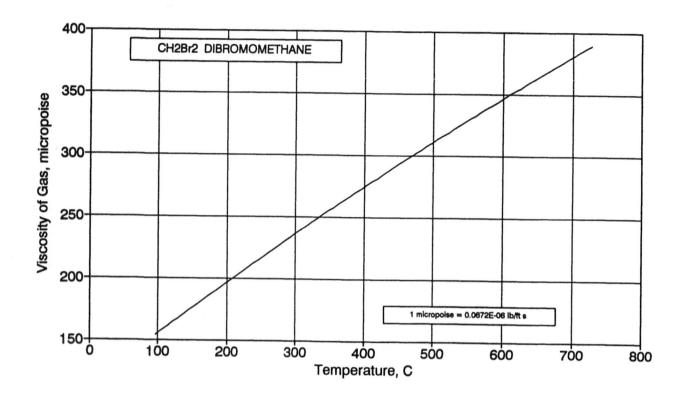

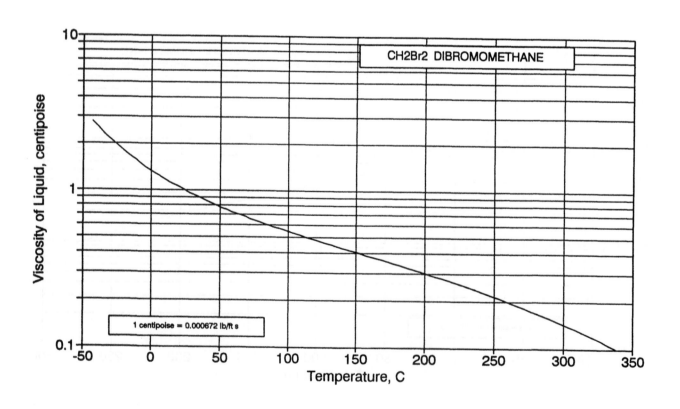

20

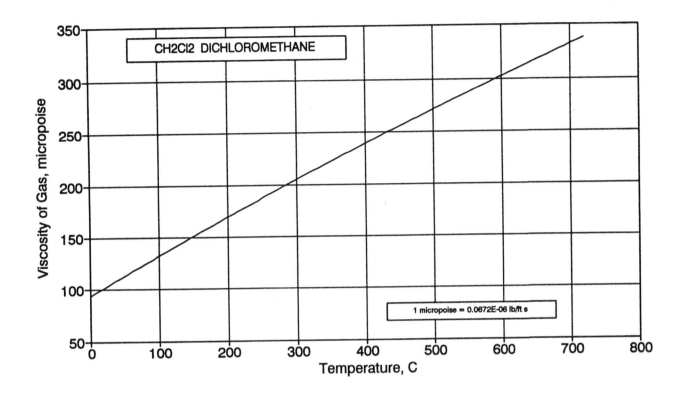

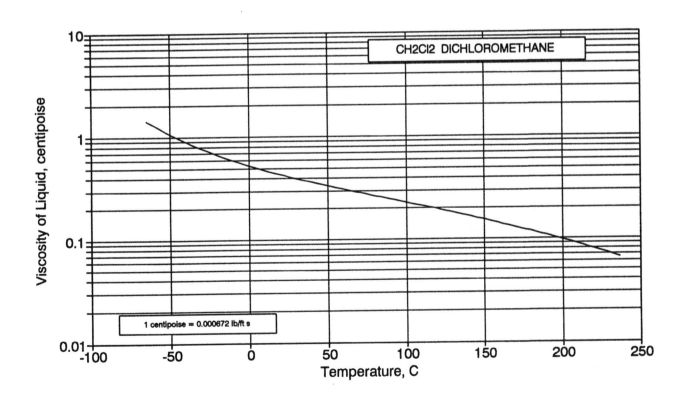

21

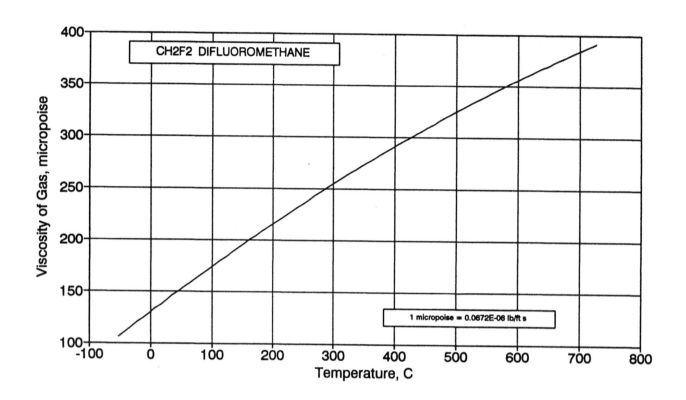

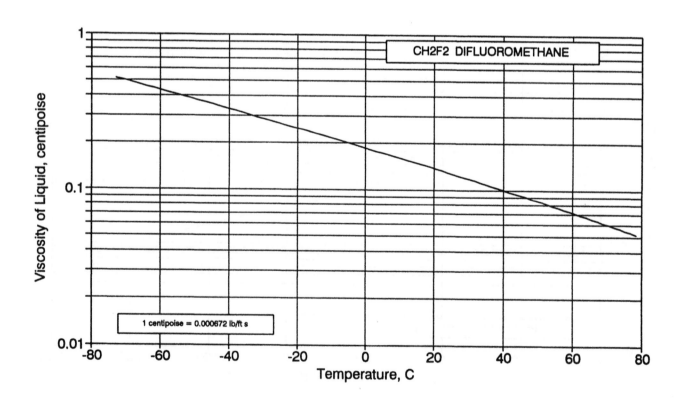

22

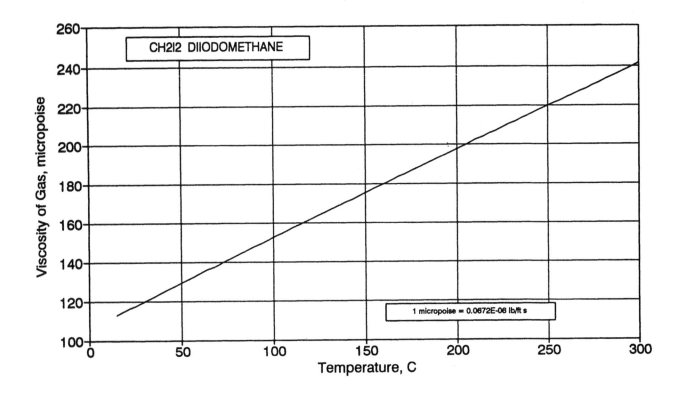

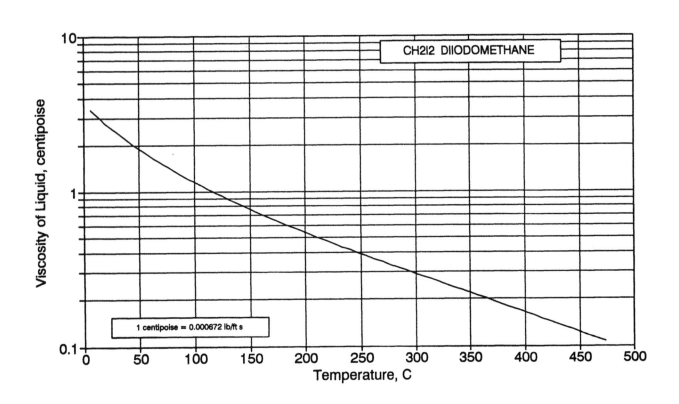

23

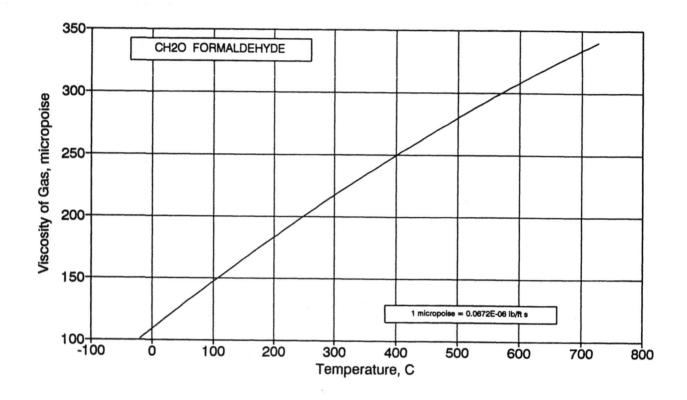

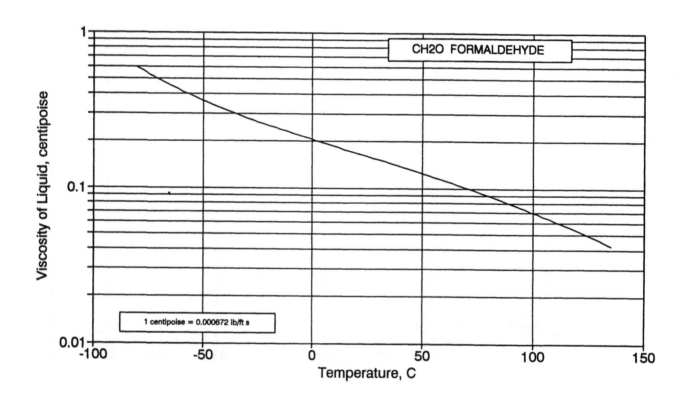

24

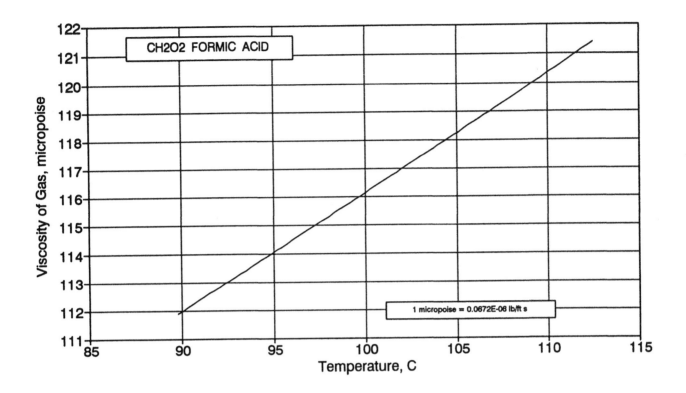

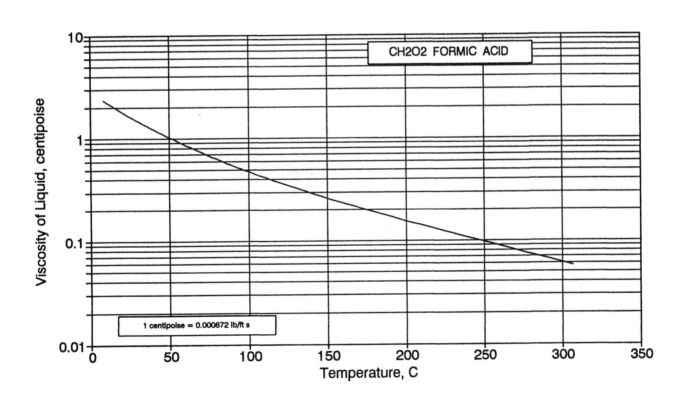

25

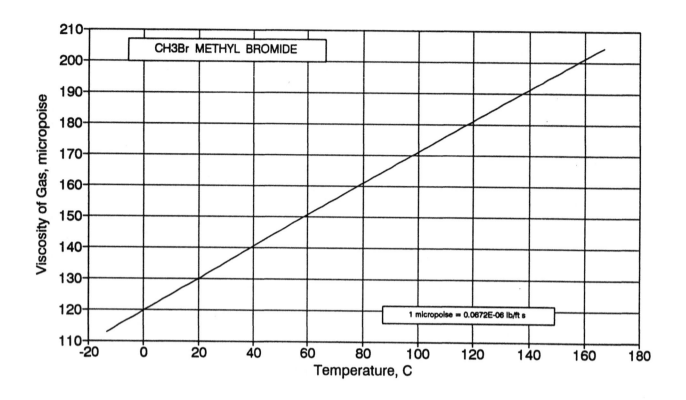

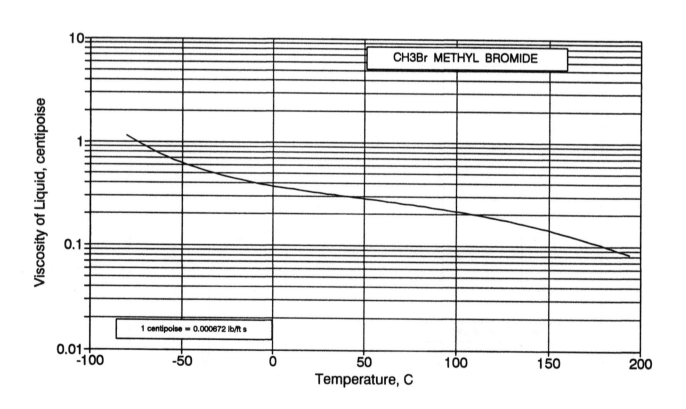

26

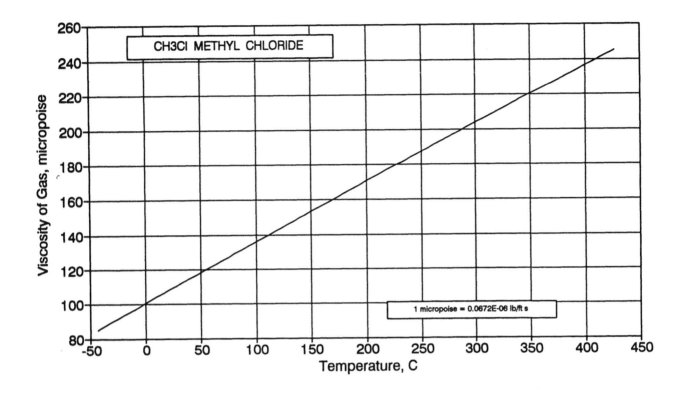

CH3Cl METHYL CHLORIDE

Viscosity of Gas, micropoise vs Temperature, C

1 micropoise = 0.0672E-06 lb/ft s

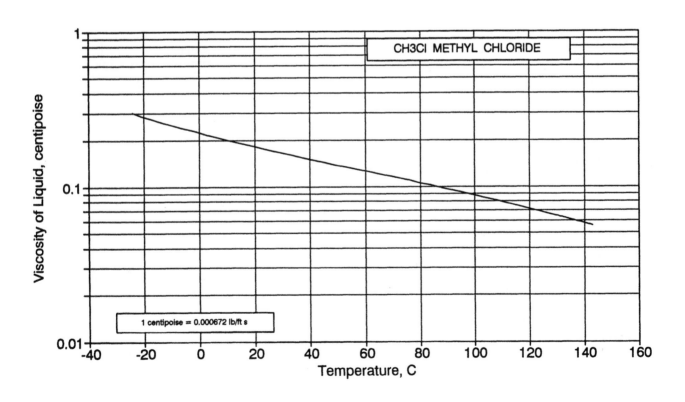

CH3Cl METHYL CHLORIDE

Viscosity of Liquid, centipoise vs Temperature, C

1 centipoise = 0.000672 lb/ft s

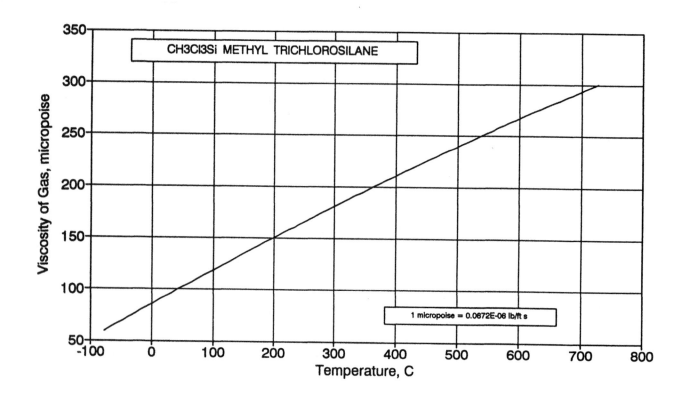

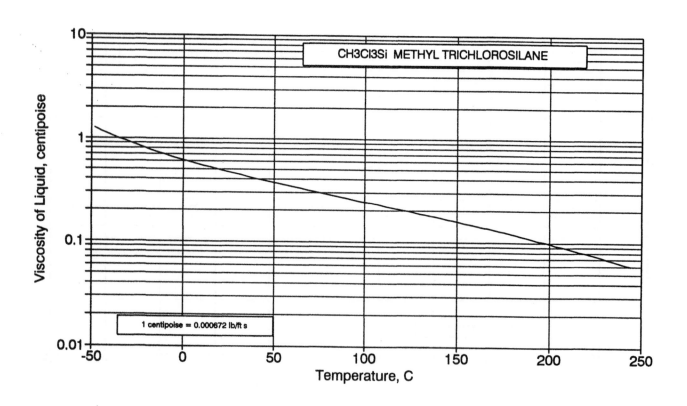

28

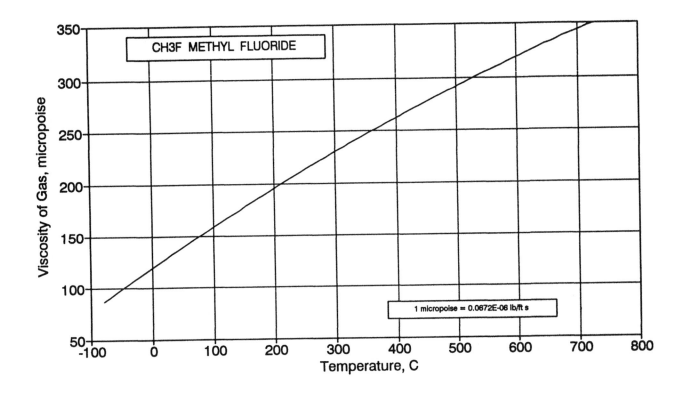

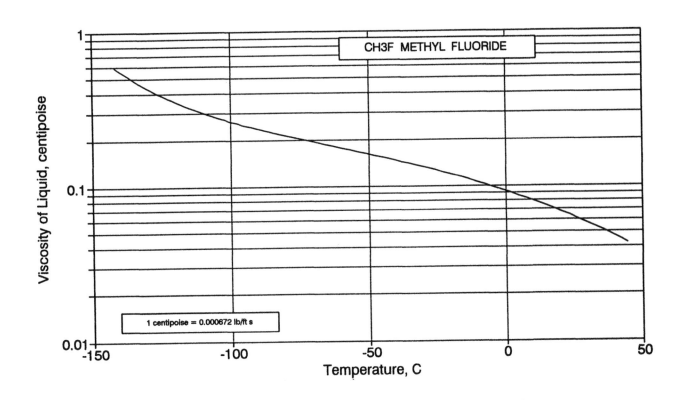

29

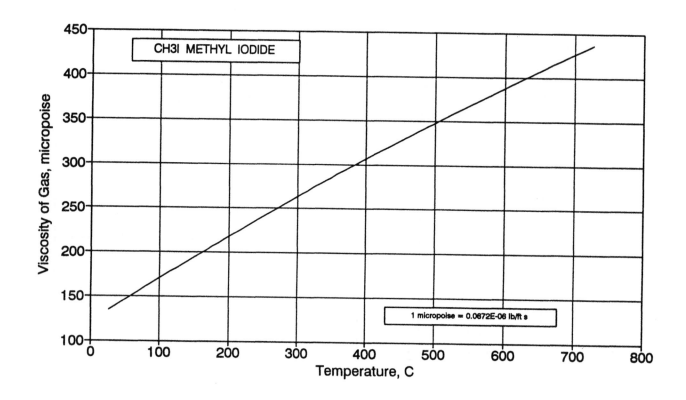

CH3I METHYL IODIDE

Viscosity of Gas, micropoise

Temperature, C

1 micropoise = 0.0672E-06 lb/ft s

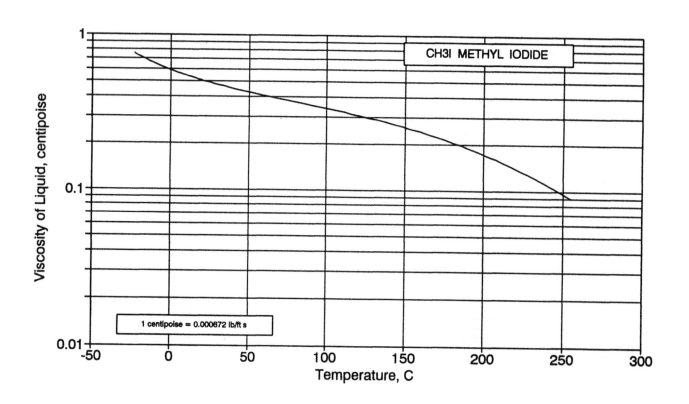

CH3I METHYL IODIDE

Viscosity of Liquid, centipoise

Temperature, C

1 centipoise = 0.000672 lb/ft s

30

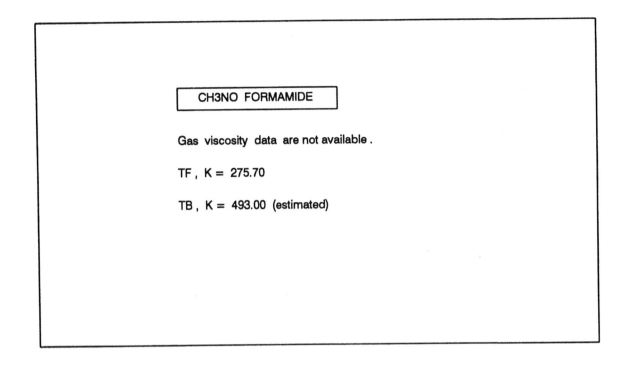

CH3NO FORMAMIDE

Gas viscosity data are not available.

TF , K = 275.70

TB , K = 493.00 (estimated)

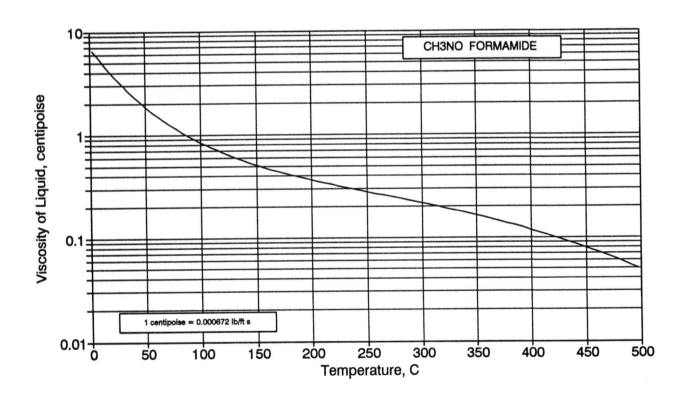

1 centipoise = 0.000672 lb/ft s

CH3NO FORMAMIDE

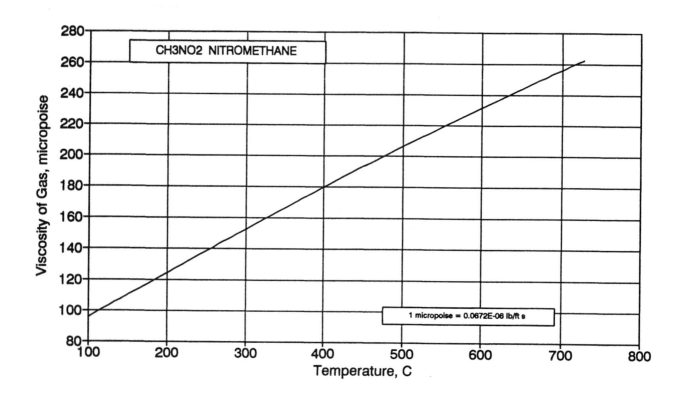

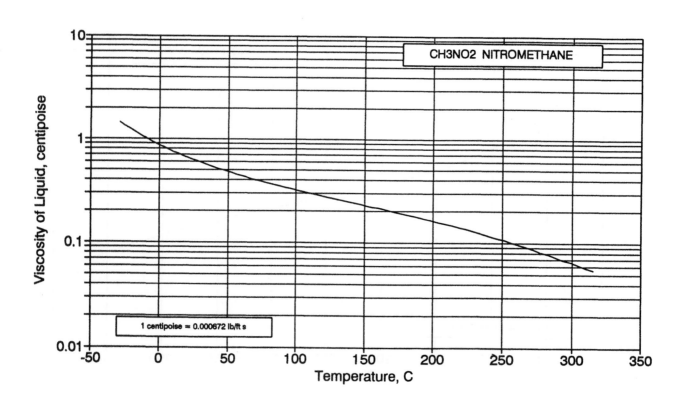

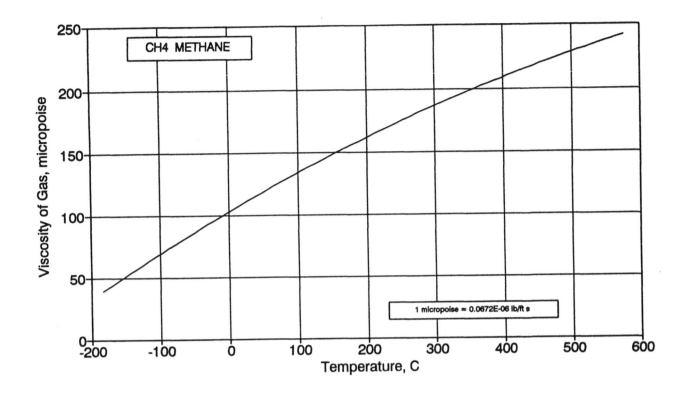

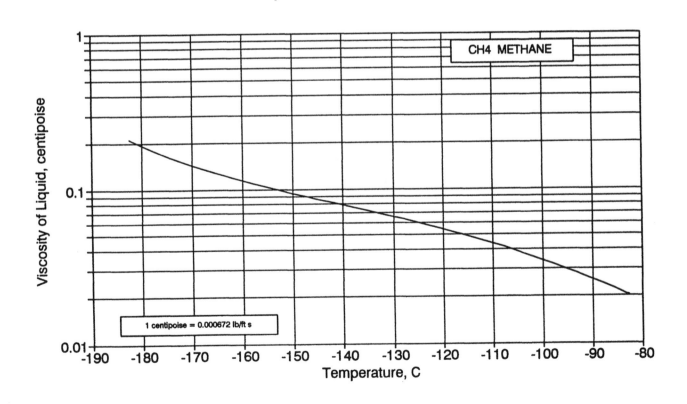

33

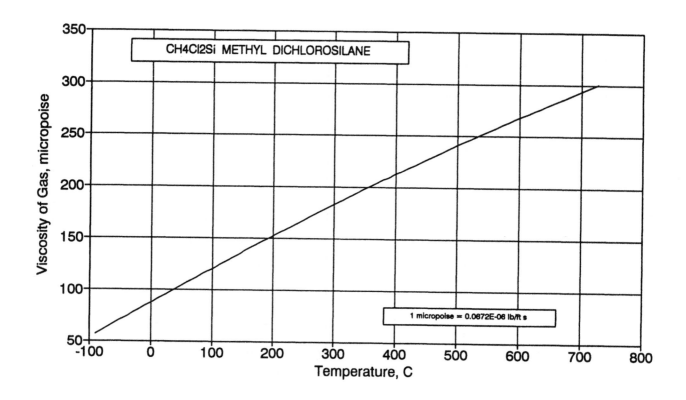

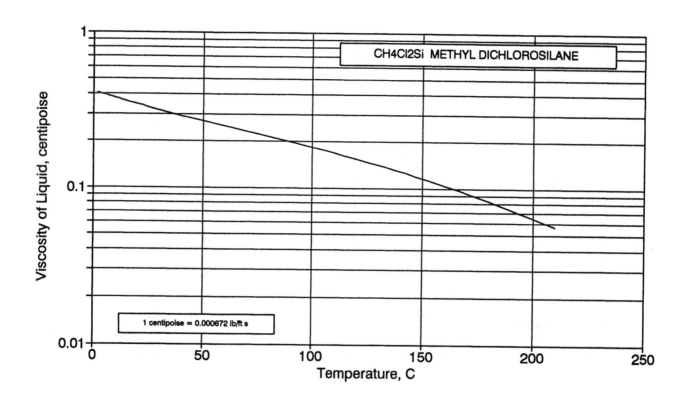

34

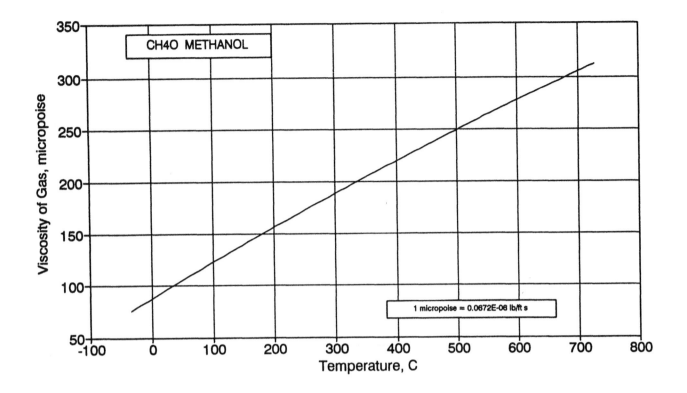

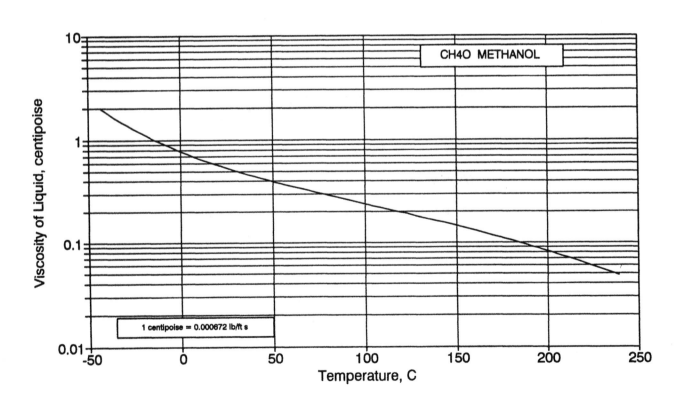

35

CH4O3S METHANESULFONIC ACID

Gas viscosity data are not available.

TF , K = 292.81

TB , K = 561 (estimated)

CH4O3S METHANESULFONIC ACID

Liquid viscosity data are not available.

TF , K = 292.81

TB , K = 561 (estimated)

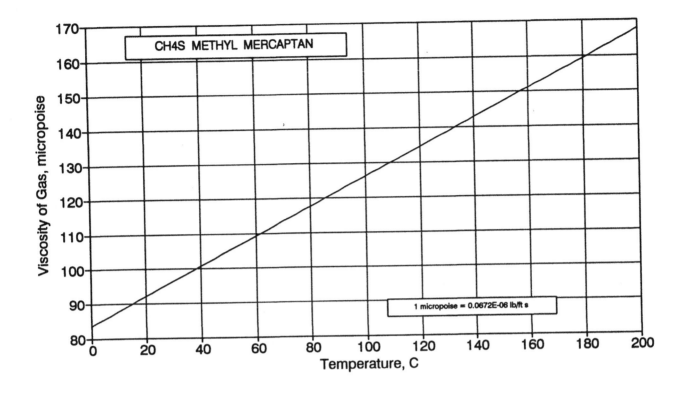

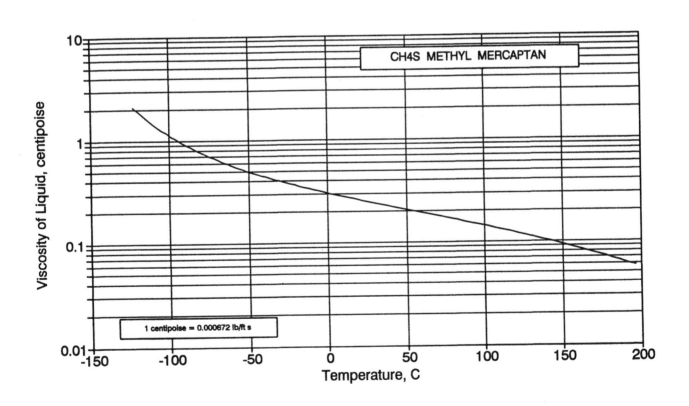

37

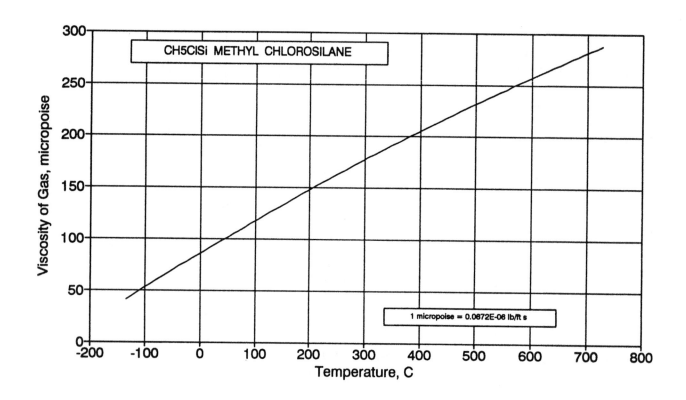

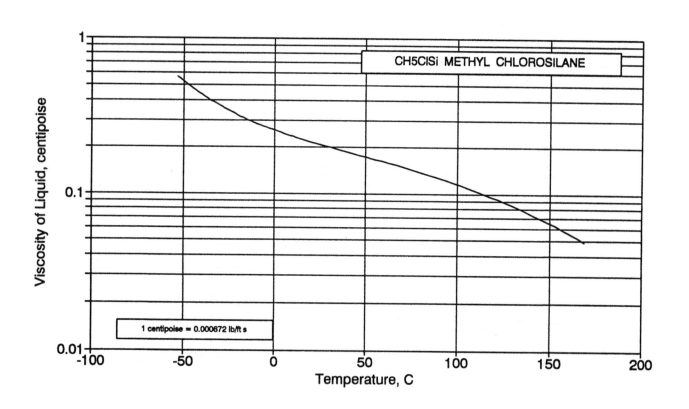

38

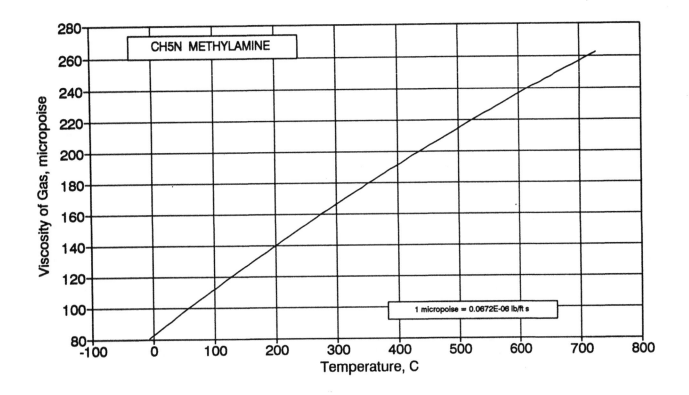

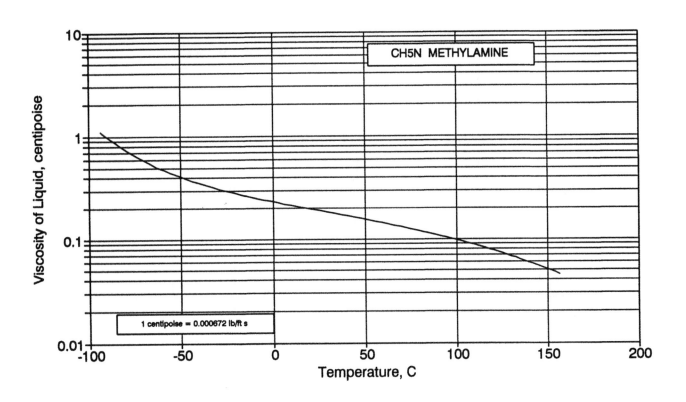

39

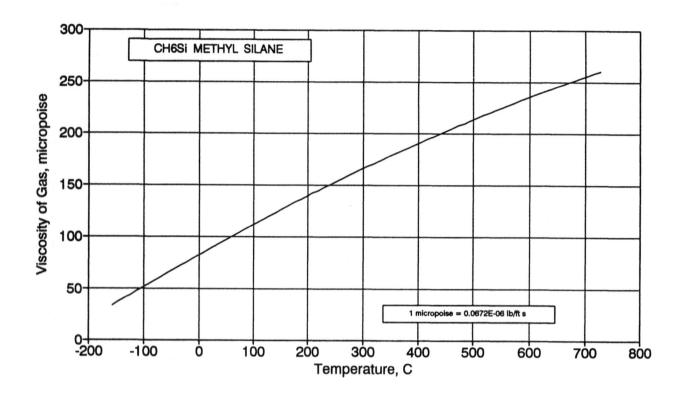

CH6Si METHYL SILANE

1 micropoise = 0.0672E-06 lb/ft s

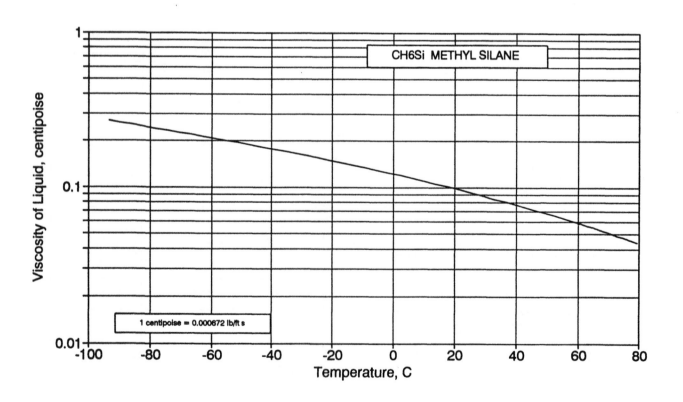

CH6Si METHYL SILANE

1 centipoise = 0.000672 lb/ft s

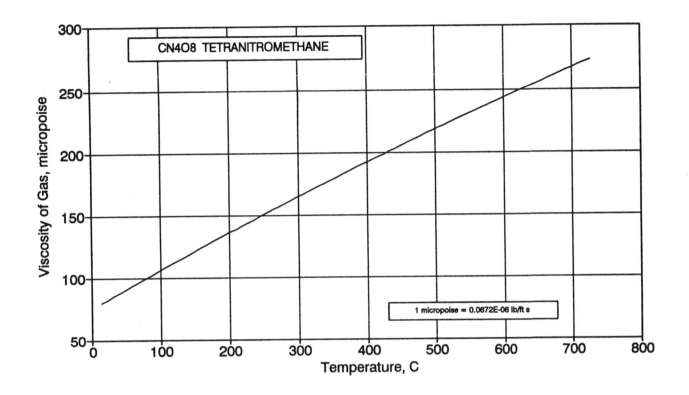

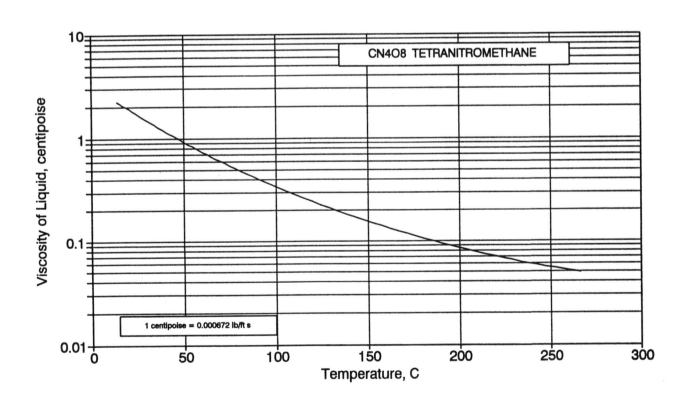

41

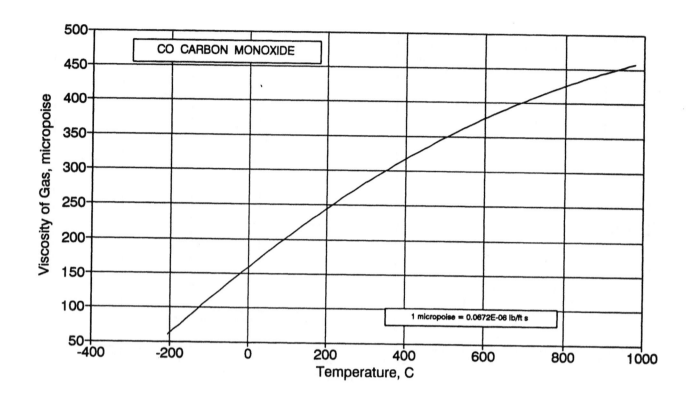

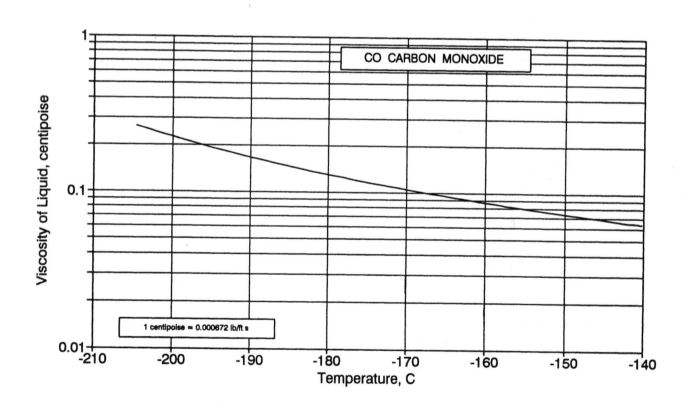

42

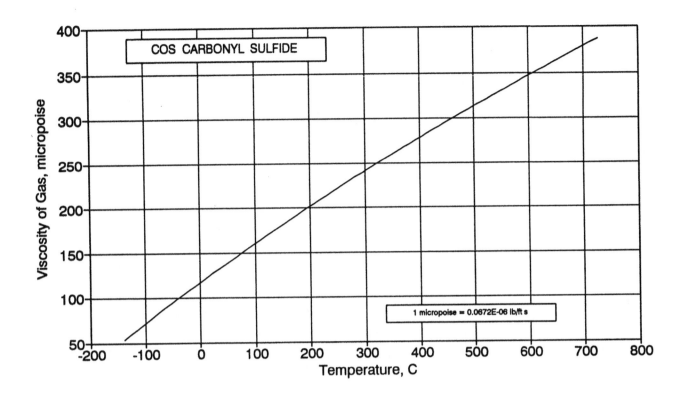

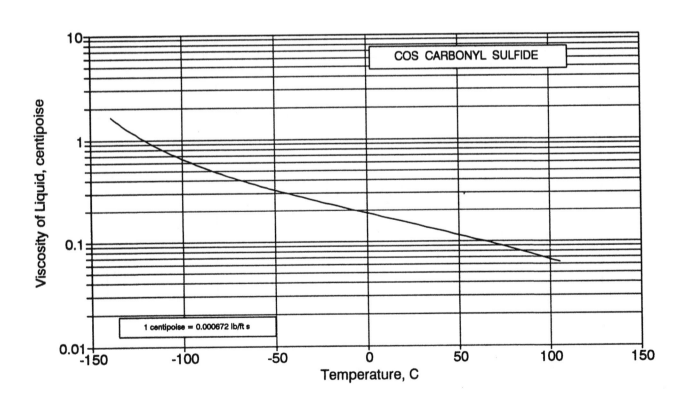

43

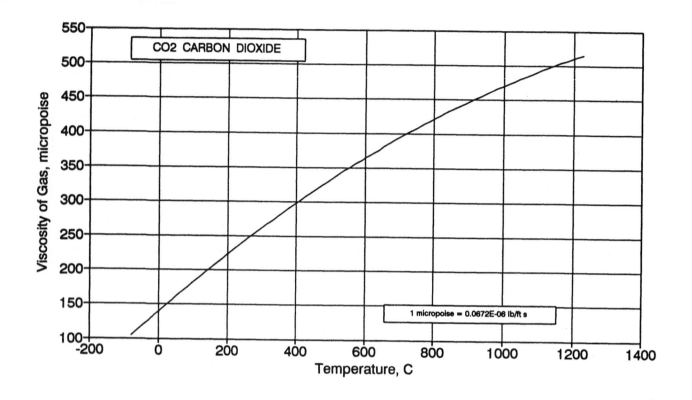

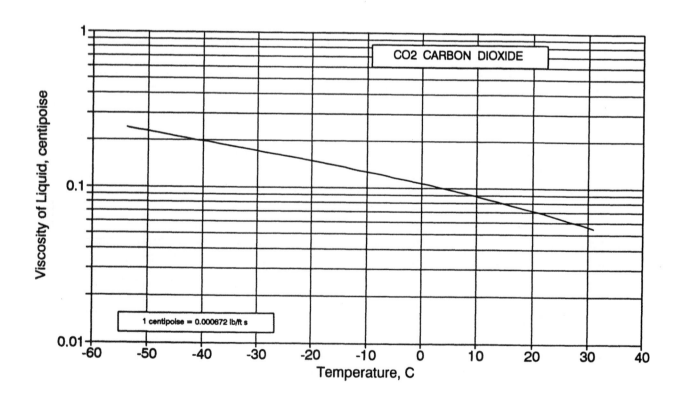

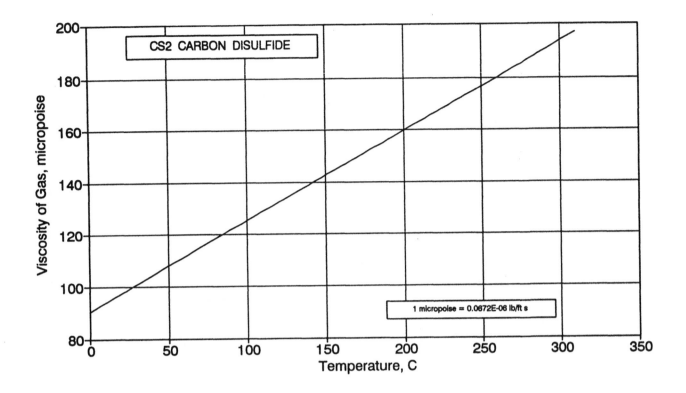

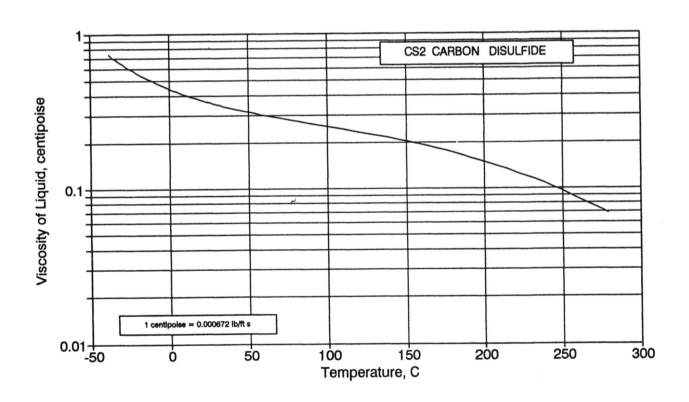

45

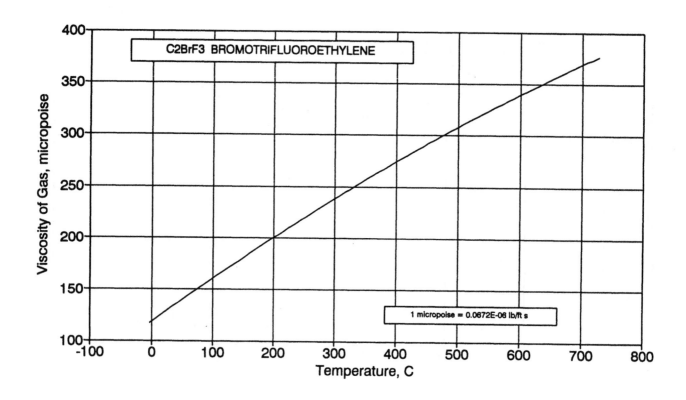

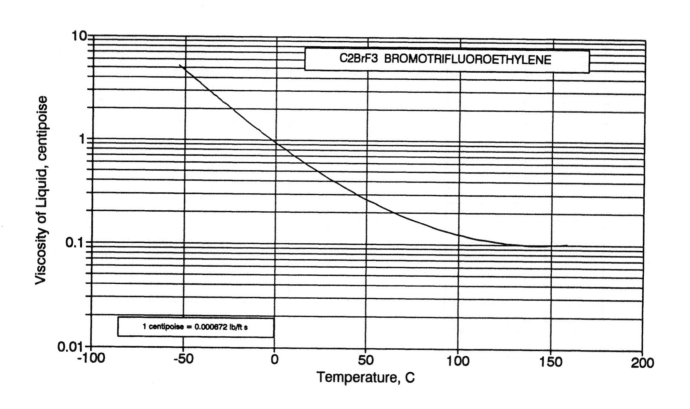

46

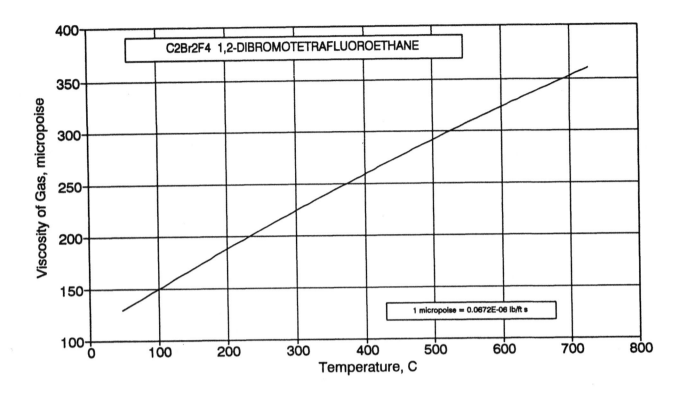

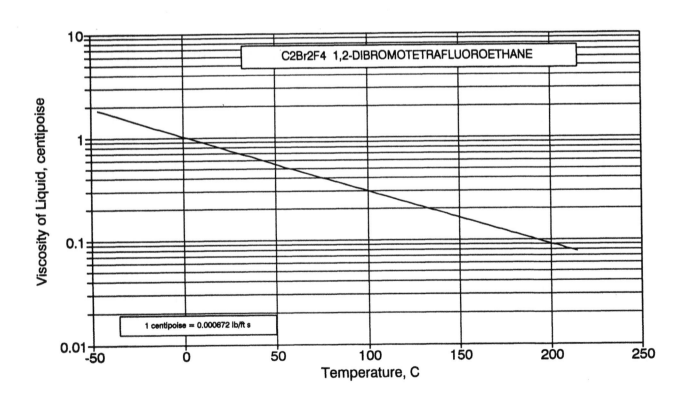

47

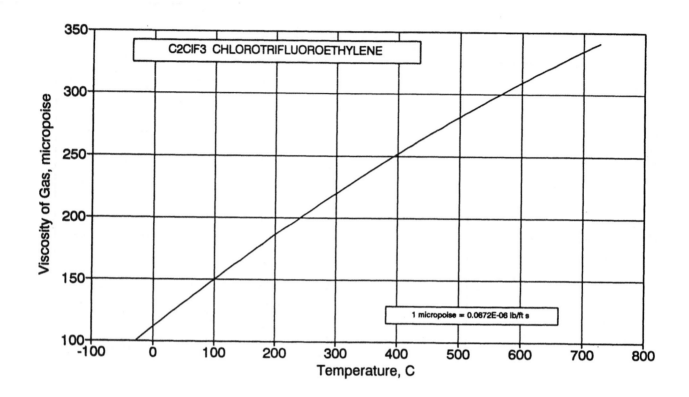

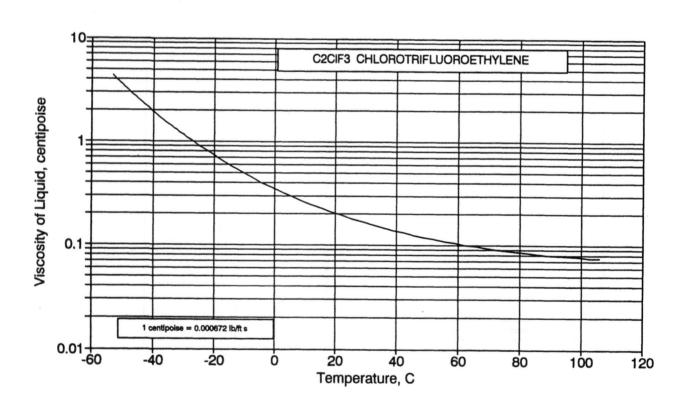

48

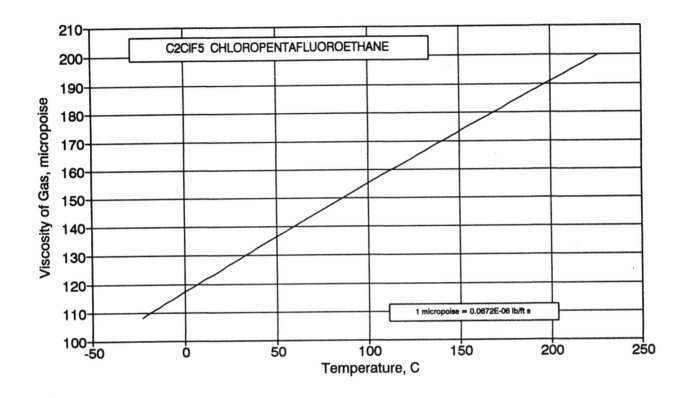

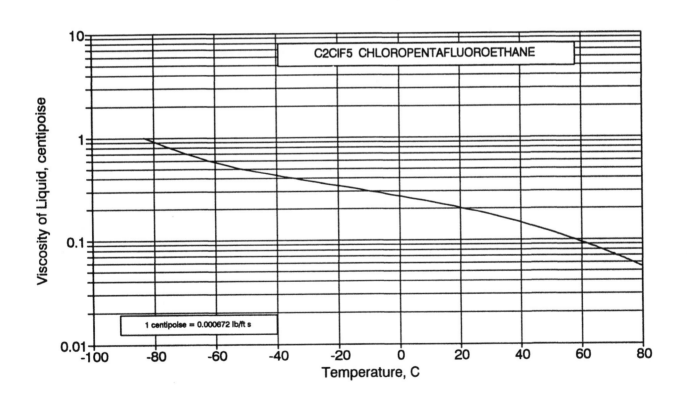

49

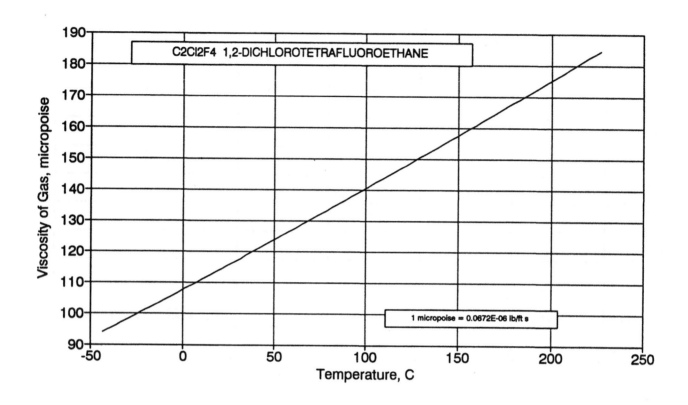

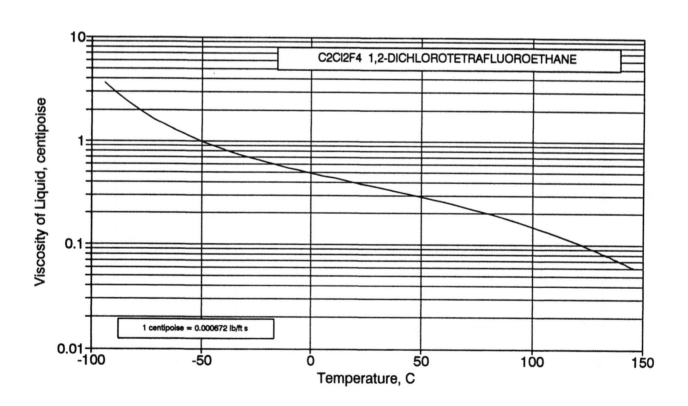

50

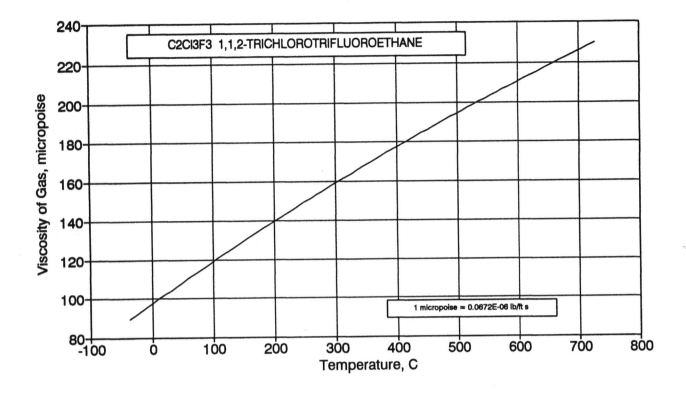

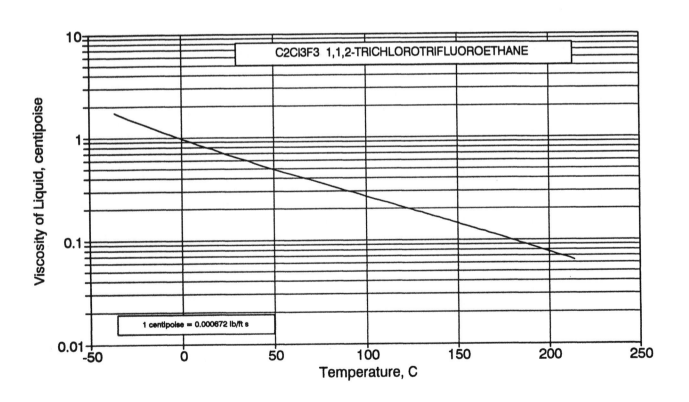

51

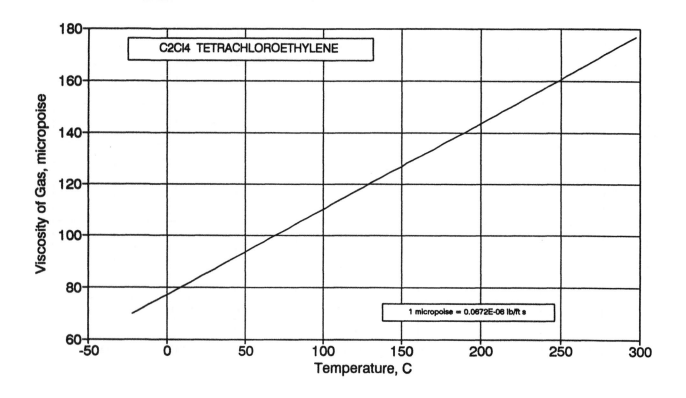

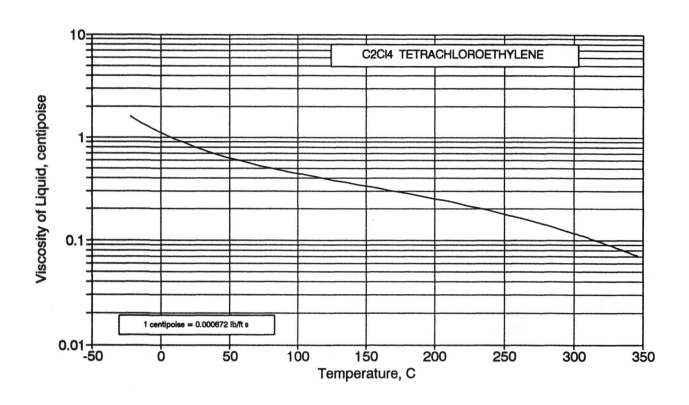

52

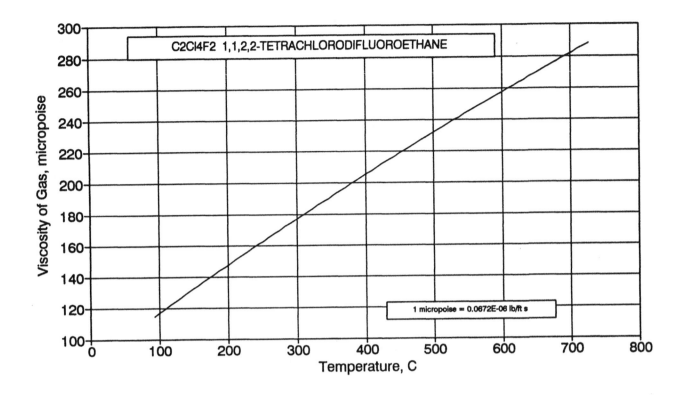

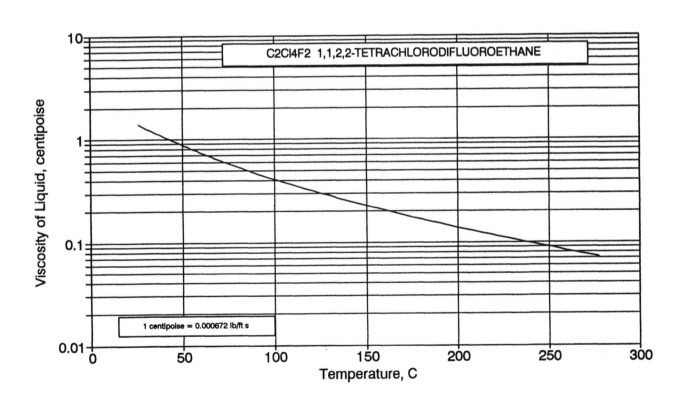

53

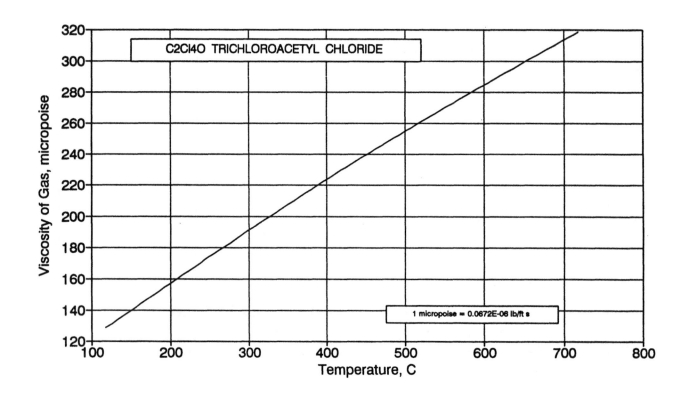

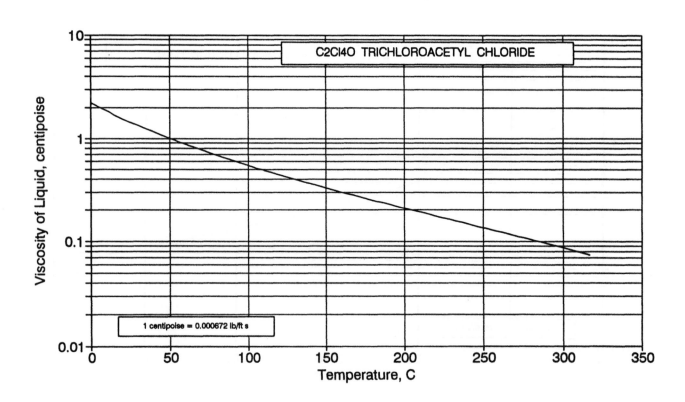

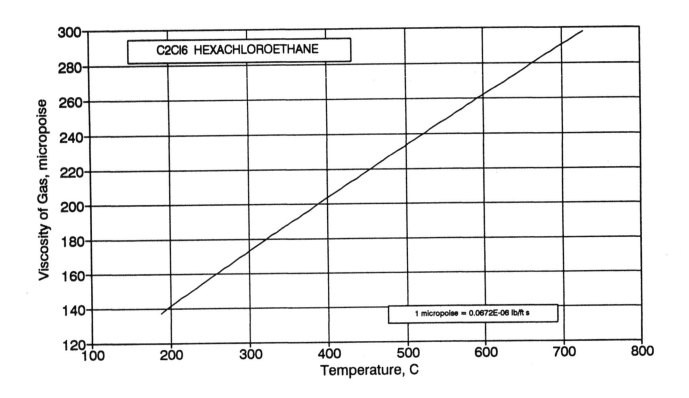

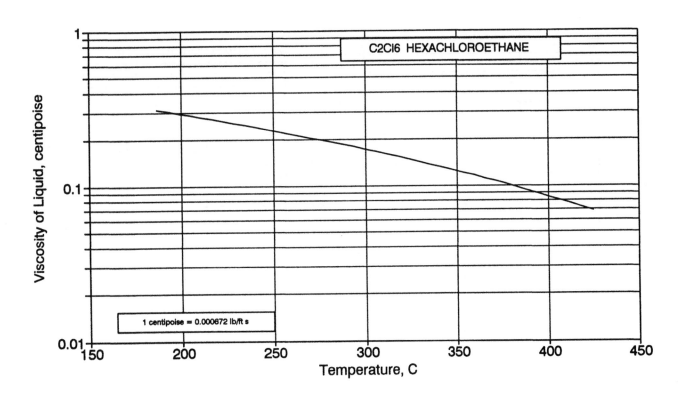

55

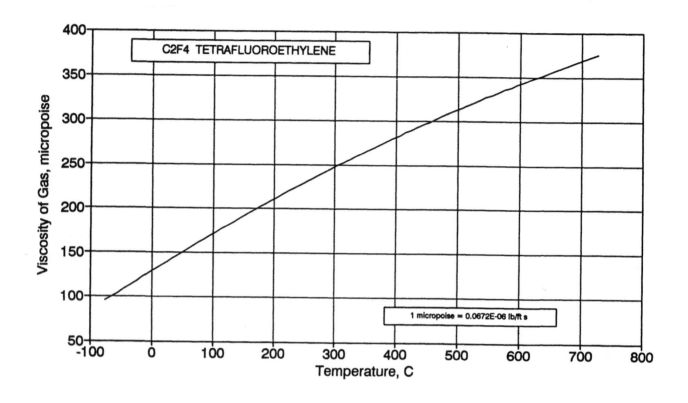

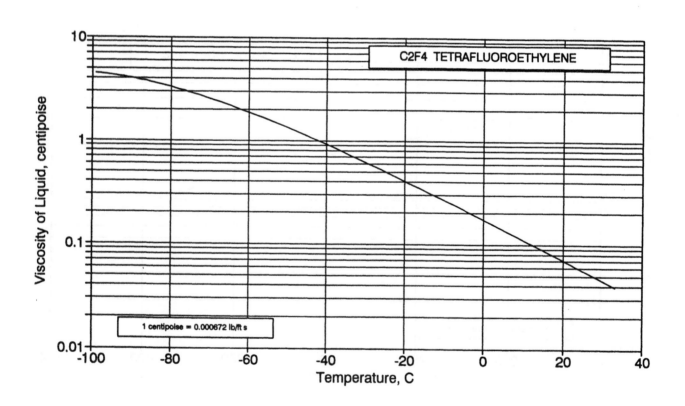

56

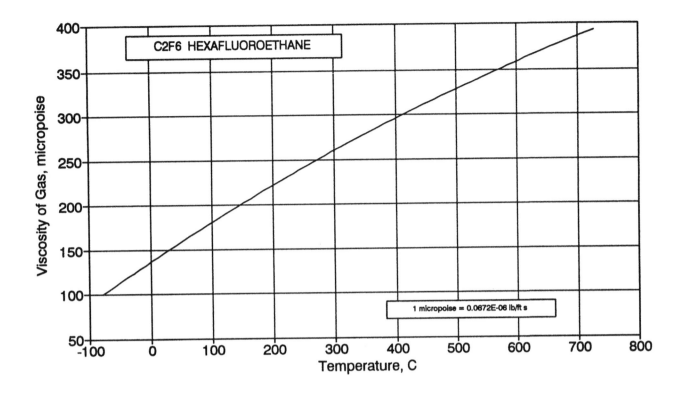

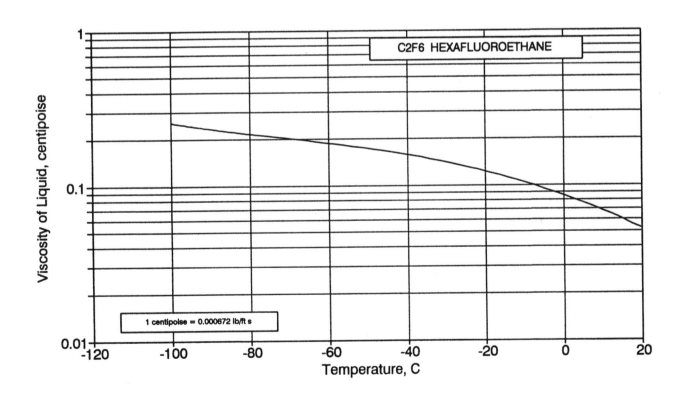

57

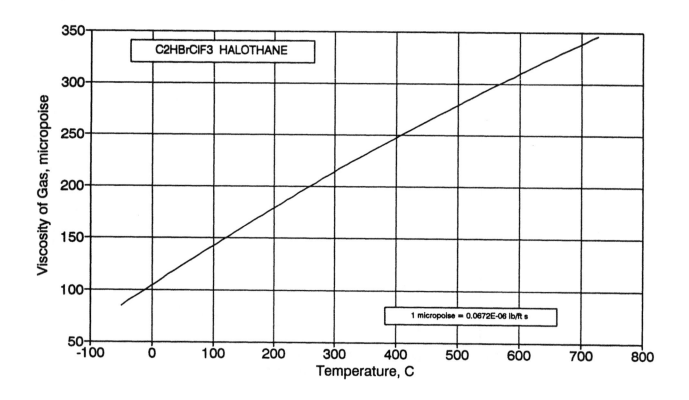

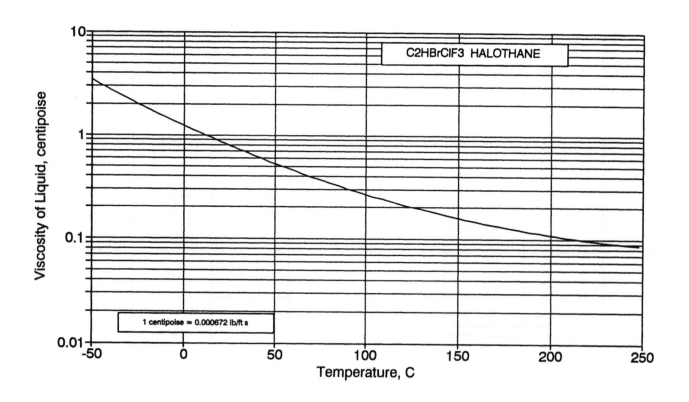

58

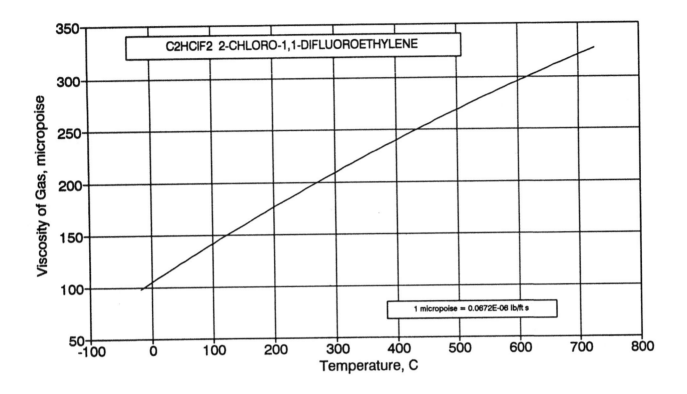

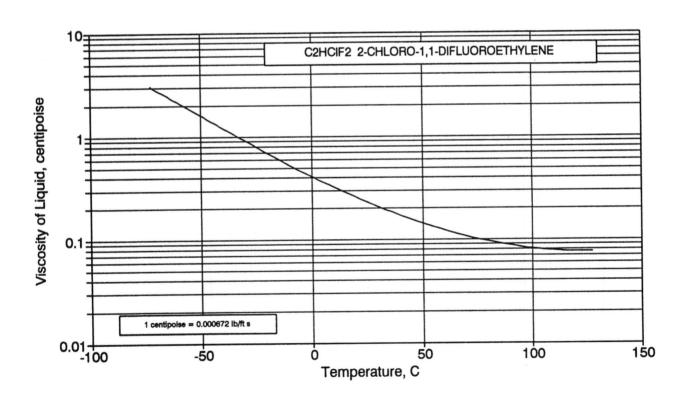

59

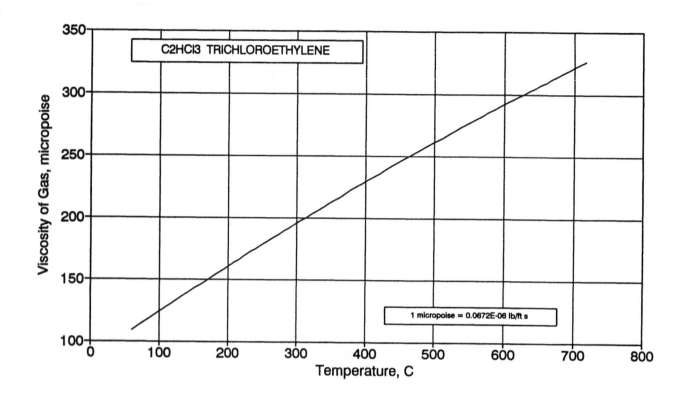

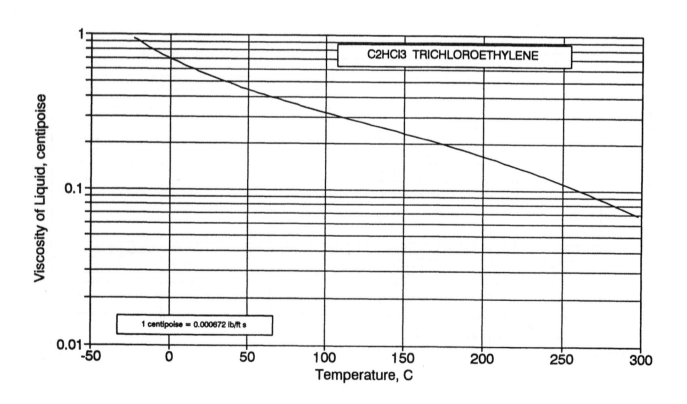

60

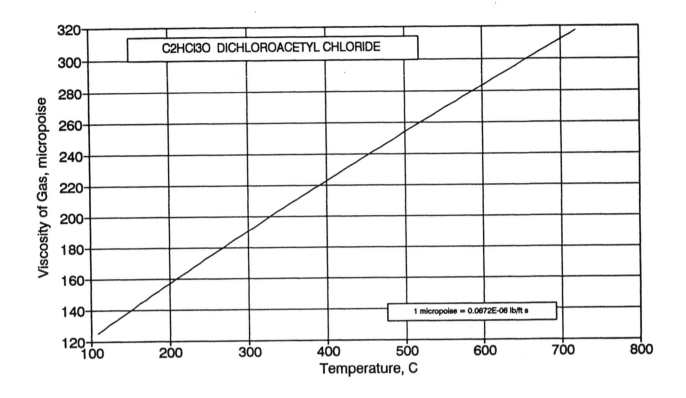

C2HCl3O DICHLOROACETYL CHLORIDE

1 micropoise = 0.0672E-06 lb/ft s

Viscosity of Gas, micropoise

Temperature, C

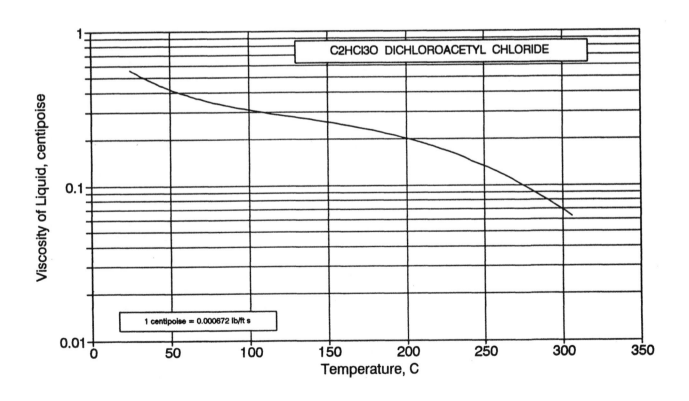

C2HCl3O DICHLOROACETYL CHLORIDE

1 centipoise = 0.000672 lb/ft s

Viscosity of Liquid, centipoise

Temperature, C

61

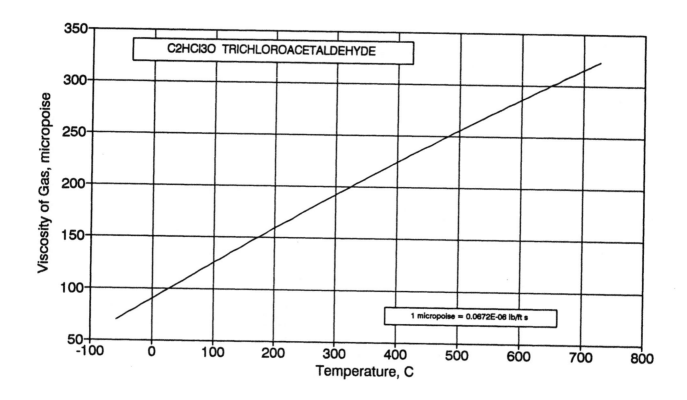

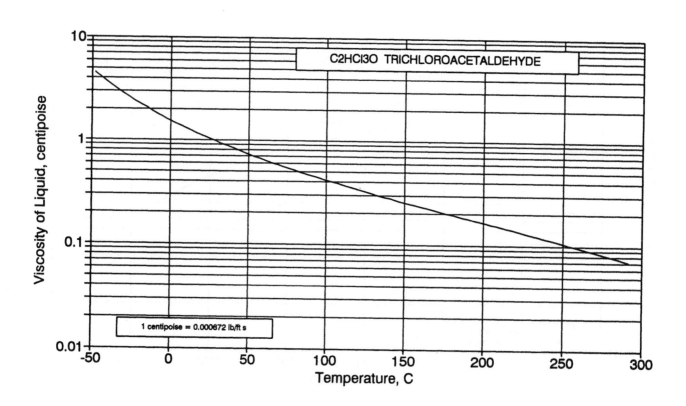

62

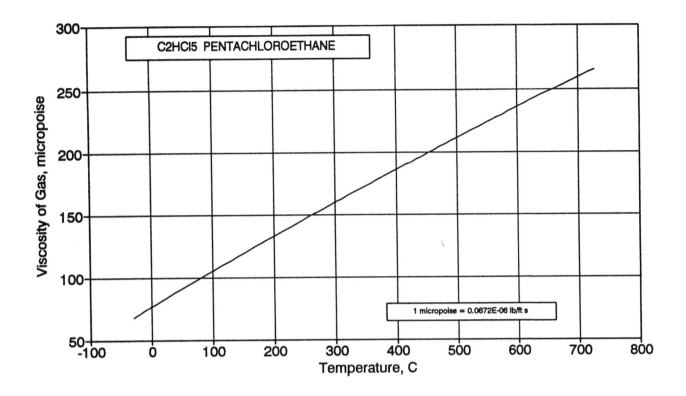

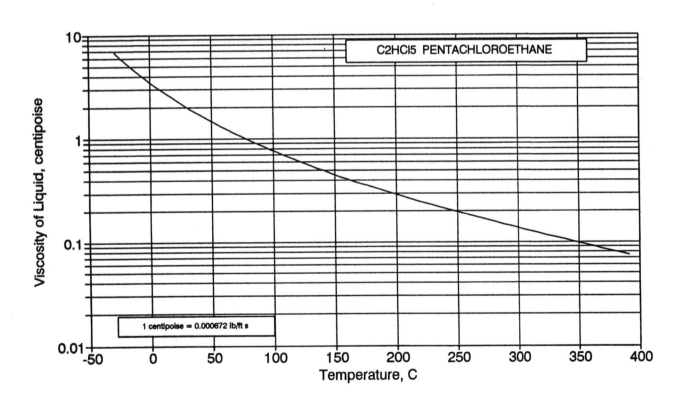

63

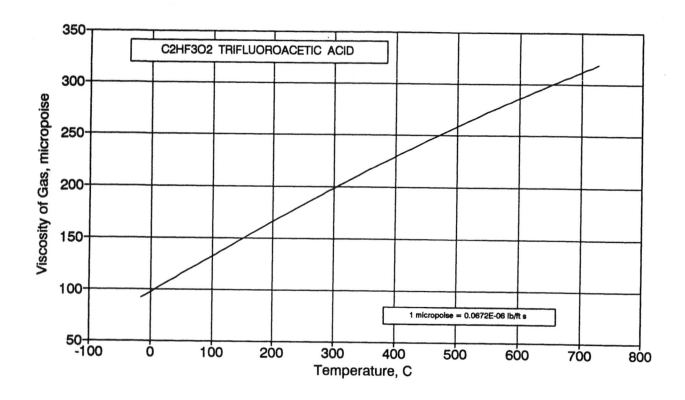

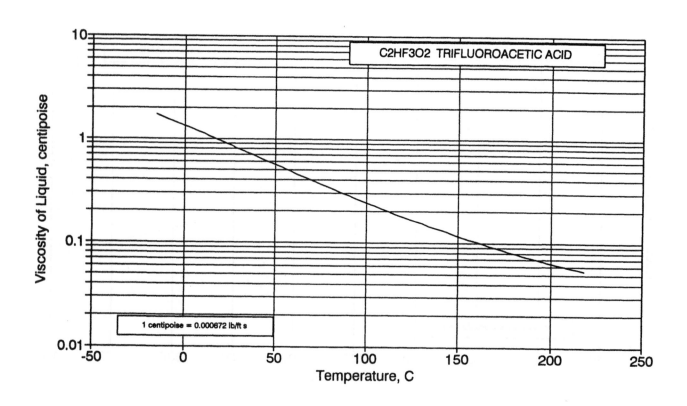

64

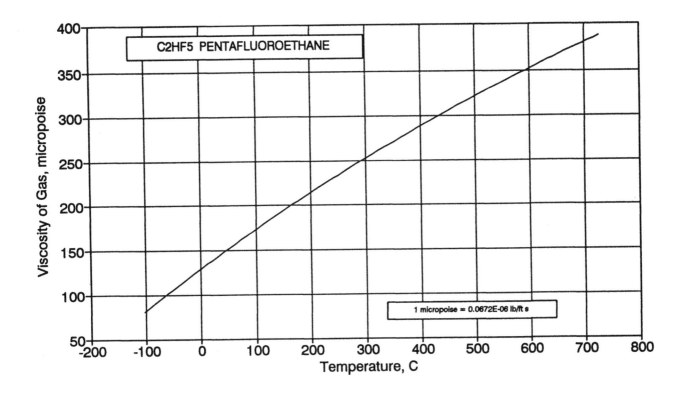

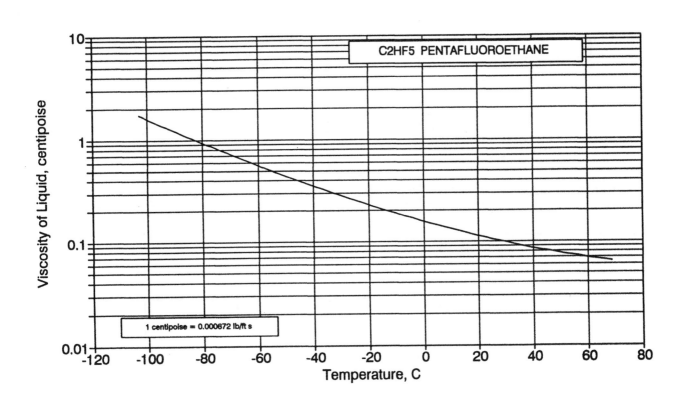

65

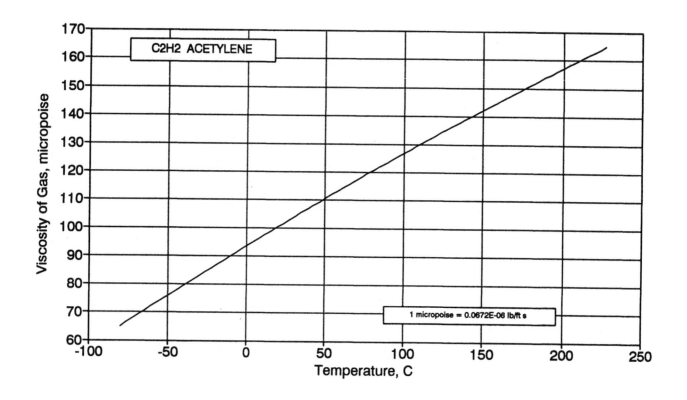

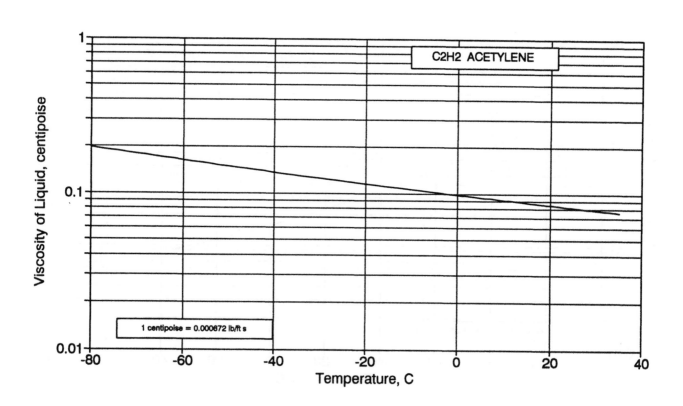

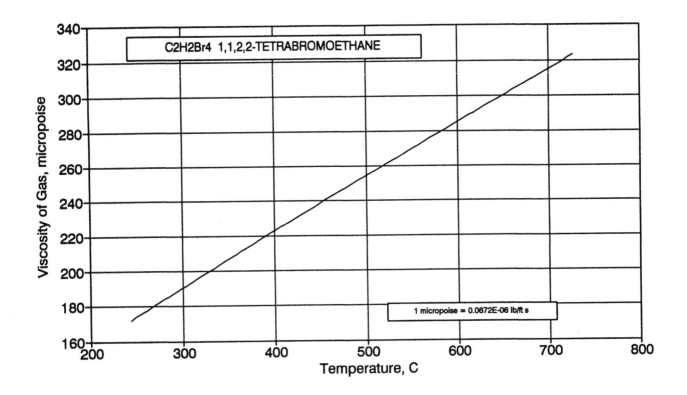

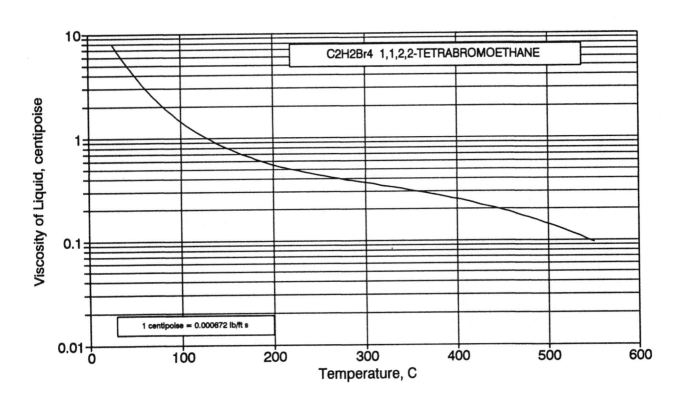

67

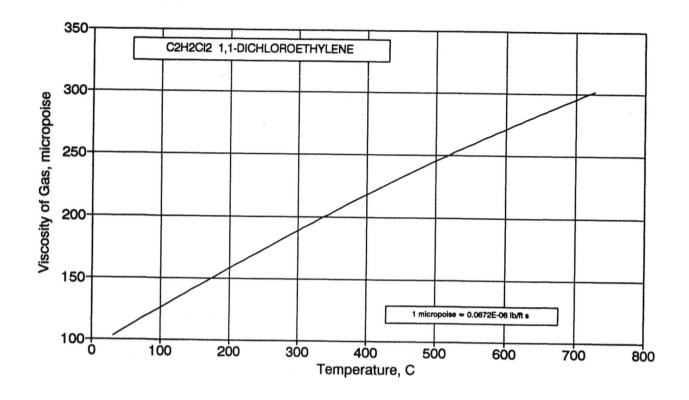

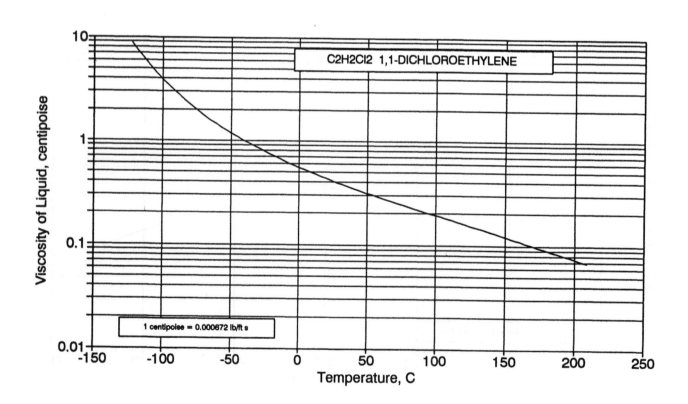

68

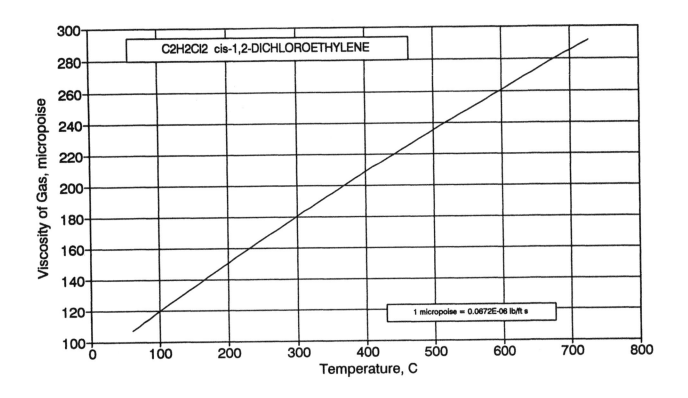

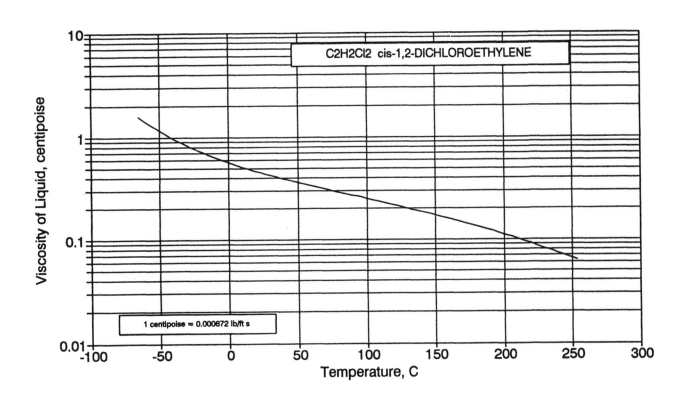

69

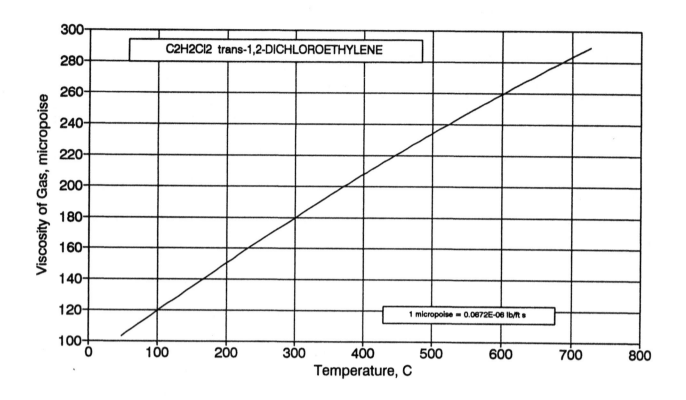

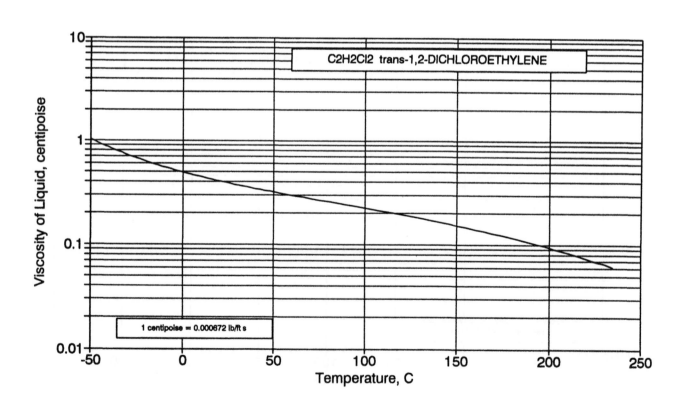

70

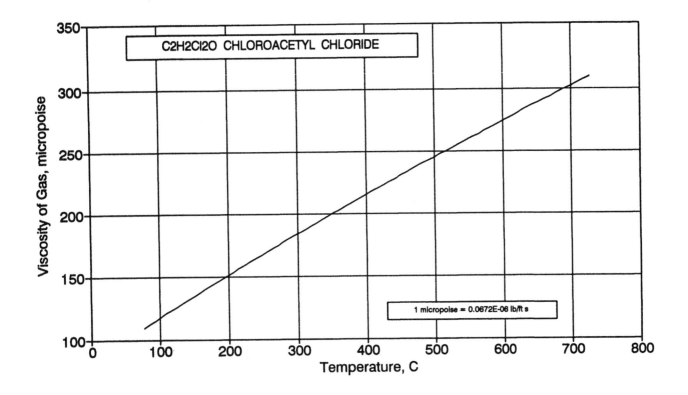

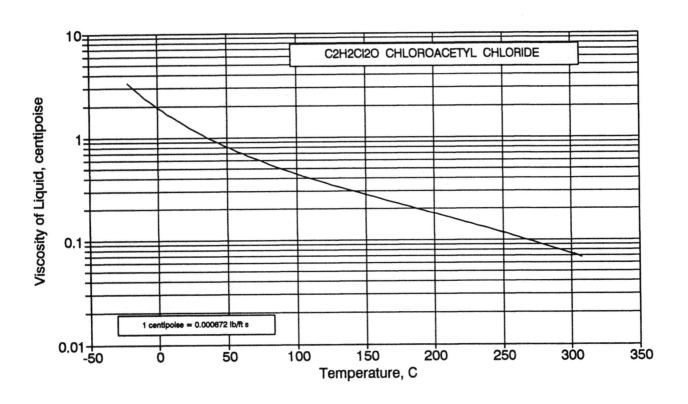

71

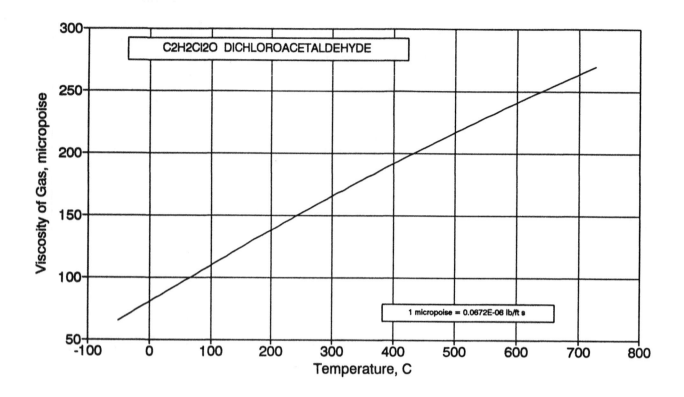

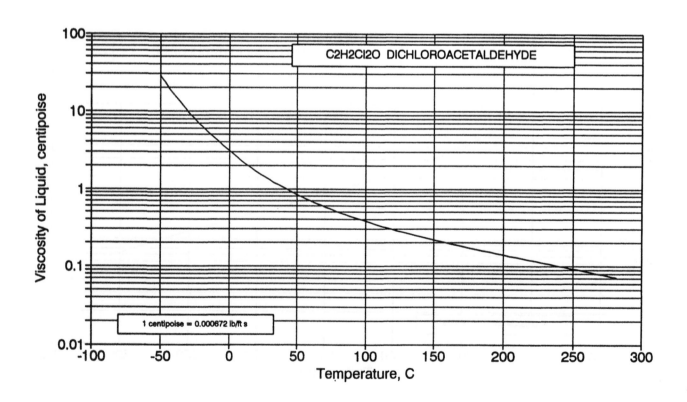

72

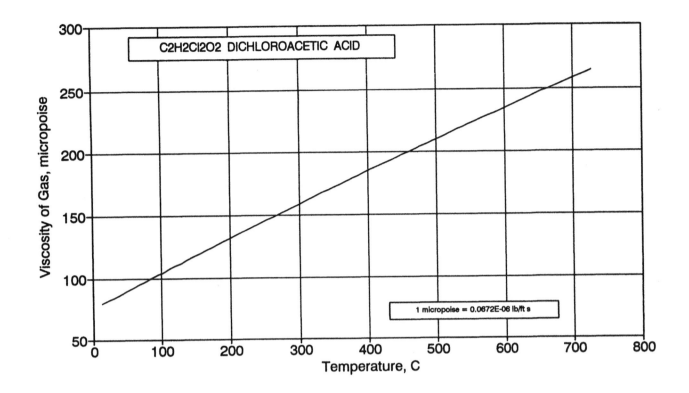

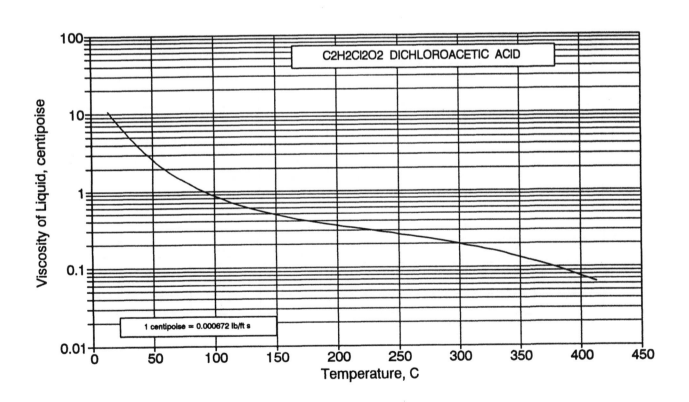

73

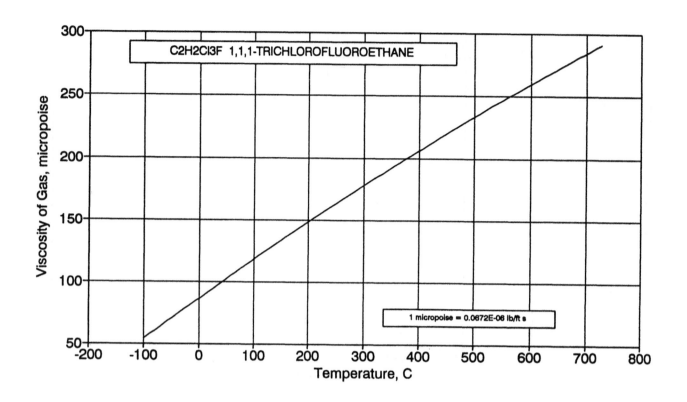

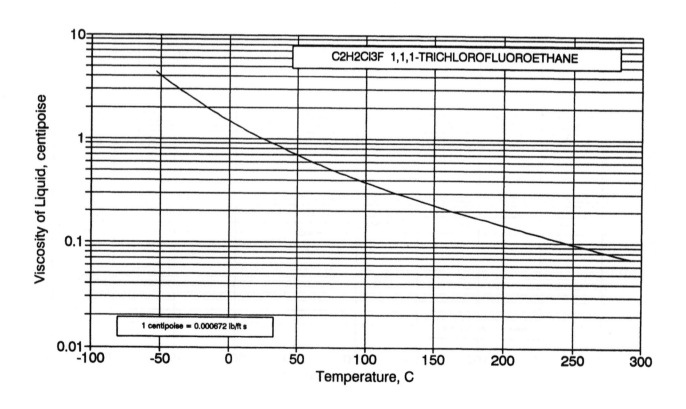

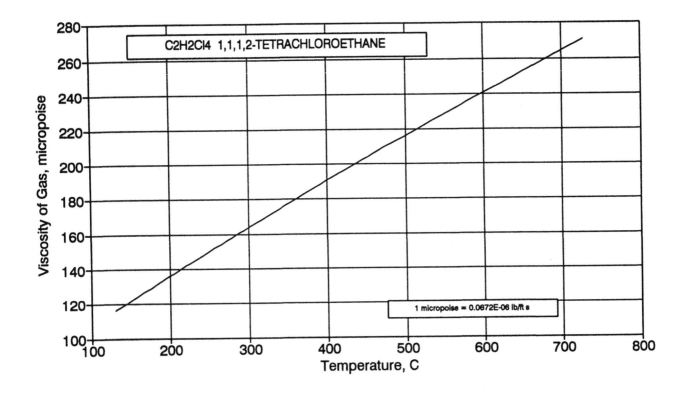

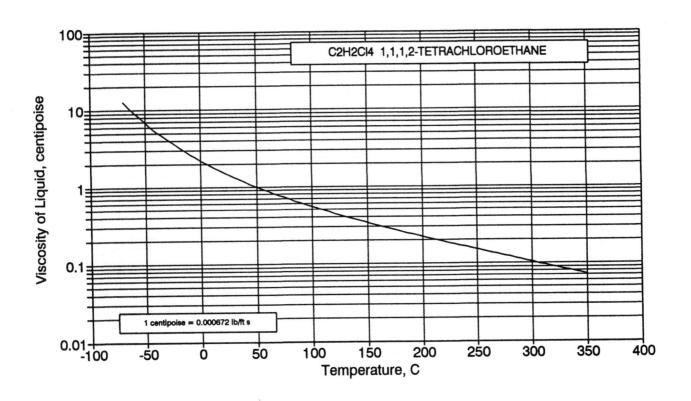

75

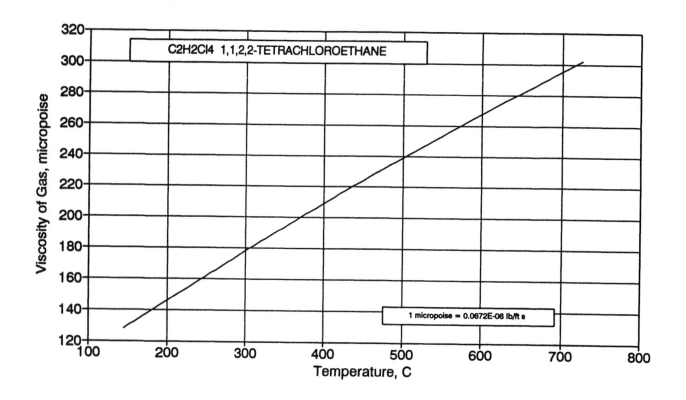

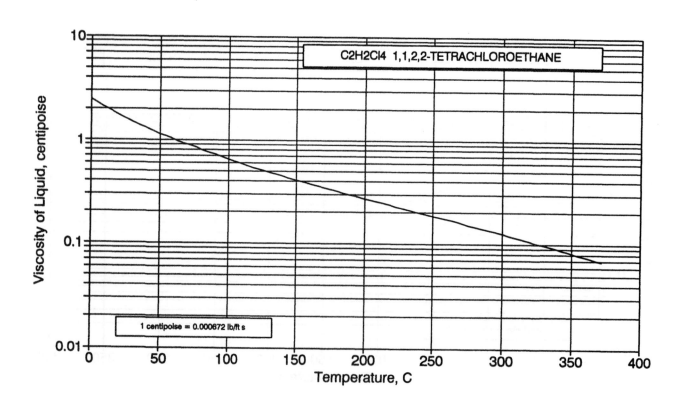

76

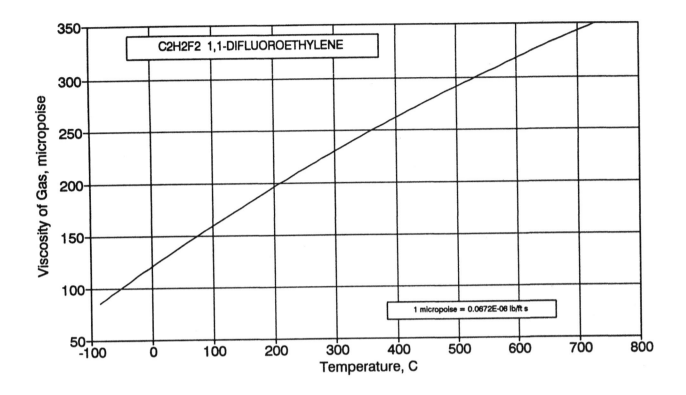

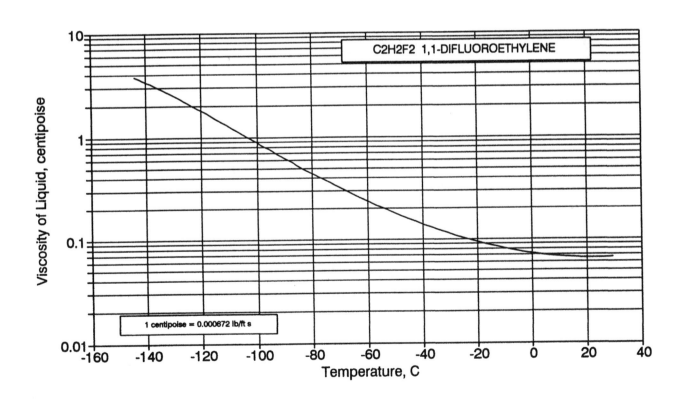

77

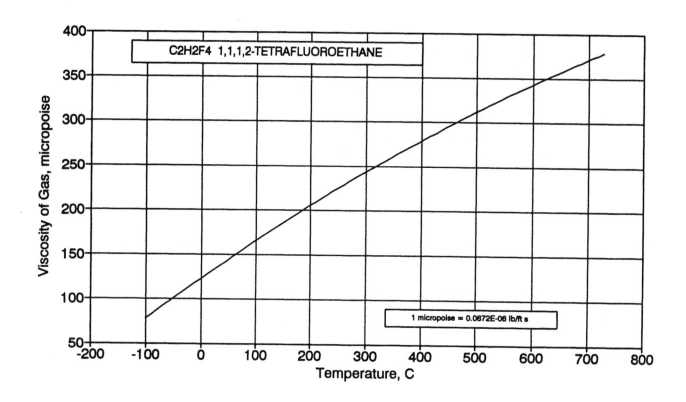

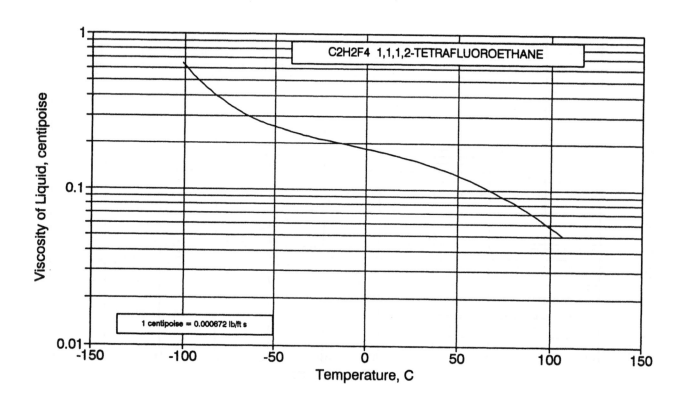

78

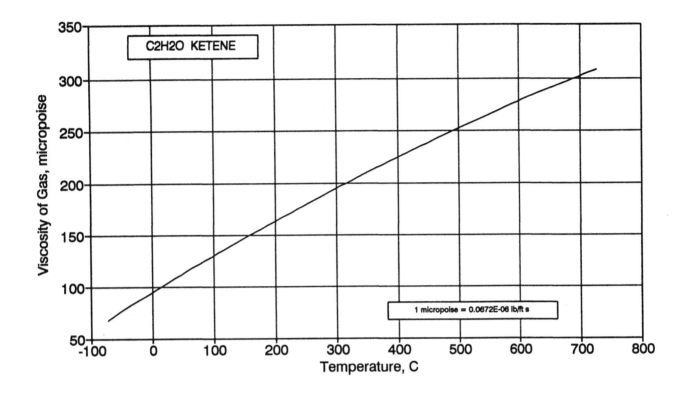

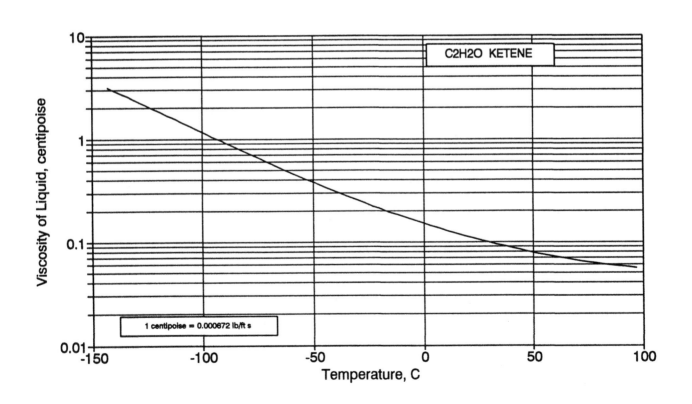

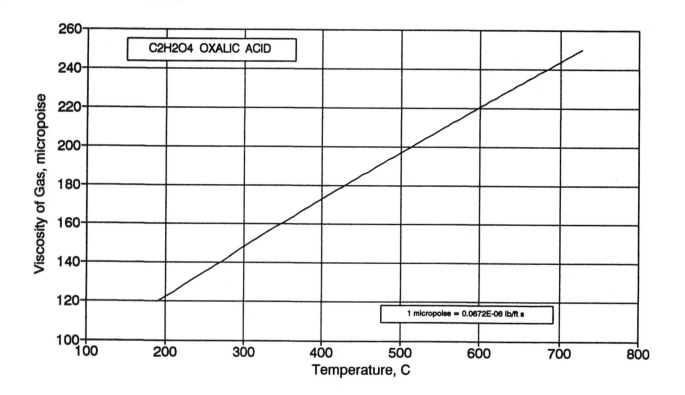

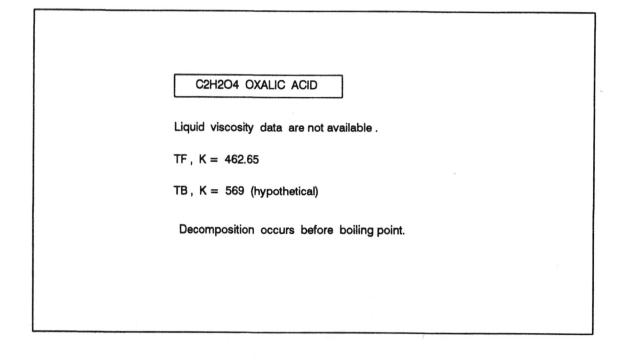

C2H2O4 OXALIC ACID

Liquid viscosity data are not available .

TF , K = 462.65

TB , K = 569 (hypothetical)

Decomposition occurs before boiling point.

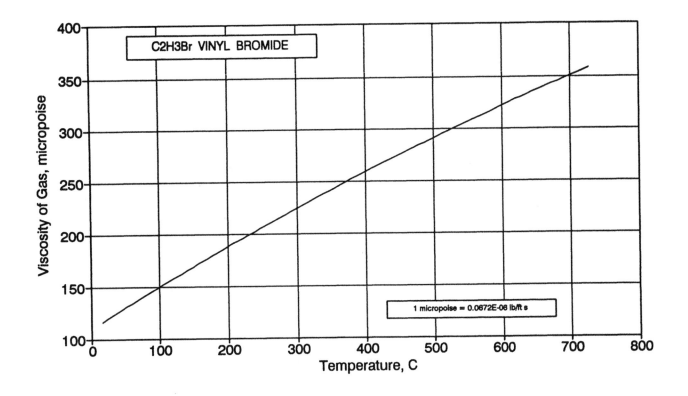

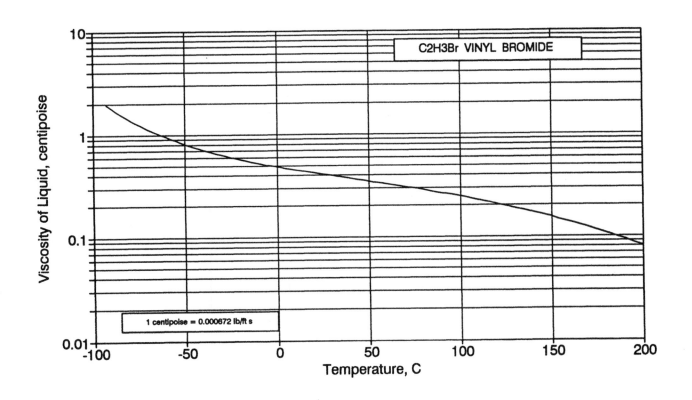

81

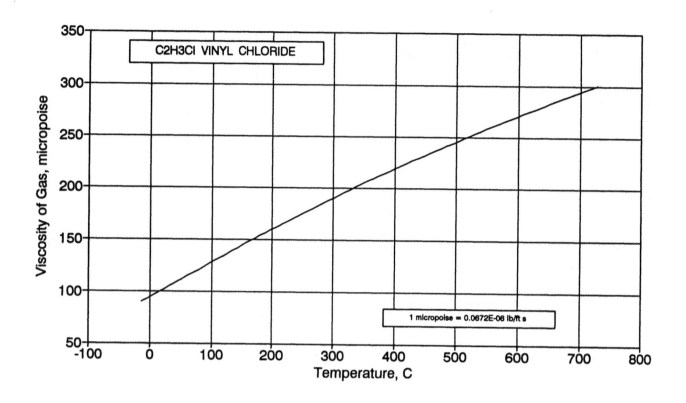

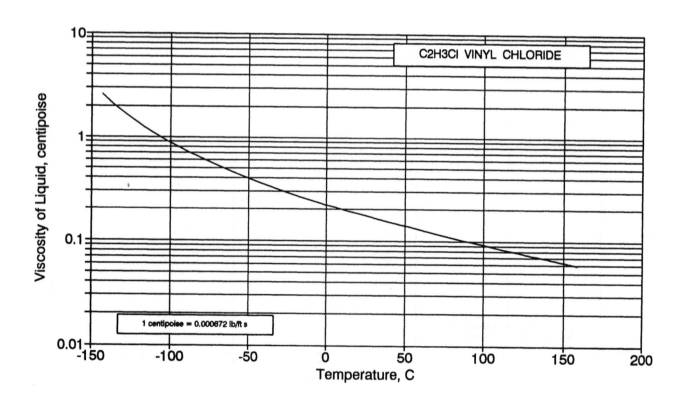

82

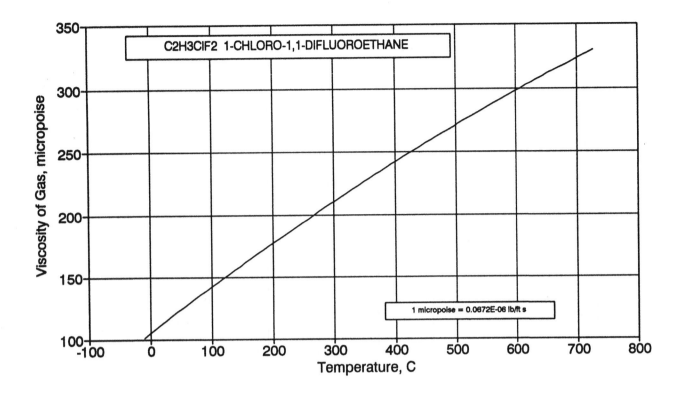

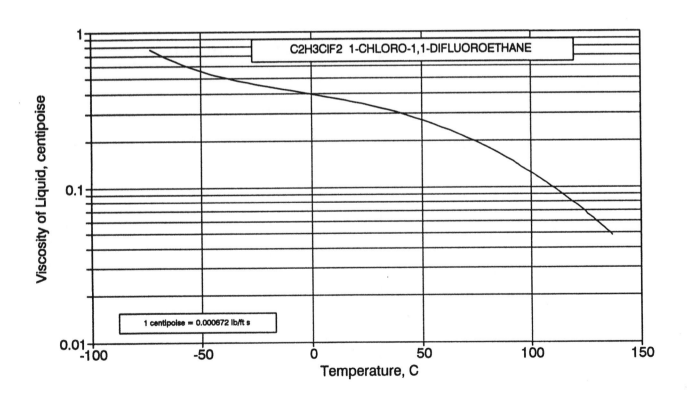

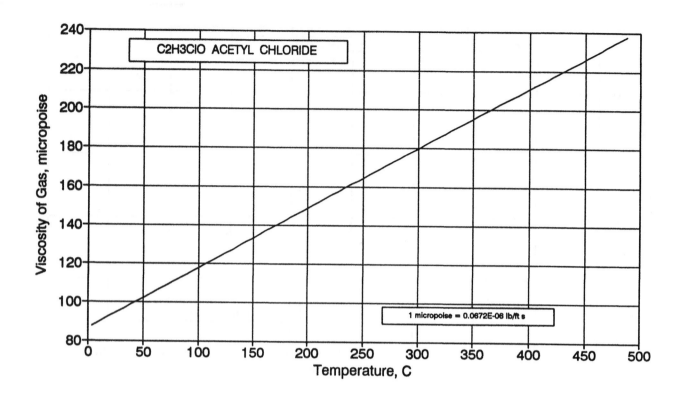

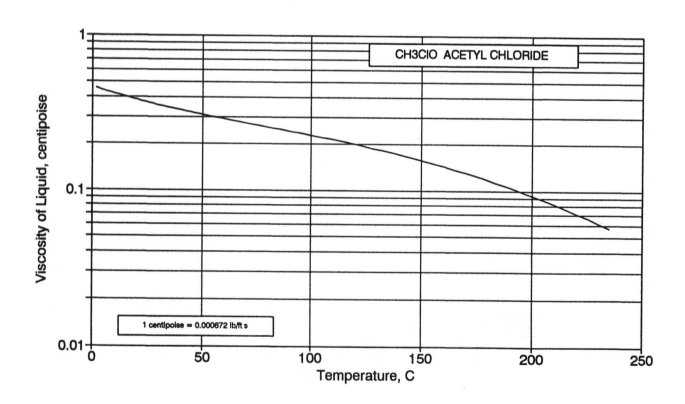

84

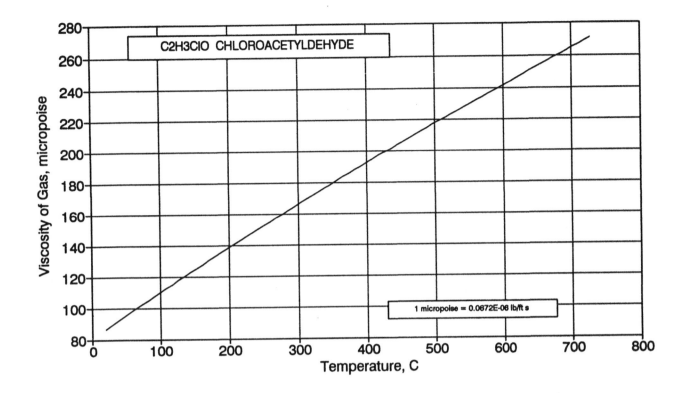

C2H3ClO CHLOROACETYLDEHYDE

Viscosity of Gas, micropoise

1 micropoise = 0.0672E-06 lb/ft s

Temperature, C

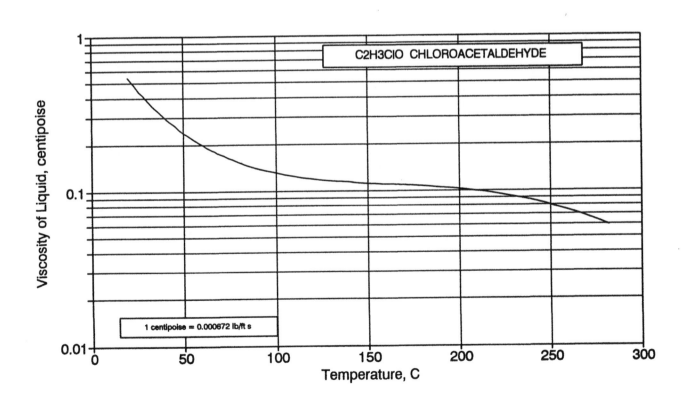

C2H3ClO CHLOROACETALDEHYDE

Viscosity of Liquid, centipoise

1 centipoise = 0.000872 lb/ft s

Temperature, C

85

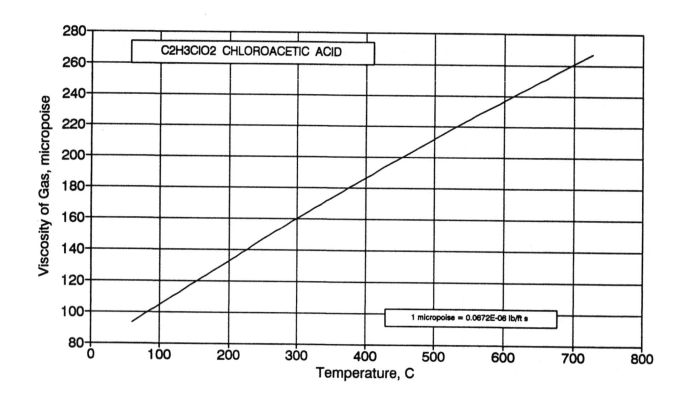

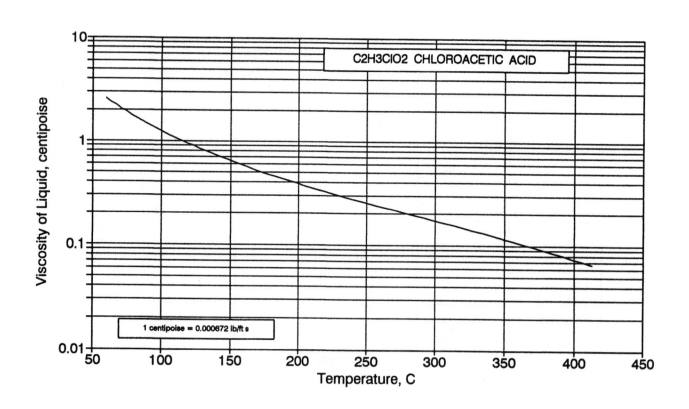

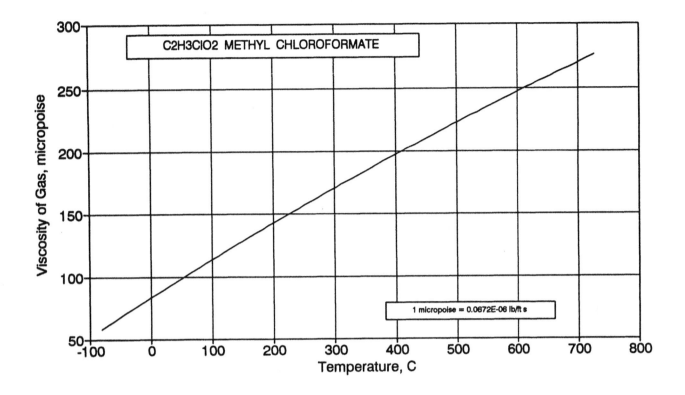

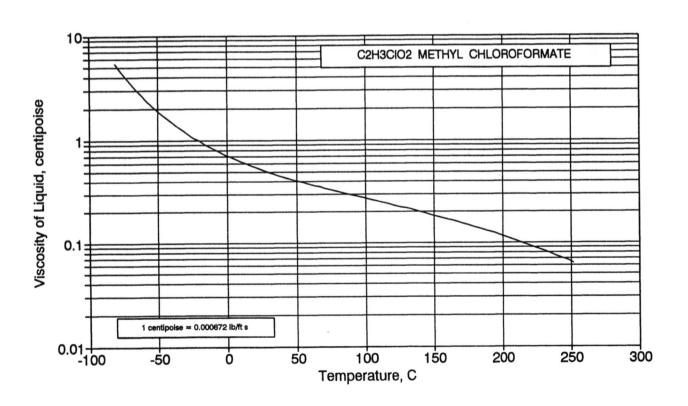

87

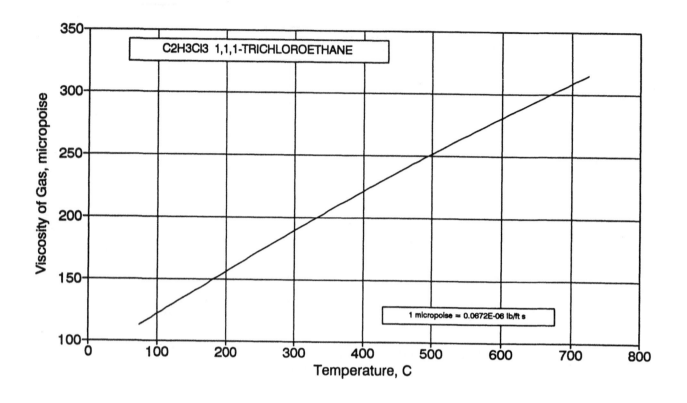

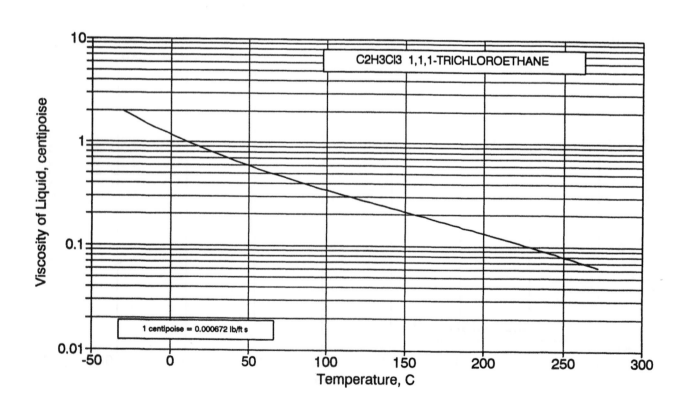

88

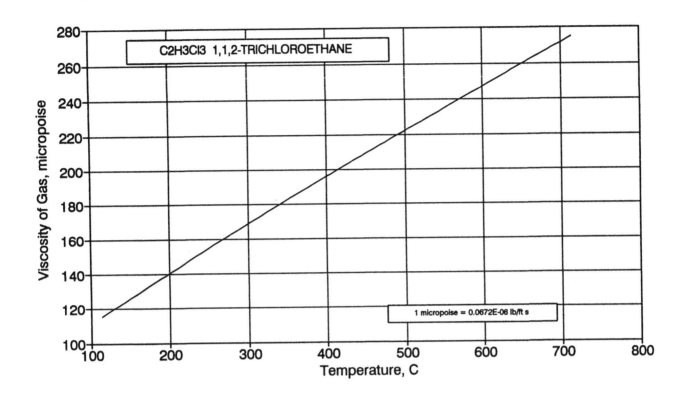

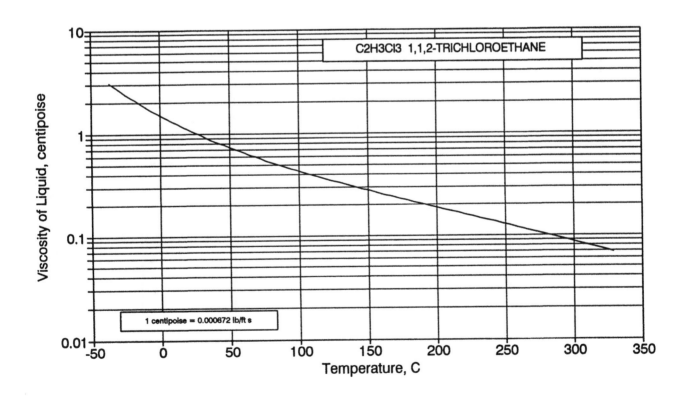

89

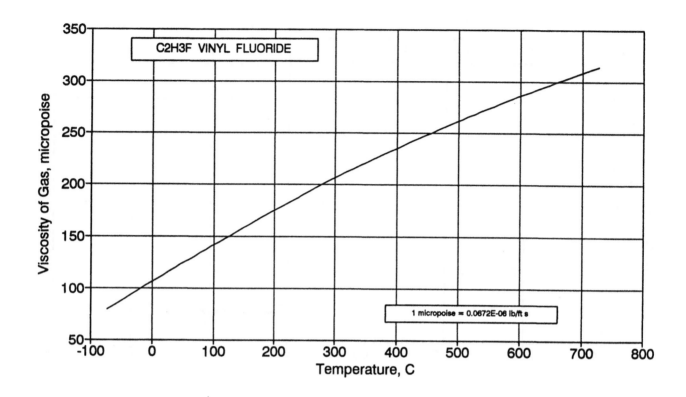

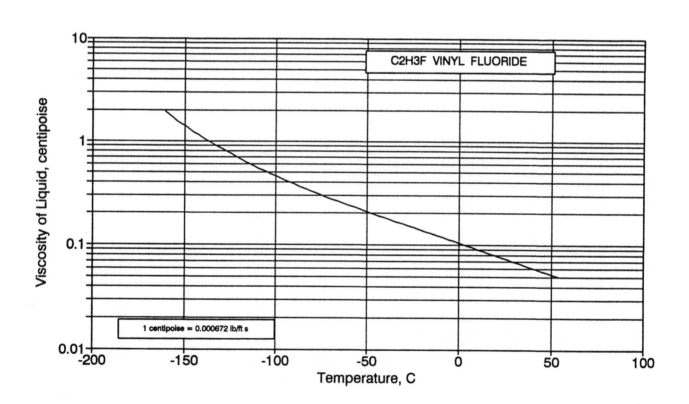

90

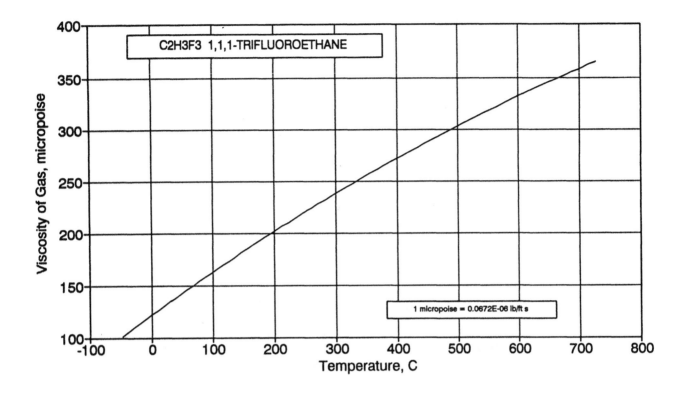

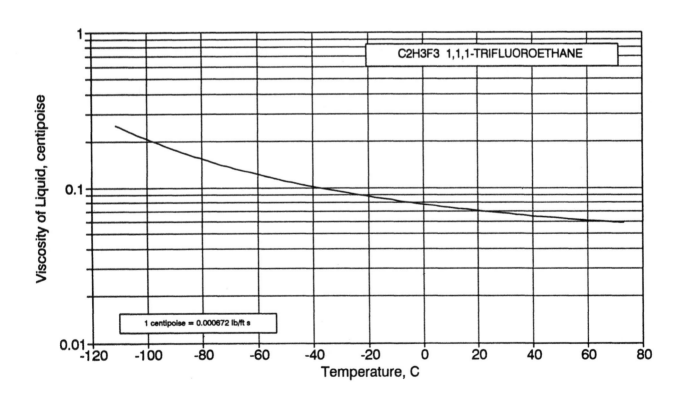

91

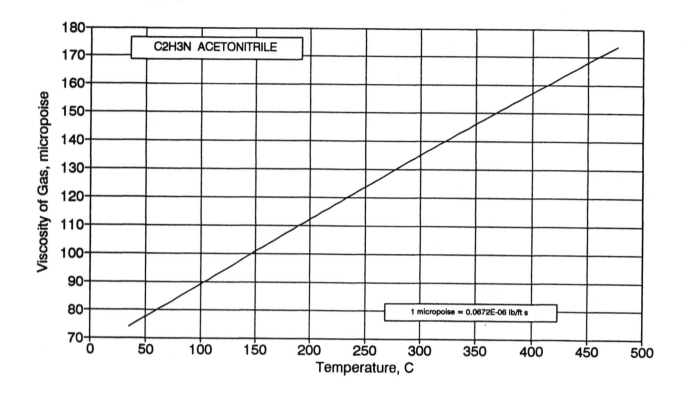

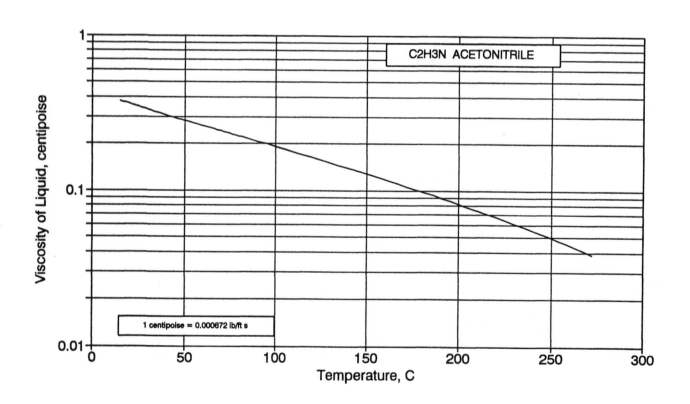

92

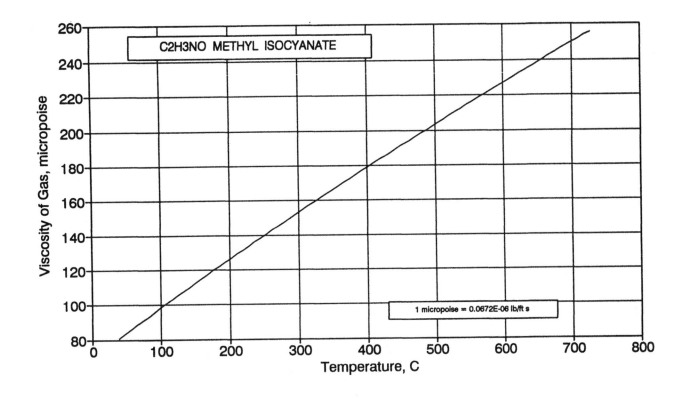

C2H3NO METHYL ISOCYANATE

1 micropoise = 0.0672E-06 lb/ft s

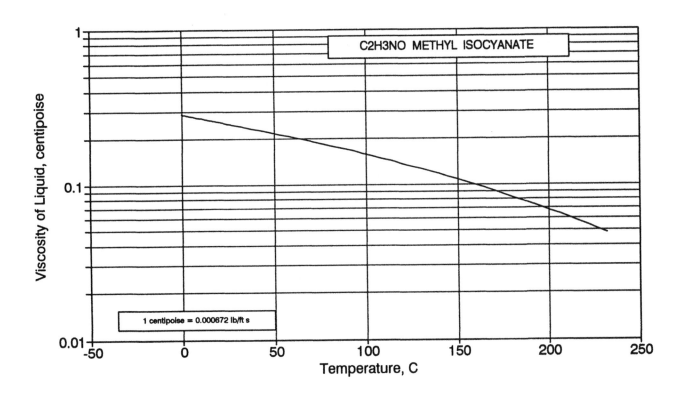

C2H3NO METHYL ISOCYANATE

1 centipoise = 0.000672 lb/ft s

93

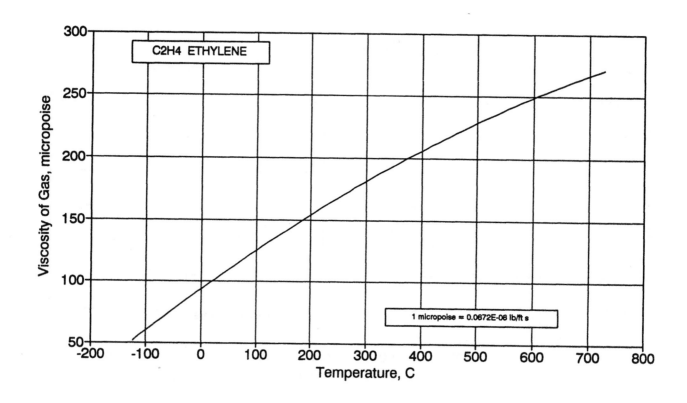

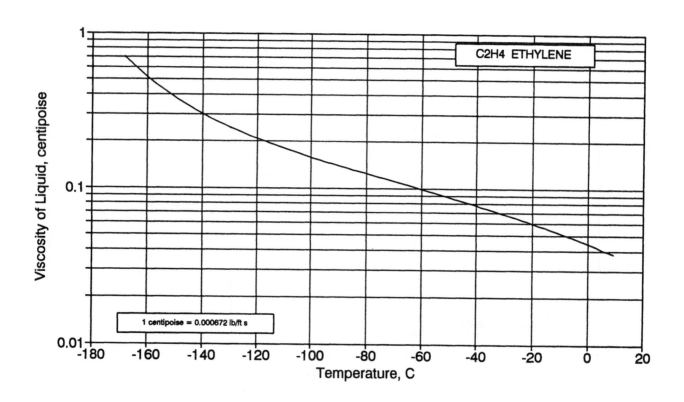

94

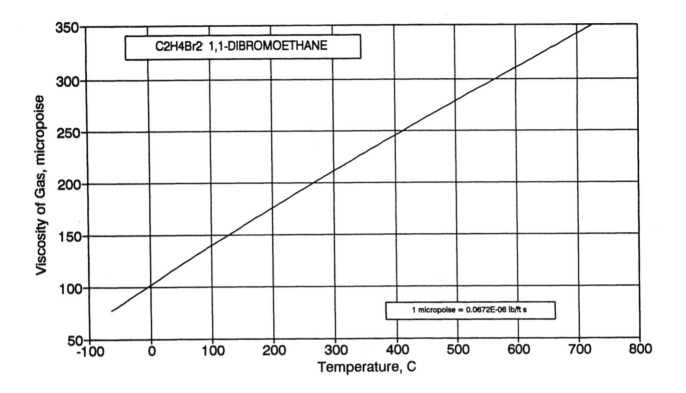

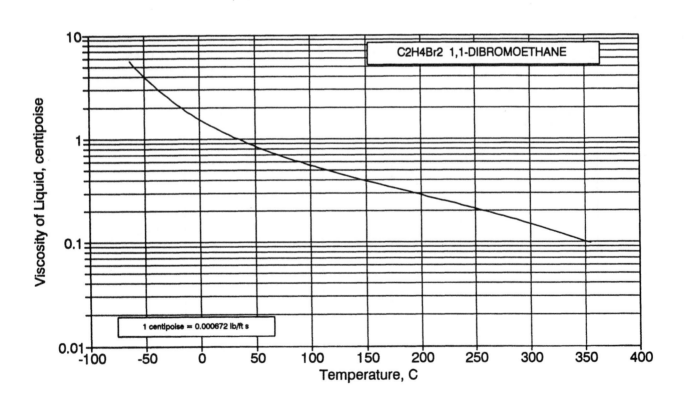

95

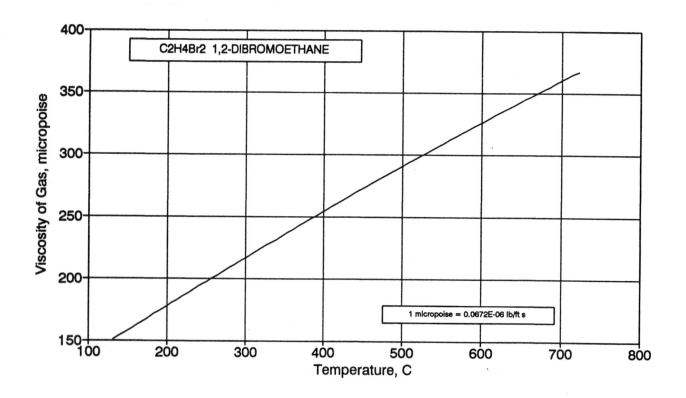

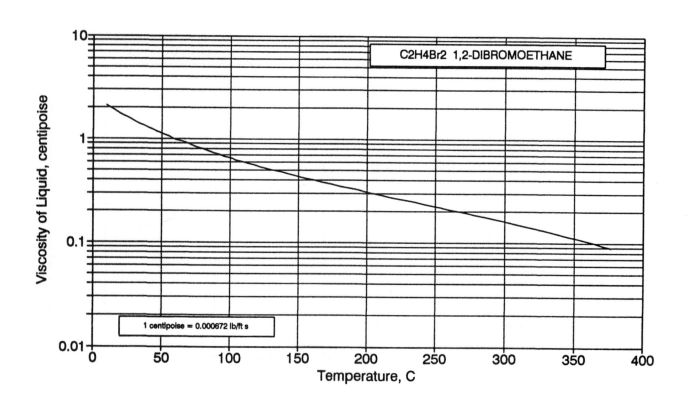

96

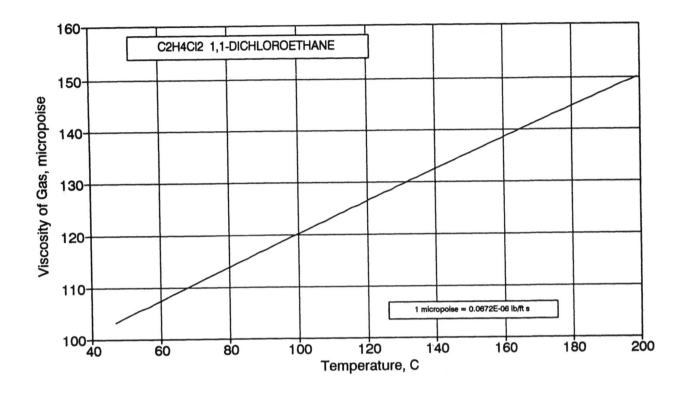

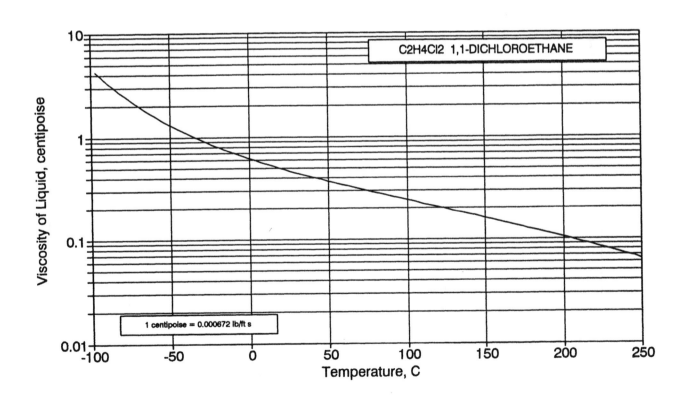

97

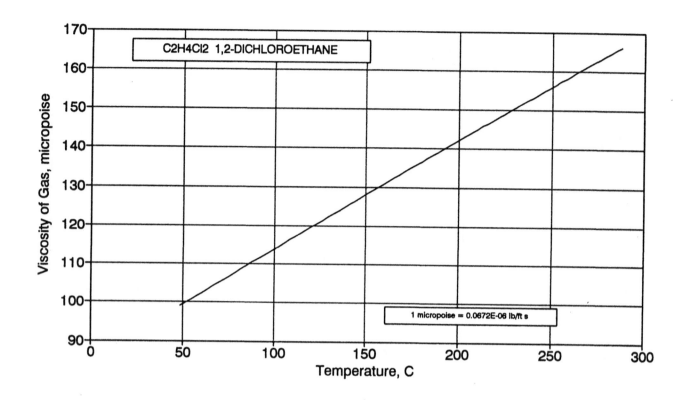

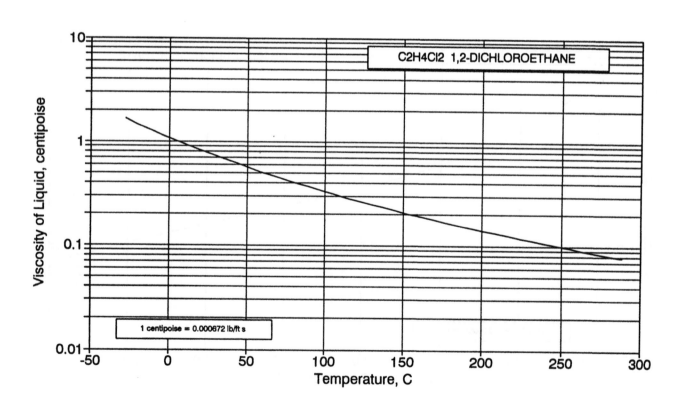

98

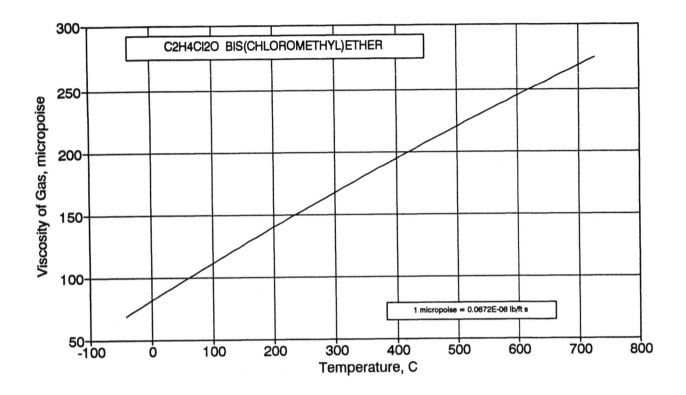

C2H4Cl2O BIS(CHLOROMETHYL)ETHER

1 micropoise = 0.0672E-06 lb/ft s

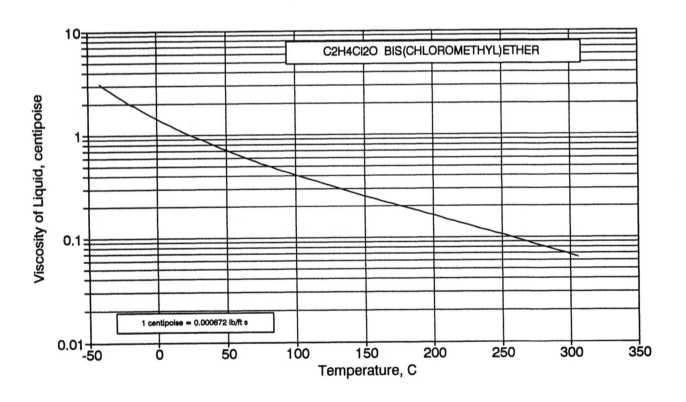

C2H4Cl2O BIS(CHLOROMETHYL)ETHER

1 centipoise = 0.000672 lb/ft s

99

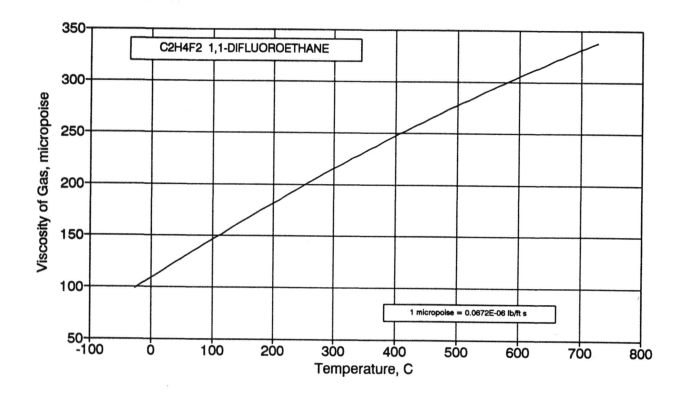

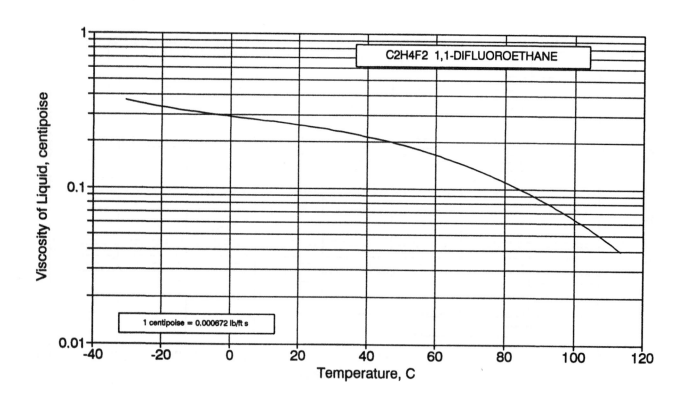

100

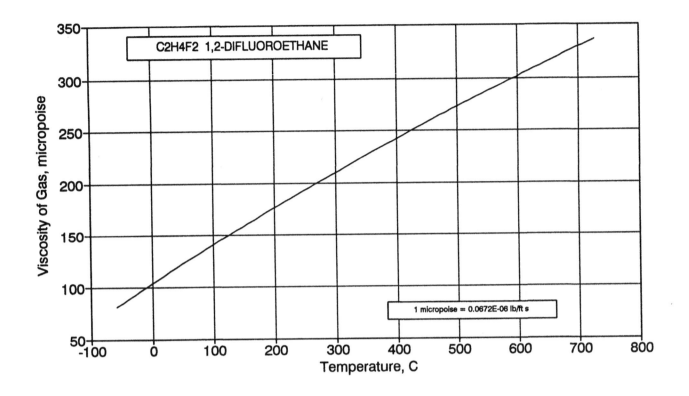

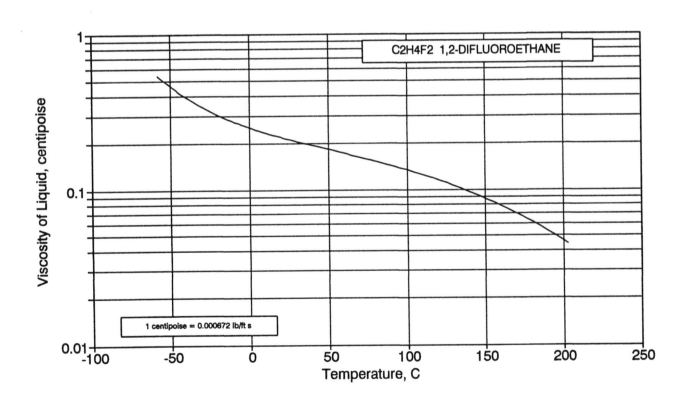

101

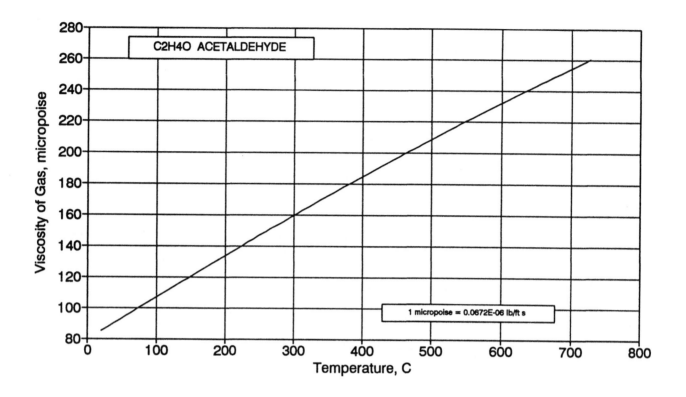

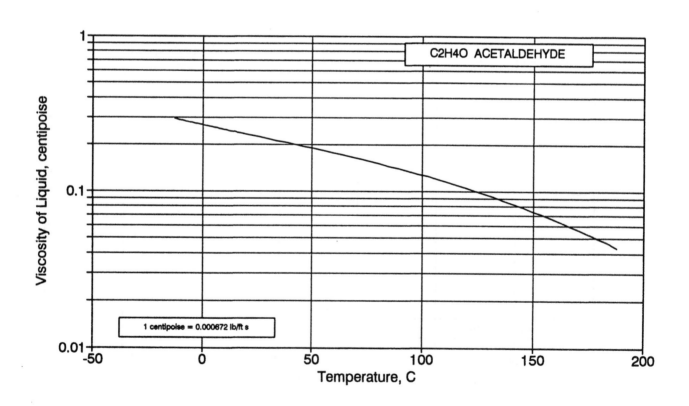

102

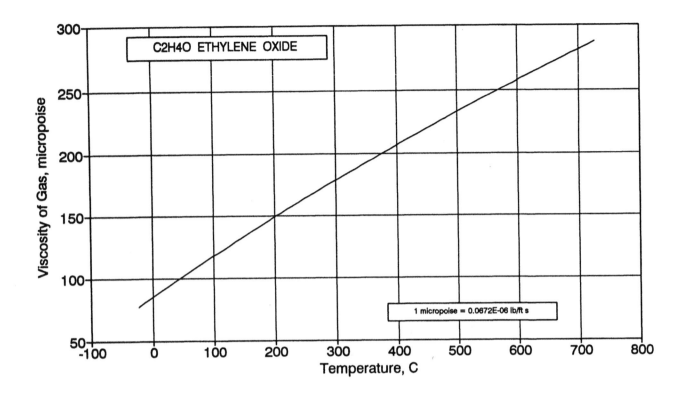

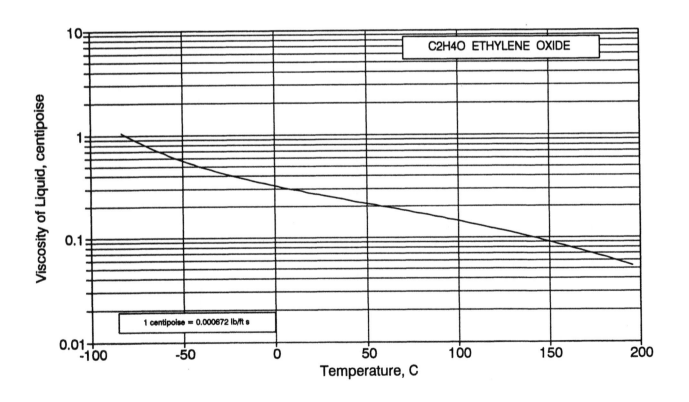

103

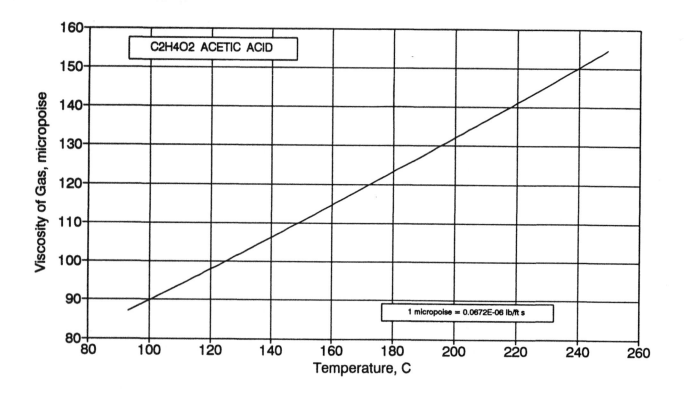

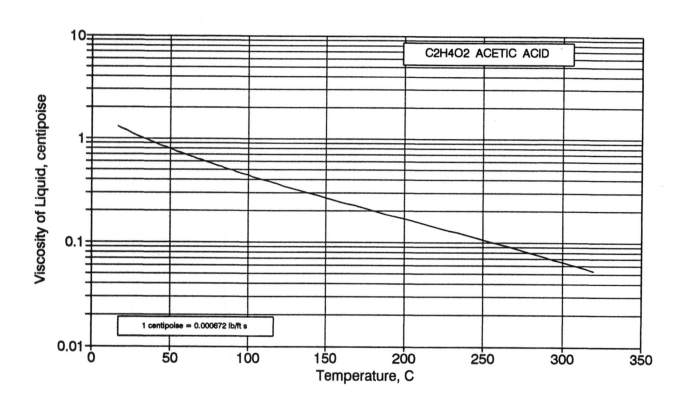

104

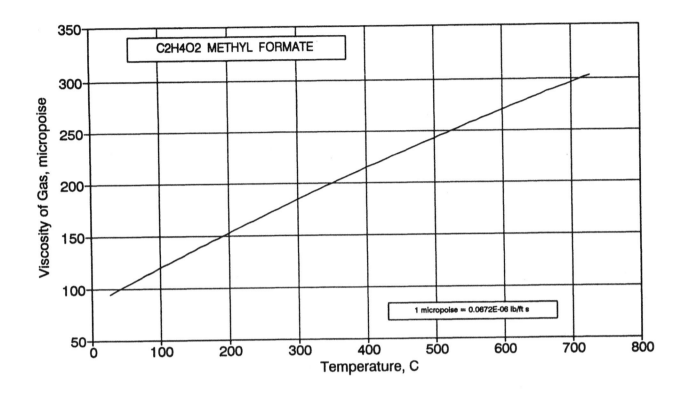

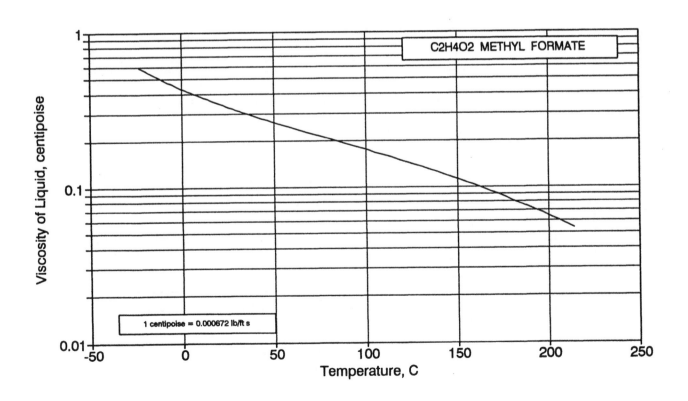

105

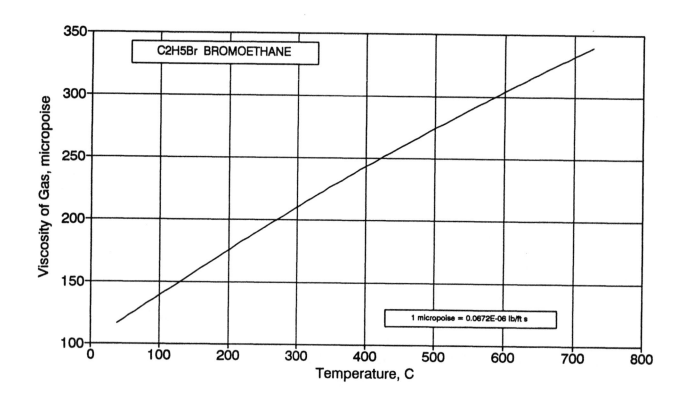

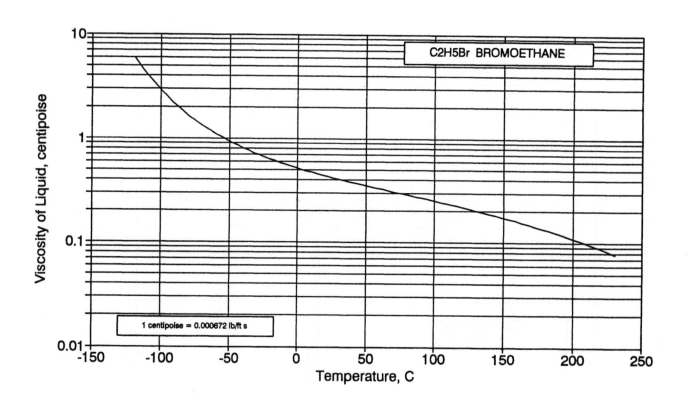

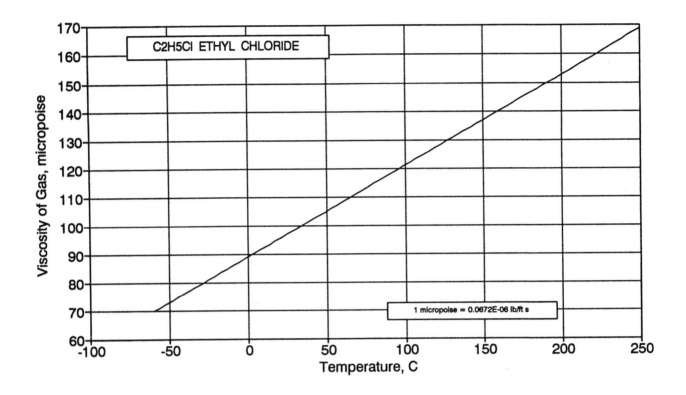

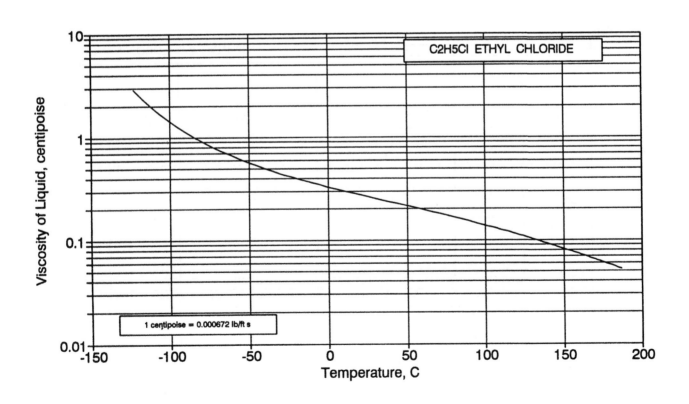

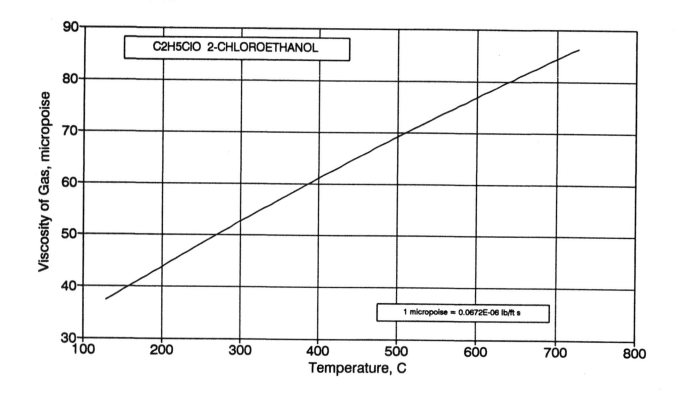

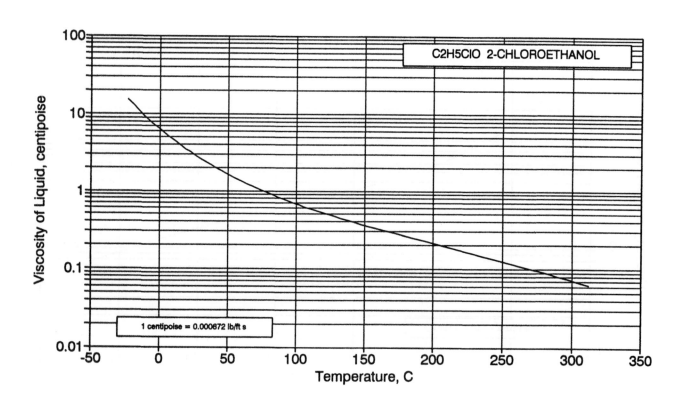

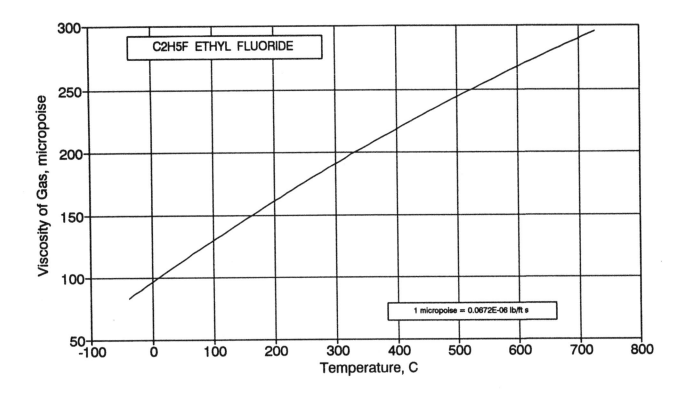

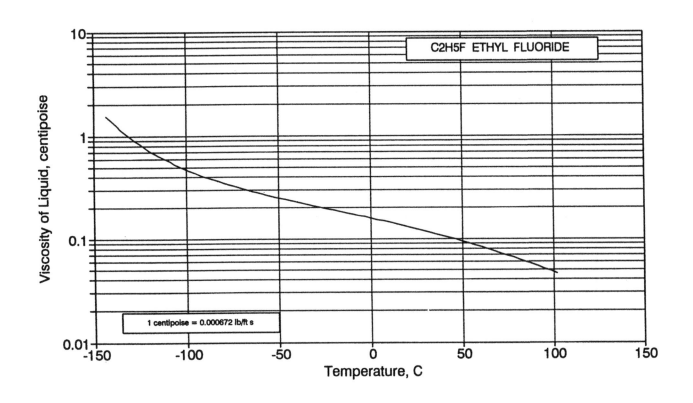

109

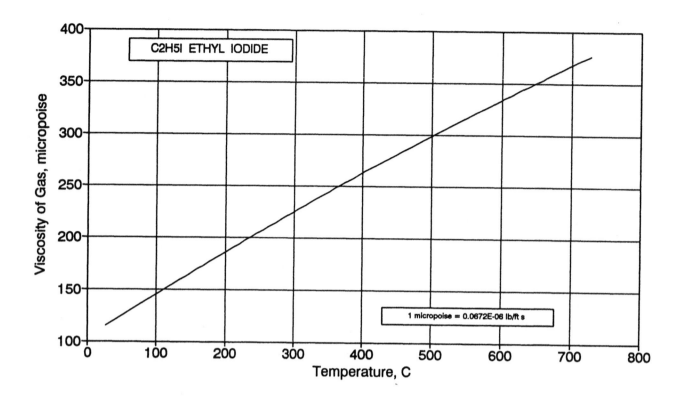

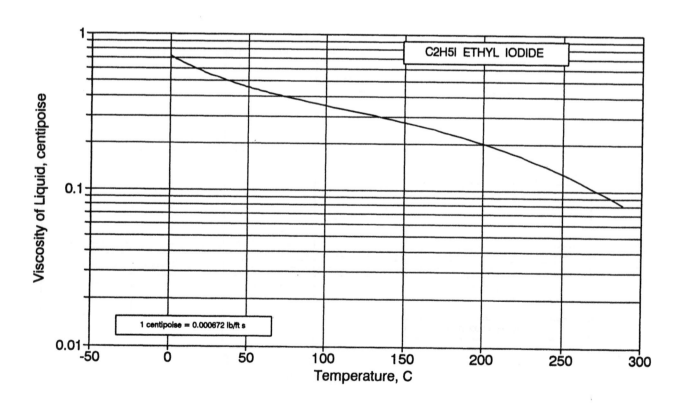

110

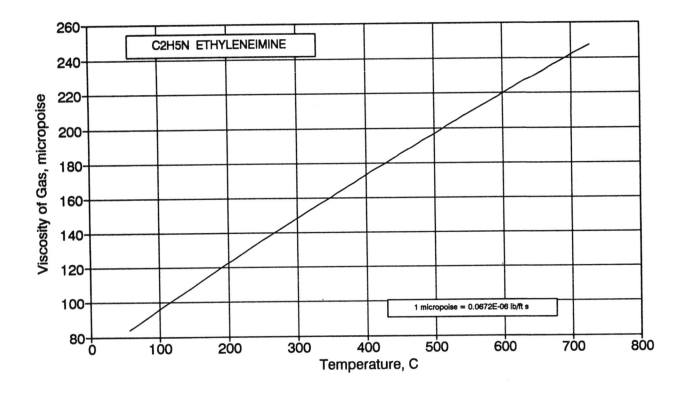

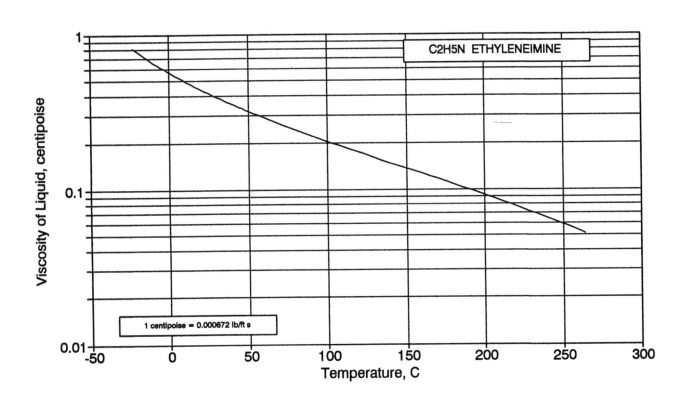

111

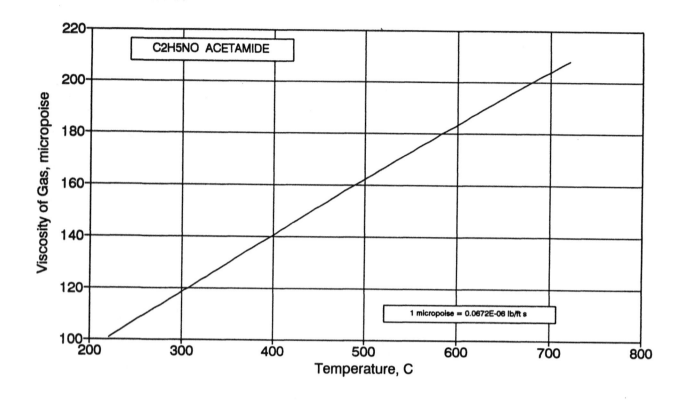

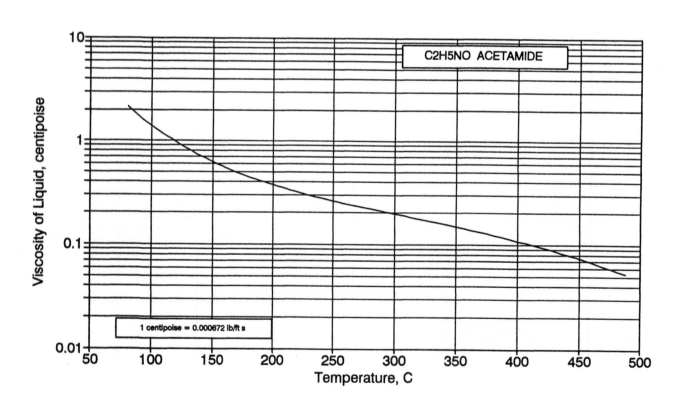

112

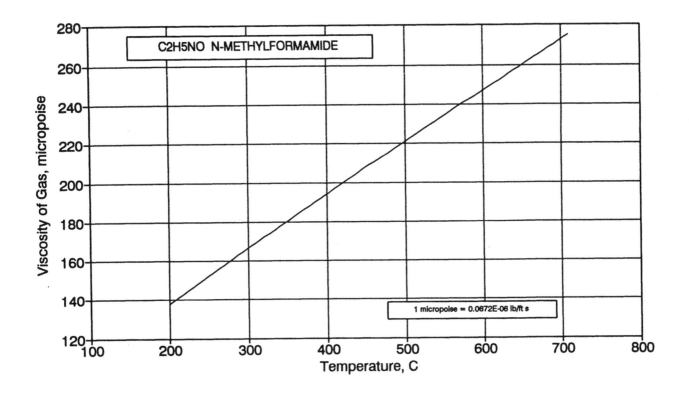

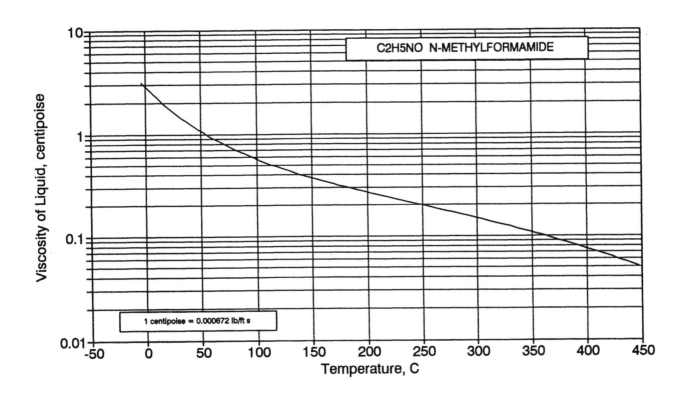

113

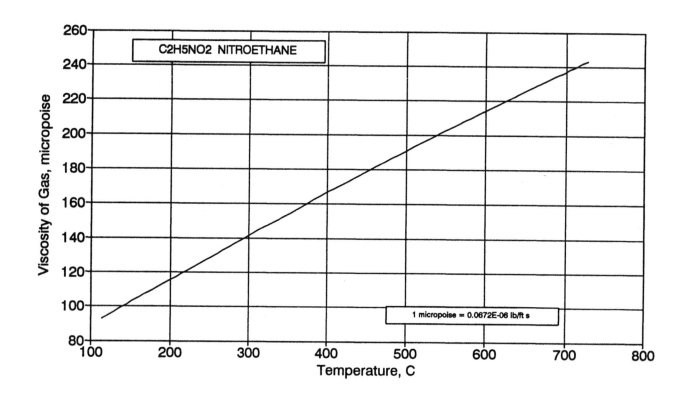

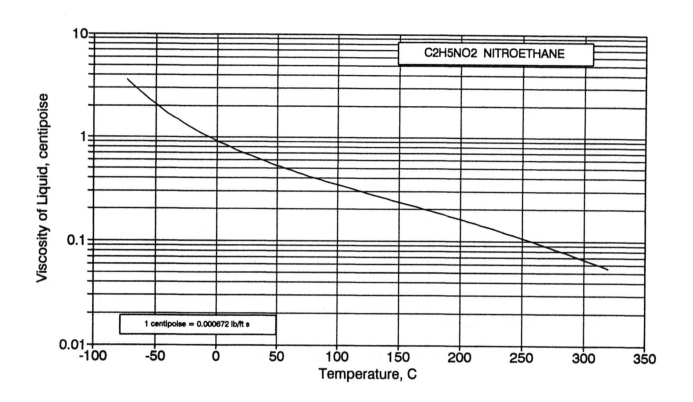

114

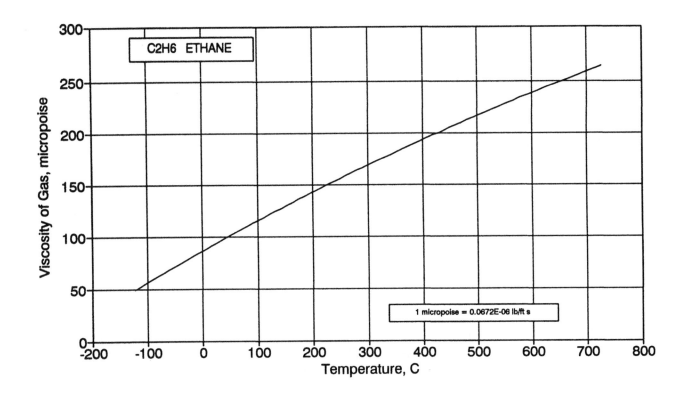

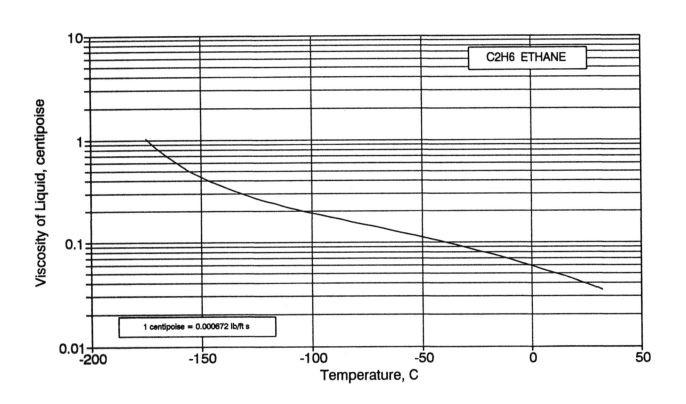

115

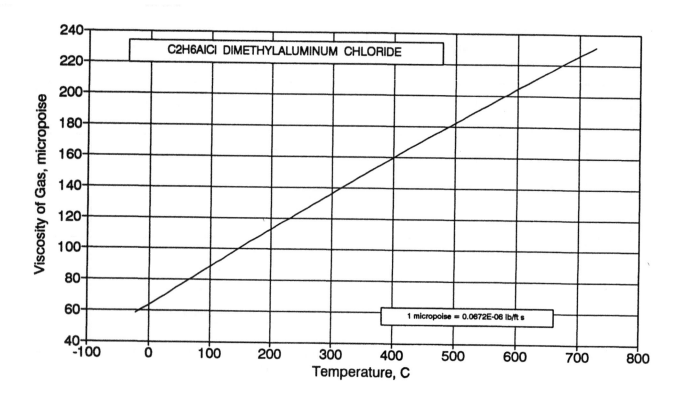

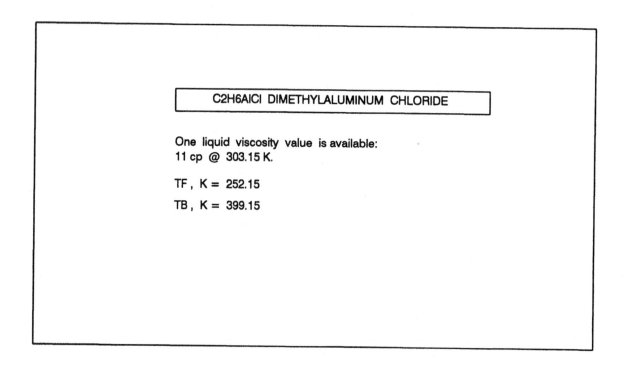

C2H6AlCl DIMETHYLALUMINUM CHLORIDE

One liquid viscosity value is available:
11 cp @ 303.15 K.

TF , K = 252.15

TB , K = 399.15

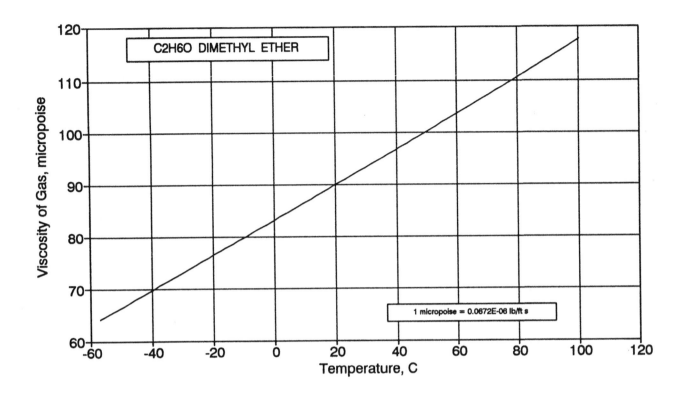

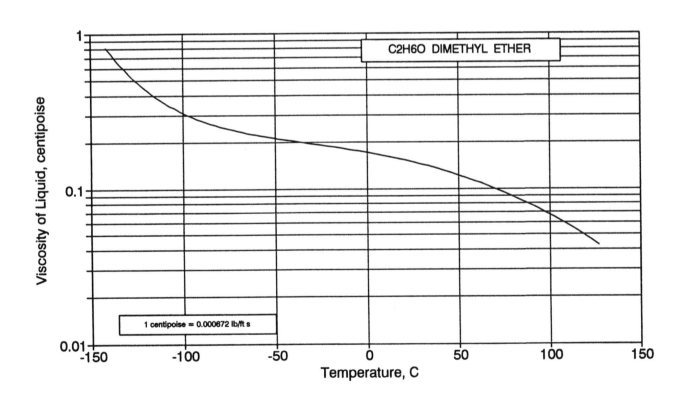

117

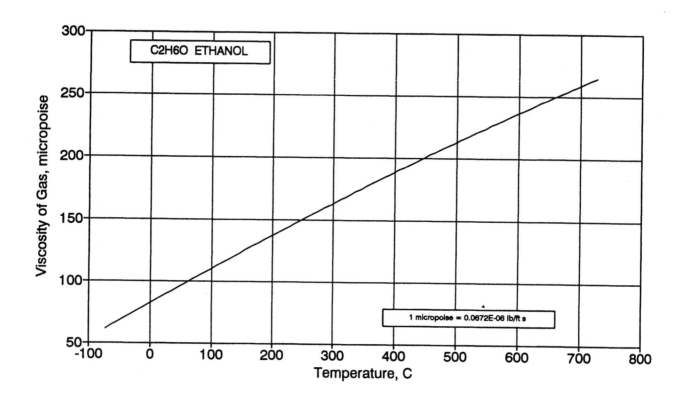

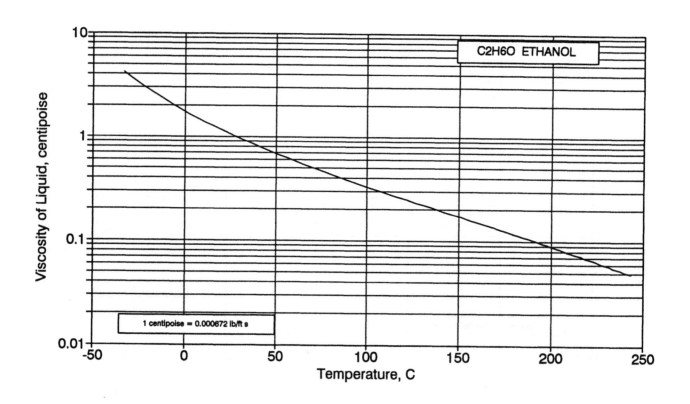

118

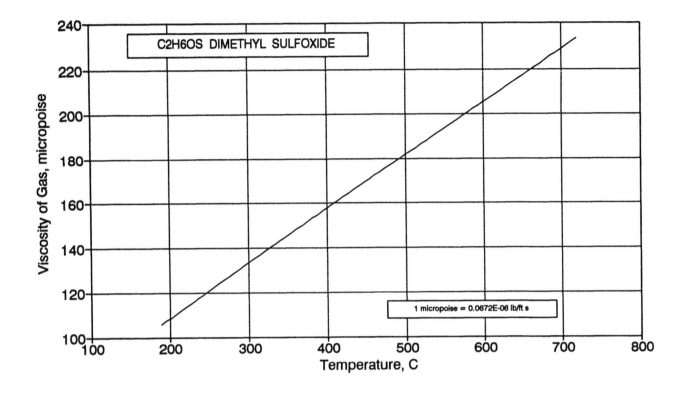

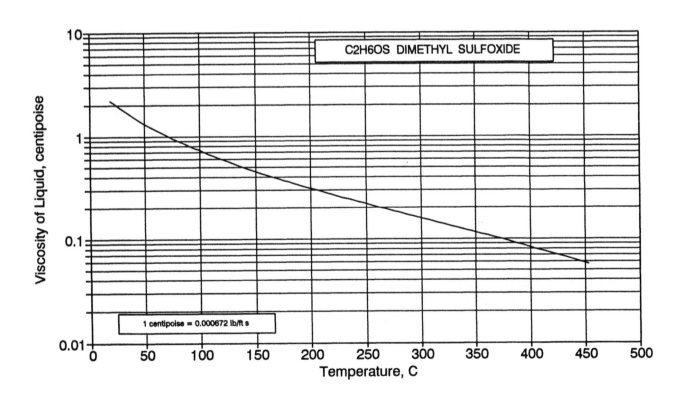

119

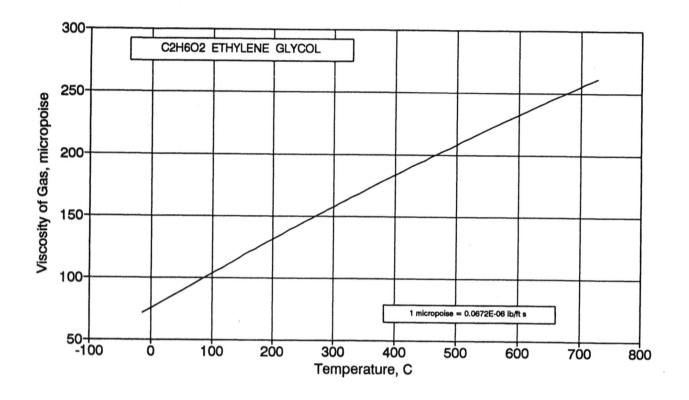

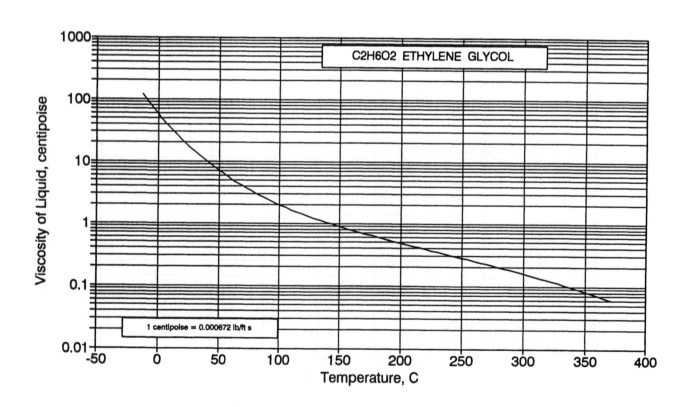

120

```
C2H6O4S  DIMETHYL SULFATE
```

Gas viscosity data are not available .

TF , K = 241.35

TB , K = 461.95

```
C2H6O4S  DIMETHYL SULFATE
```

Liquid viscosity data are not available.

TF , K = 241.35

TB , K = 461.95

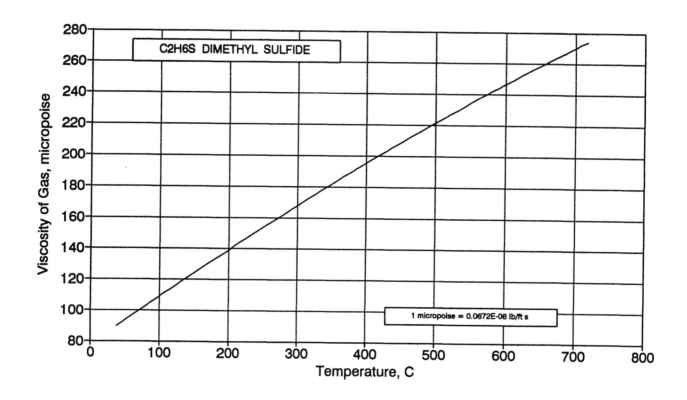

C2H6S DIMETHYL SULFIDE

1 micropoise = 0.0672E-06 lb/ft s

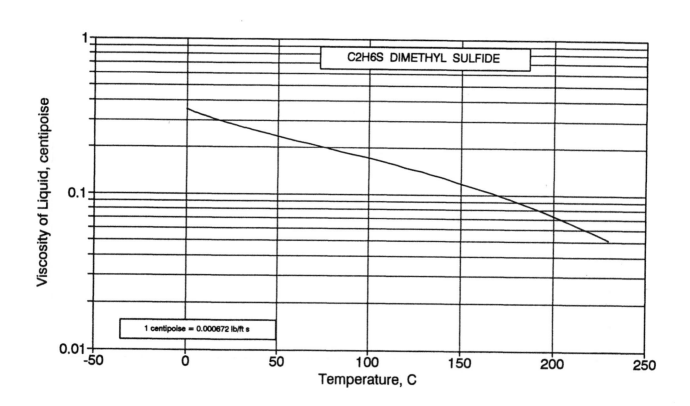

C2H6S DIMETHYL SULFIDE

1 centipoise = 0.000672 lb/ft s

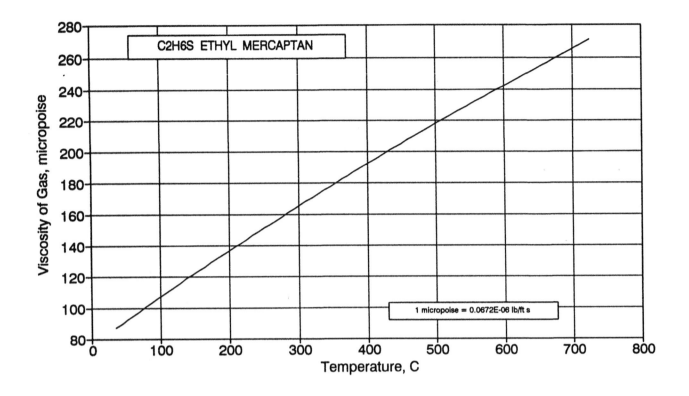

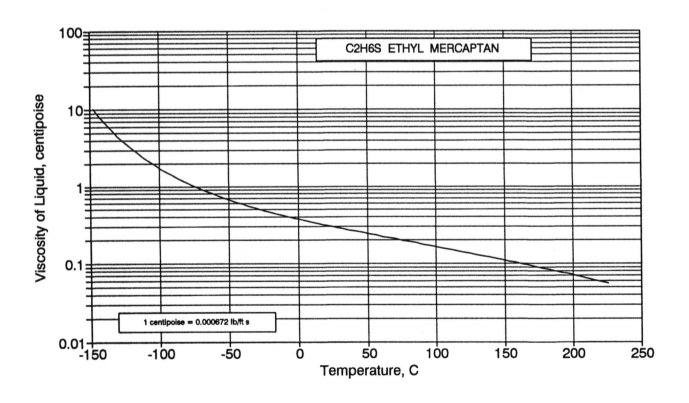

123

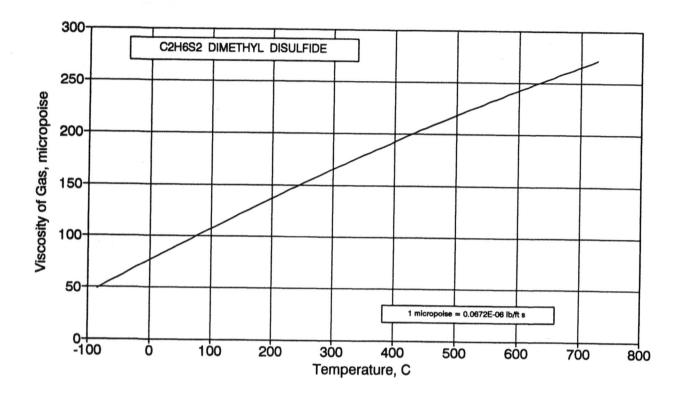

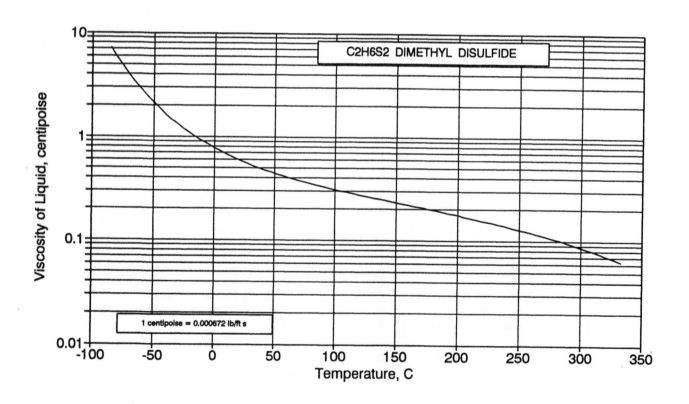

124

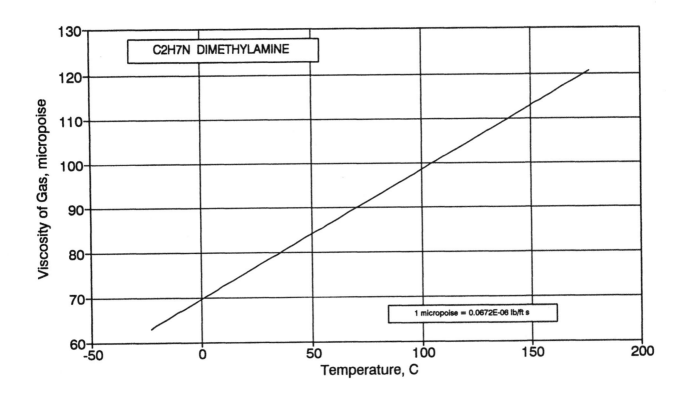

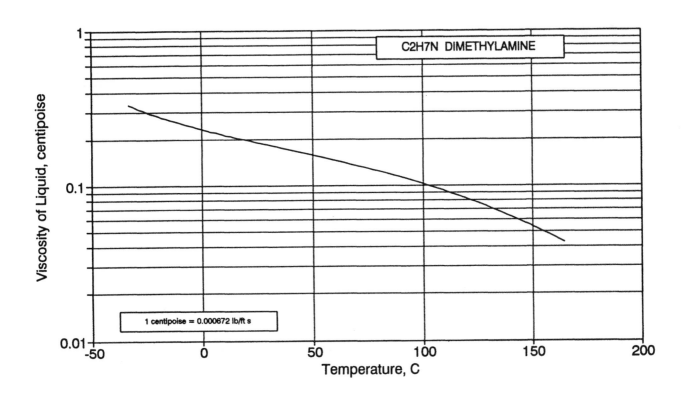

125

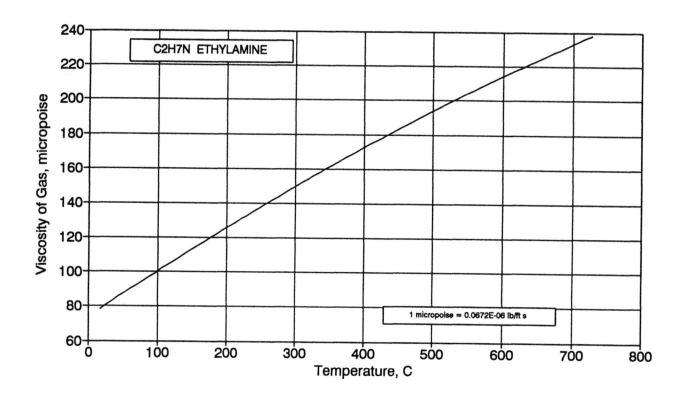

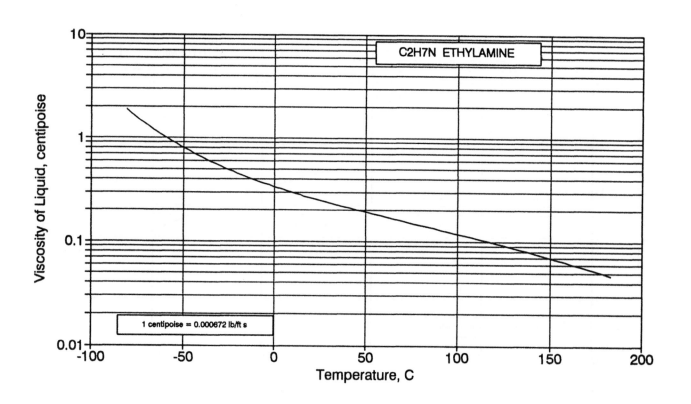

126

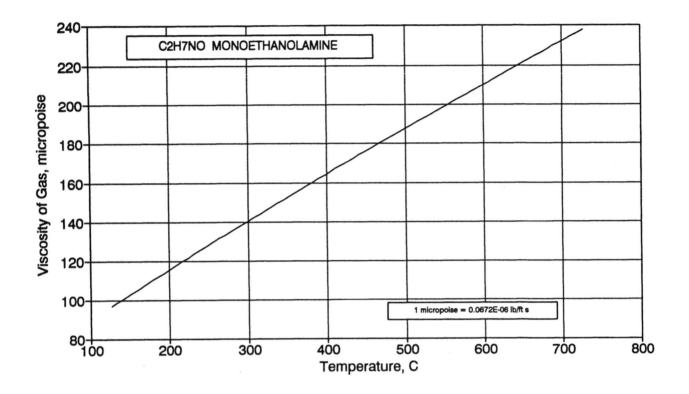

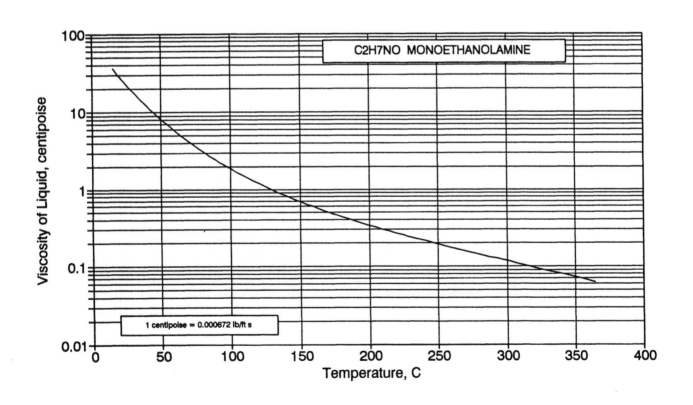

127

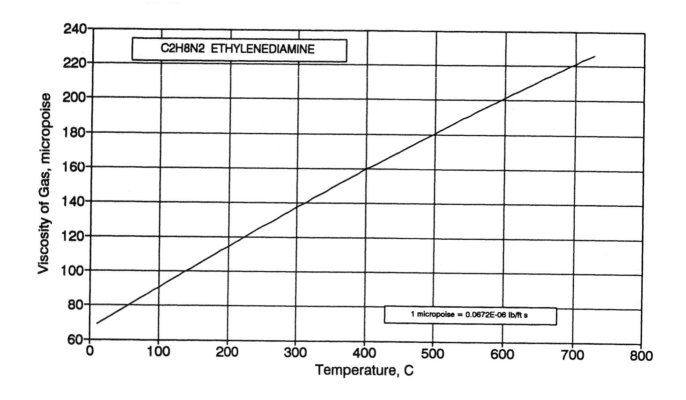

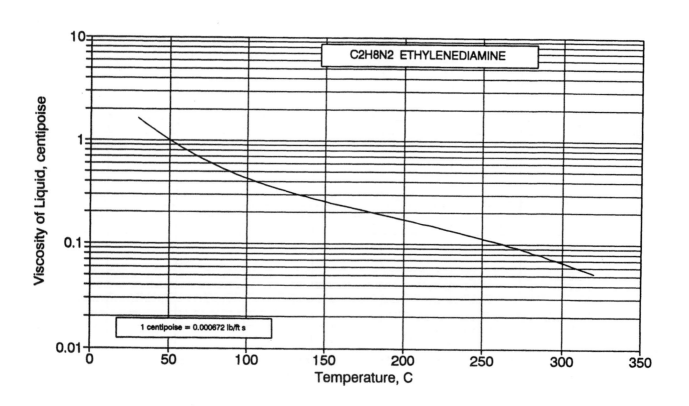

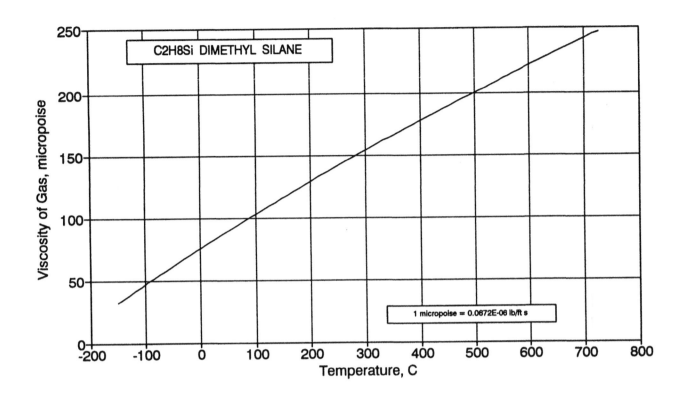

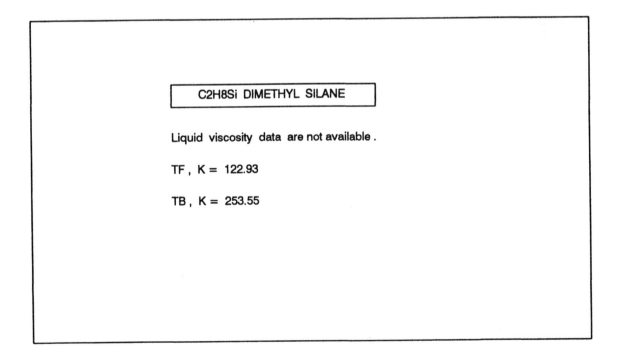

C2H8Si DIMETHYL SILANE

Liquid viscosity data are not available .

TF , K = 122.93

TB , K = 253.55

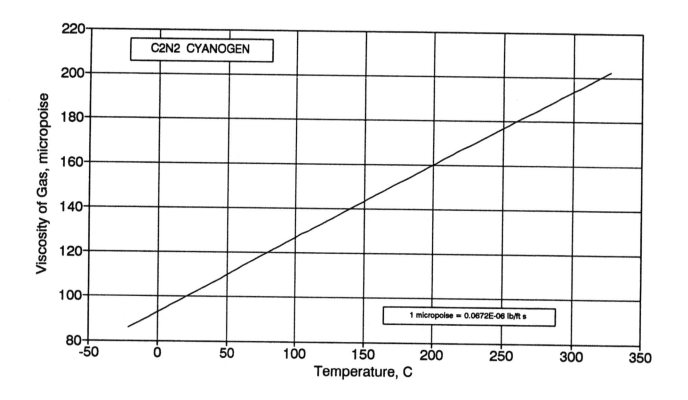

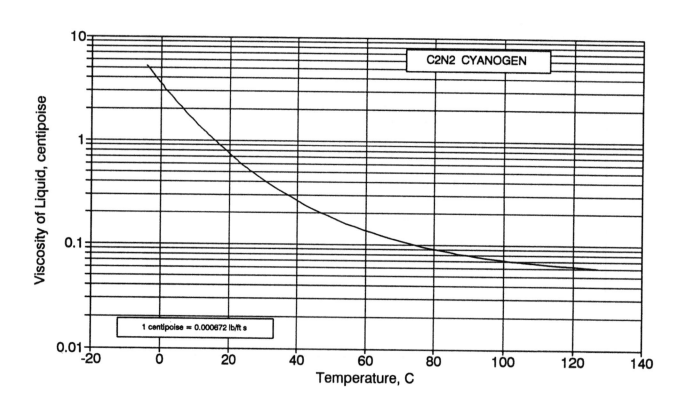

130

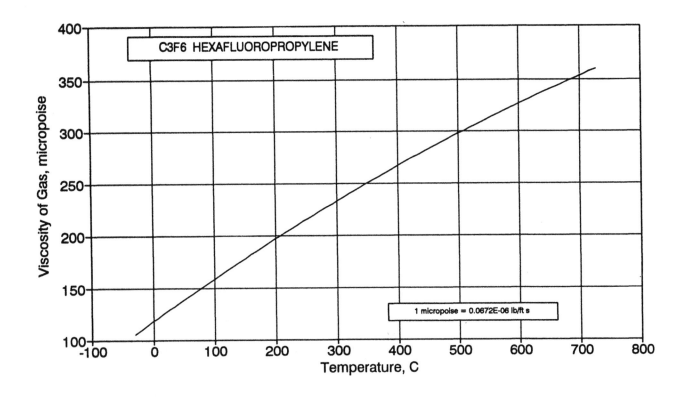

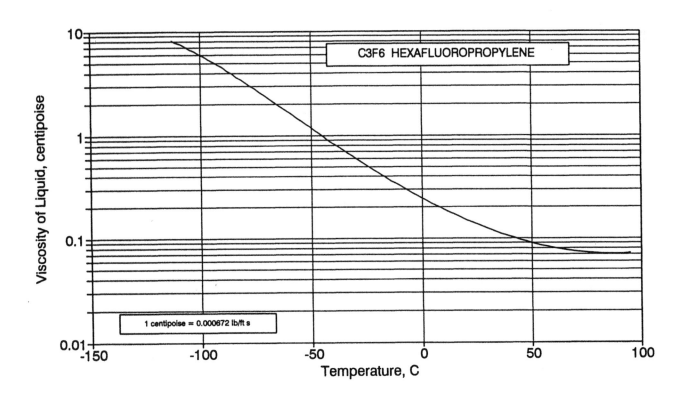

131

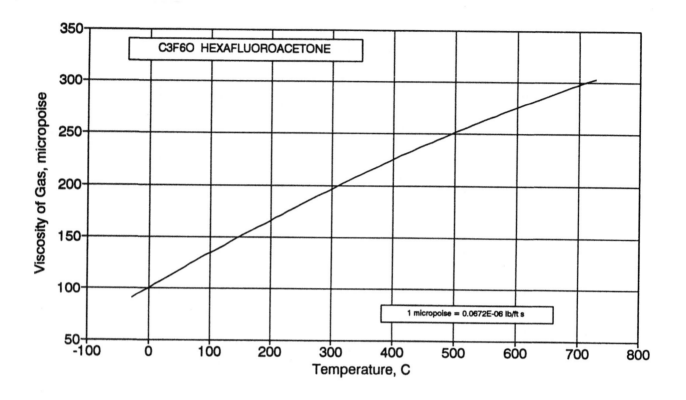

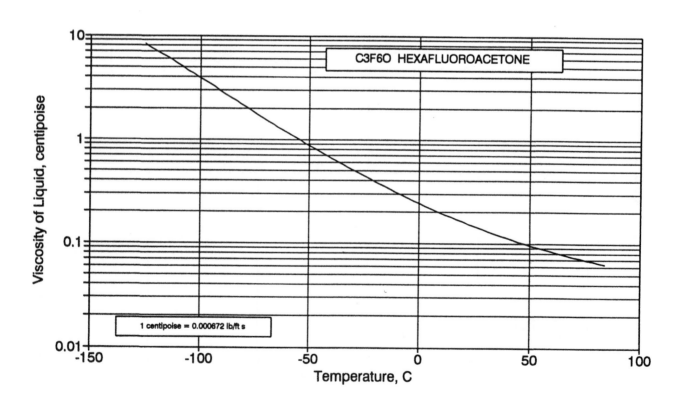

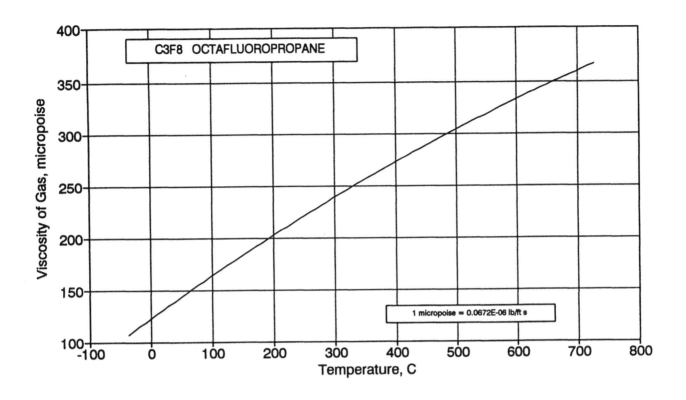

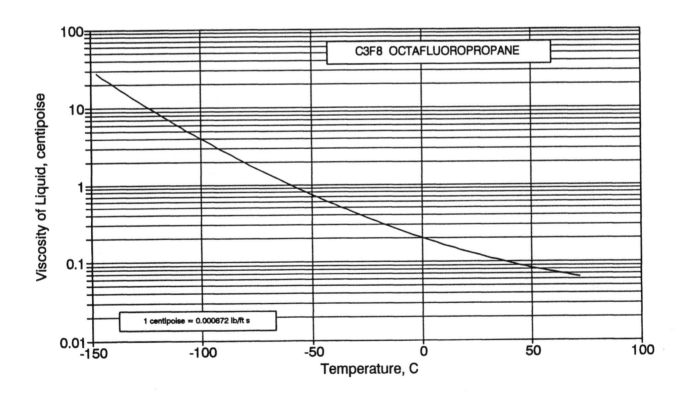

133

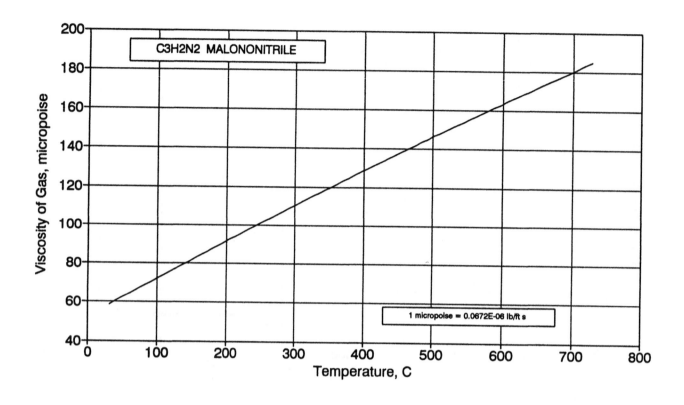

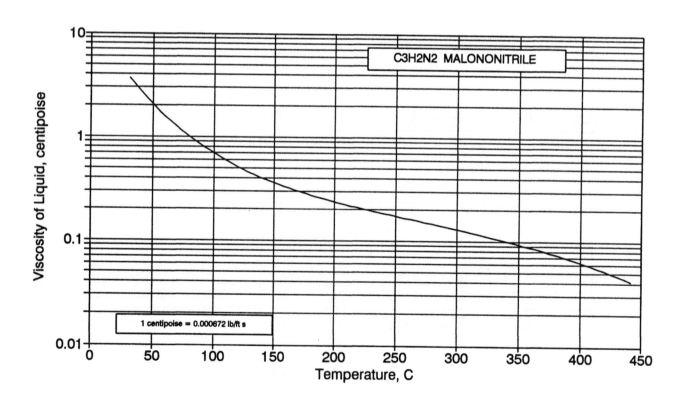

134

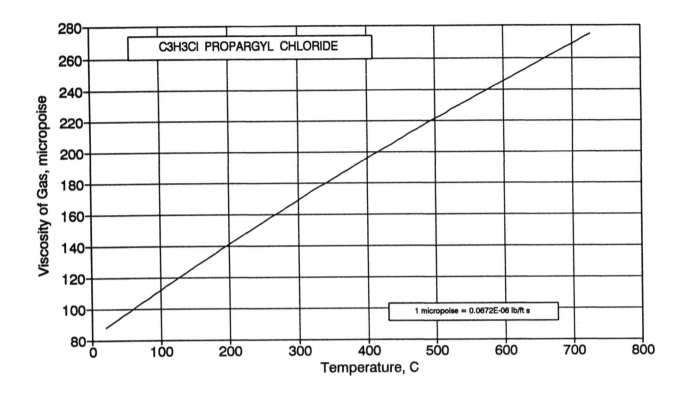

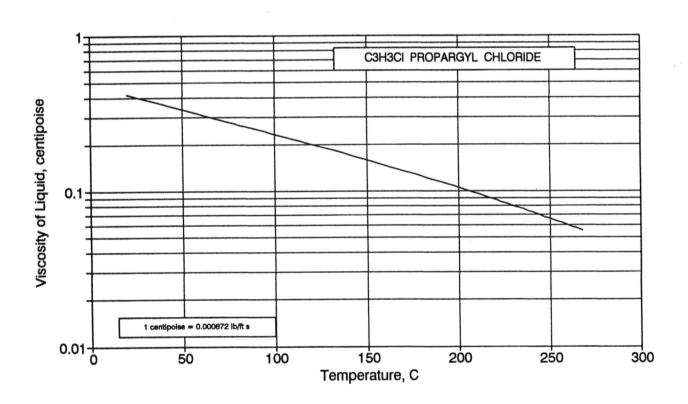

135

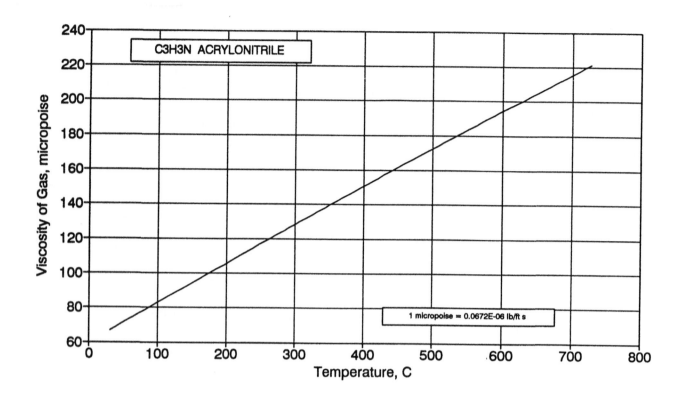

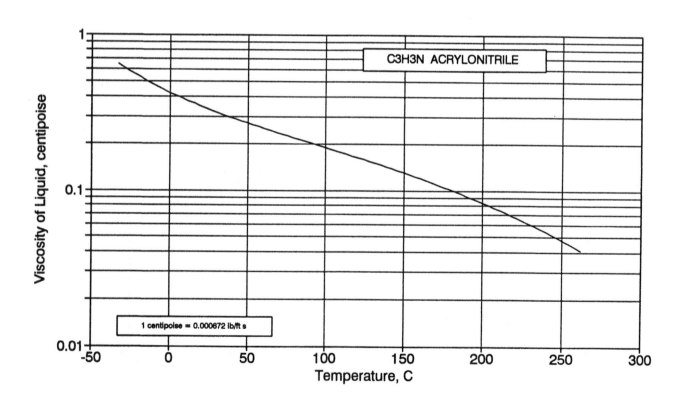

136

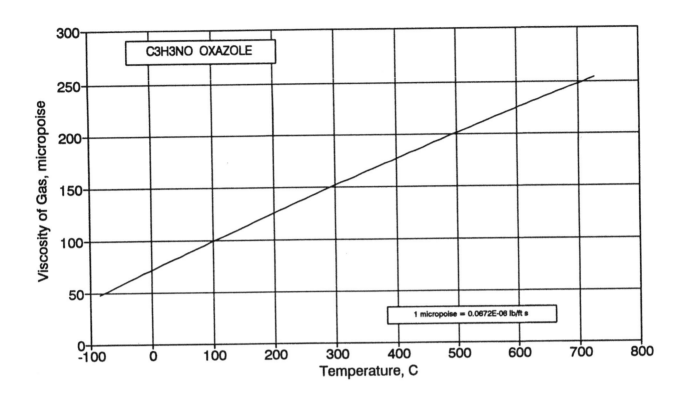

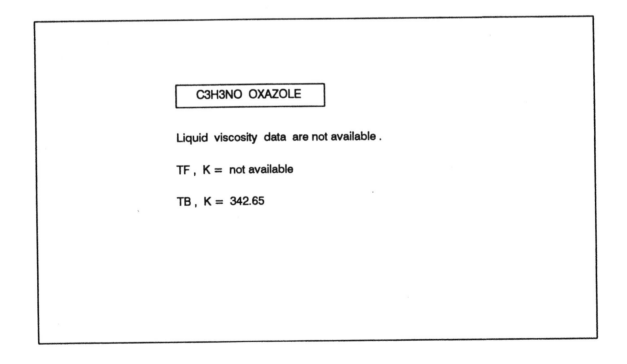

137

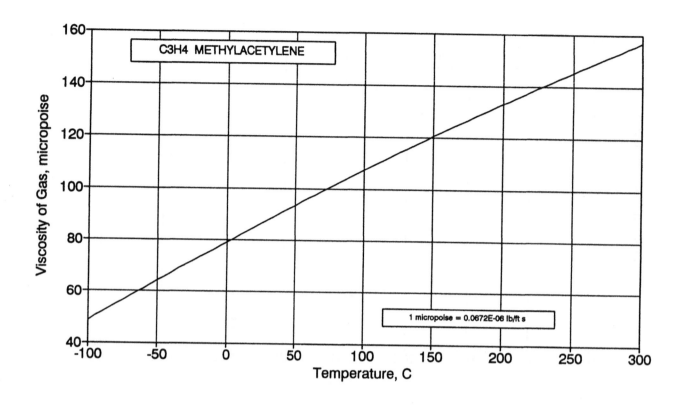

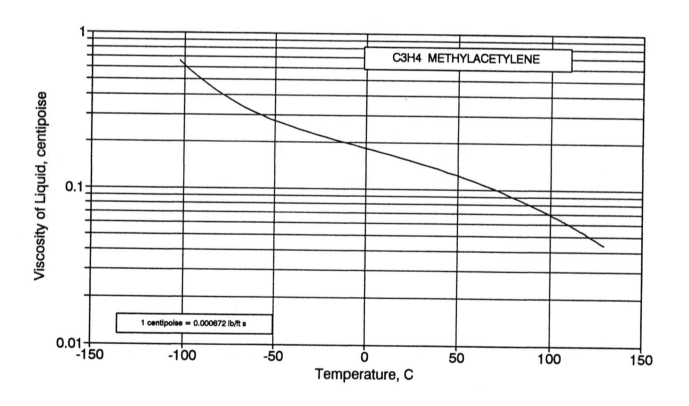

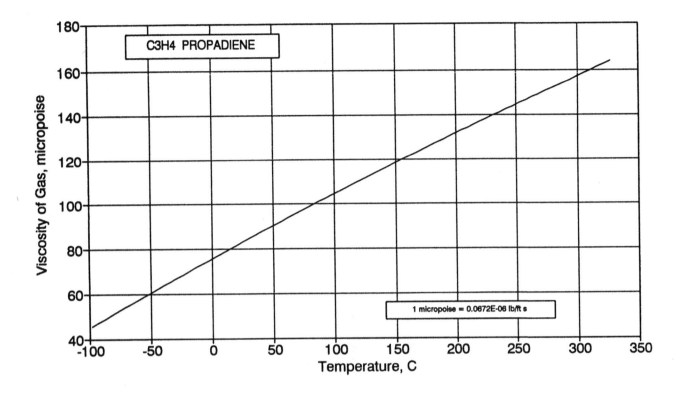

1 micropoise = 0.0672E-06 lb/ft s

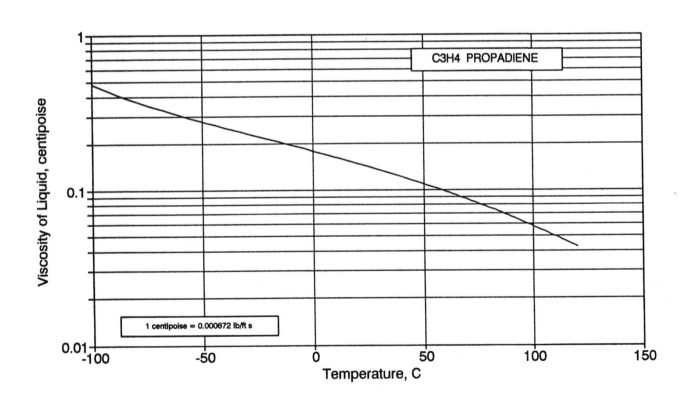

1 centipoise = 0.000672 lb/ft s

139

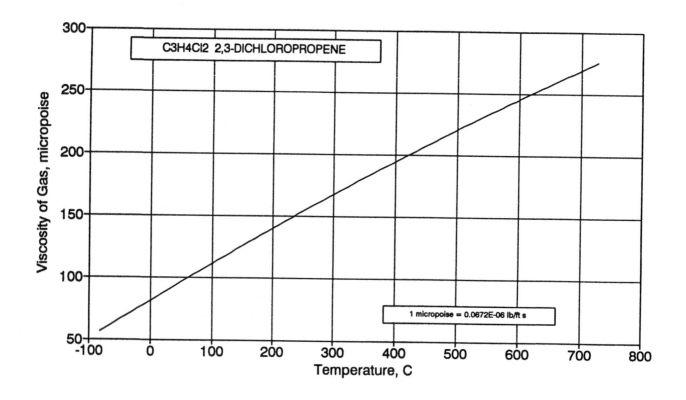

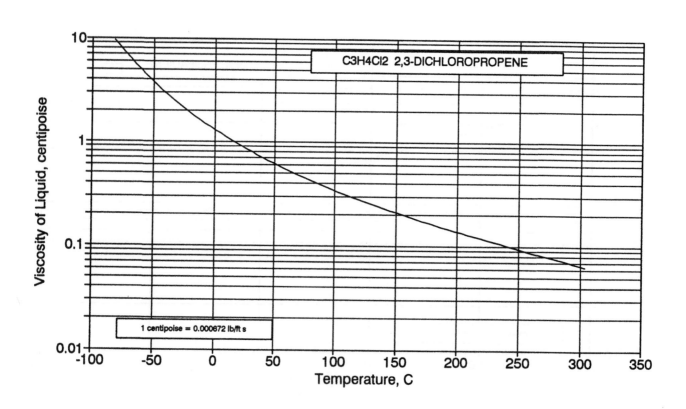

140

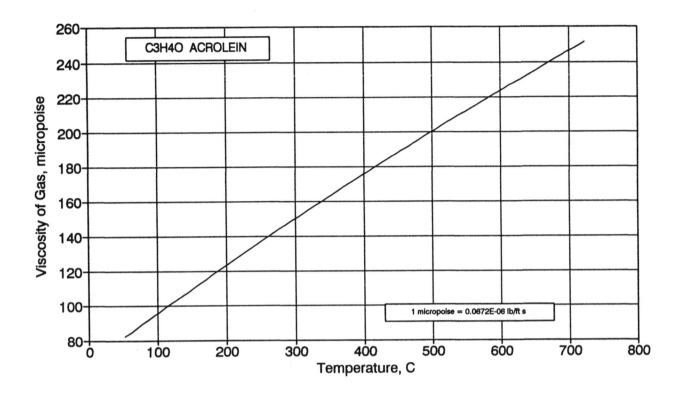

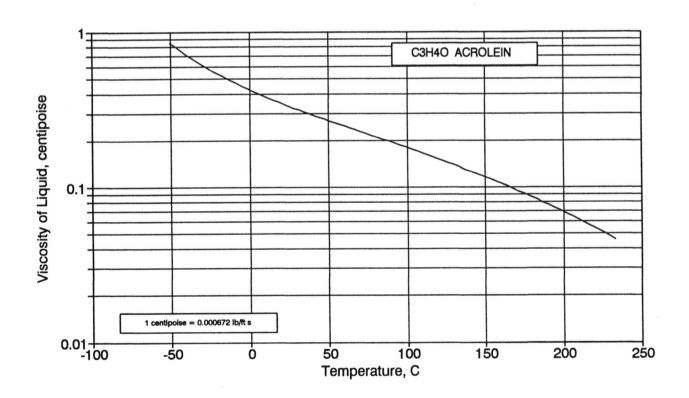

141

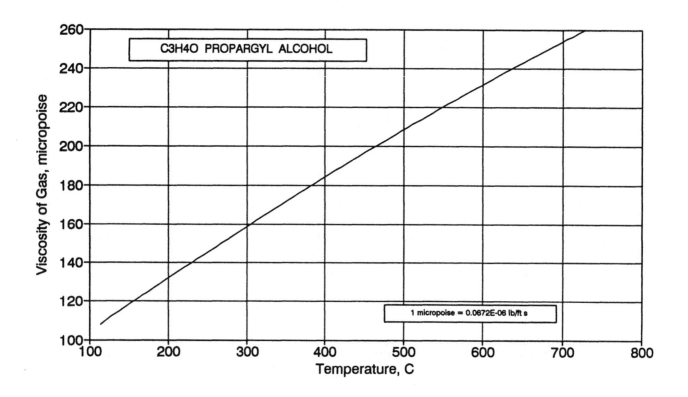

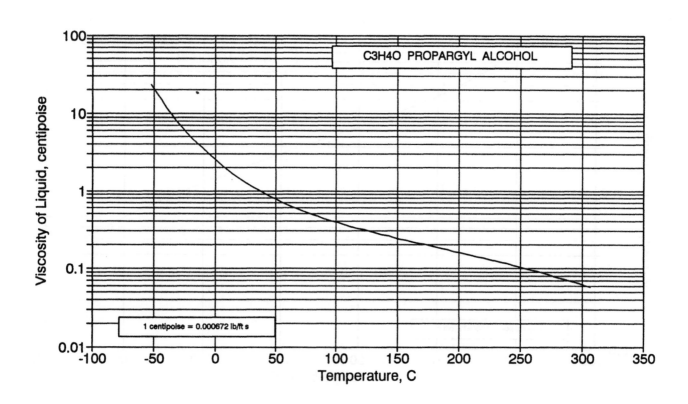

142

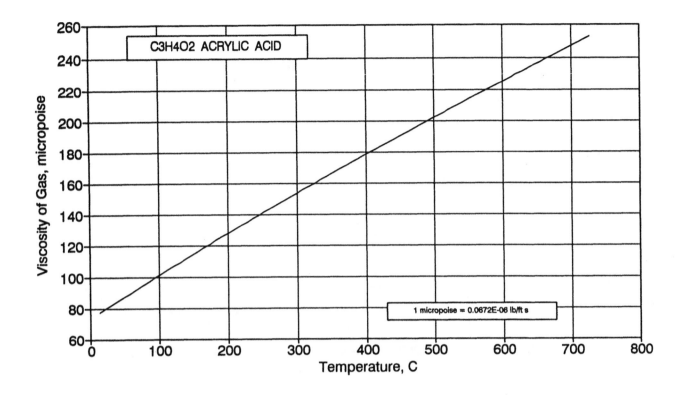

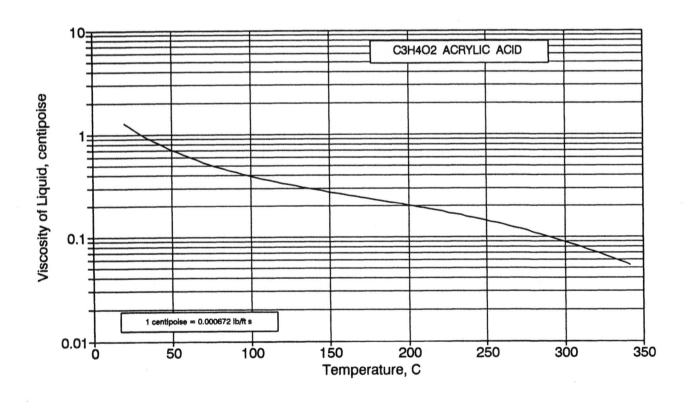

143

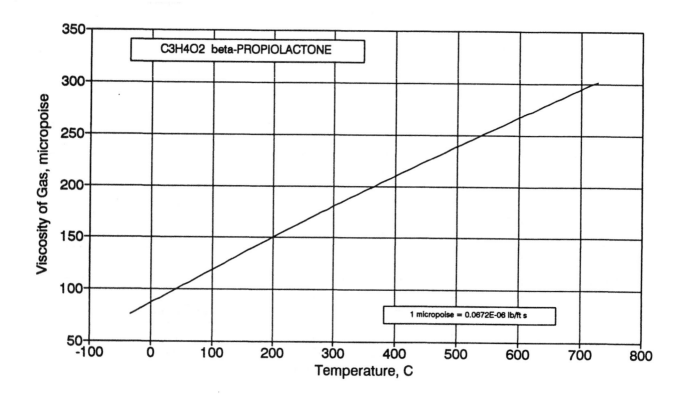

C3H4O2 beta-PROPIOLACTONE

1 micropoise = 0.0672E-06 lb/ft s

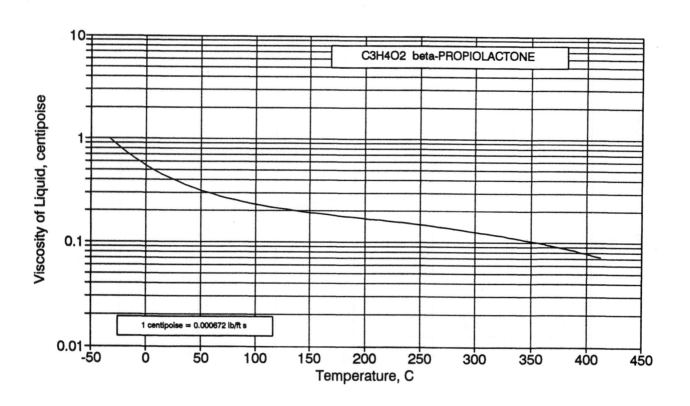

C3H4O2 beta-PROPIOLACTONE

1 centipoise = 0.000672 lb/ft s

144

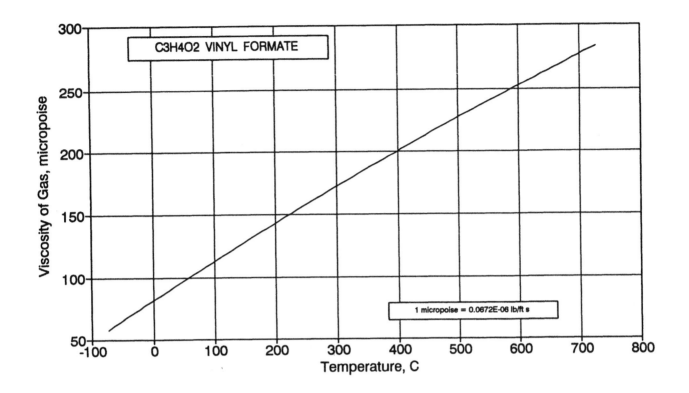

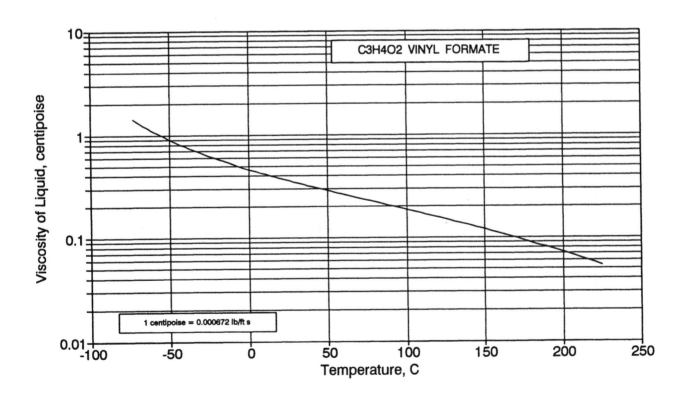

145

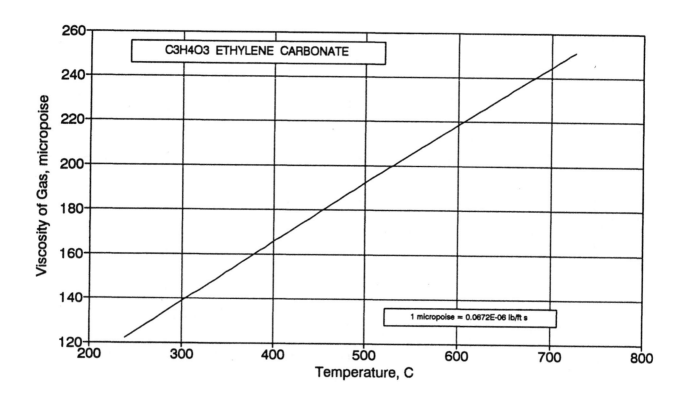

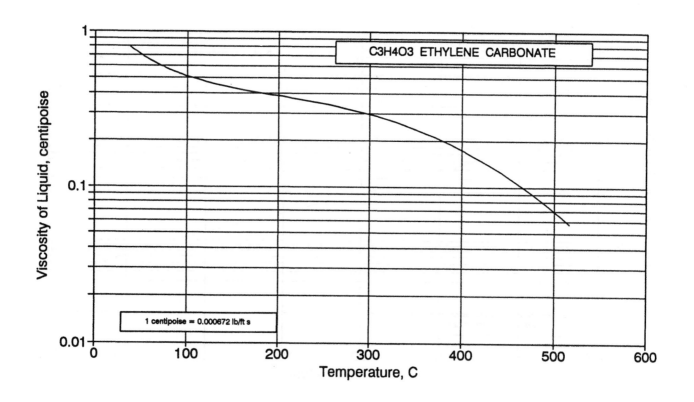

146

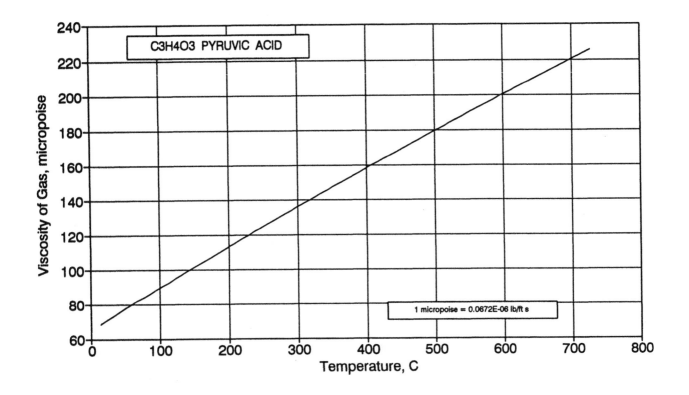

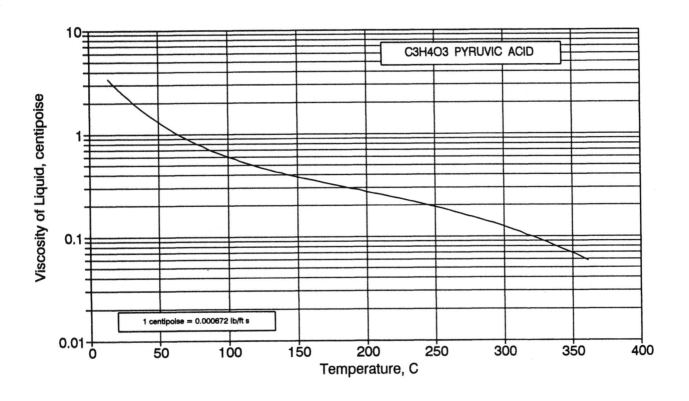

147

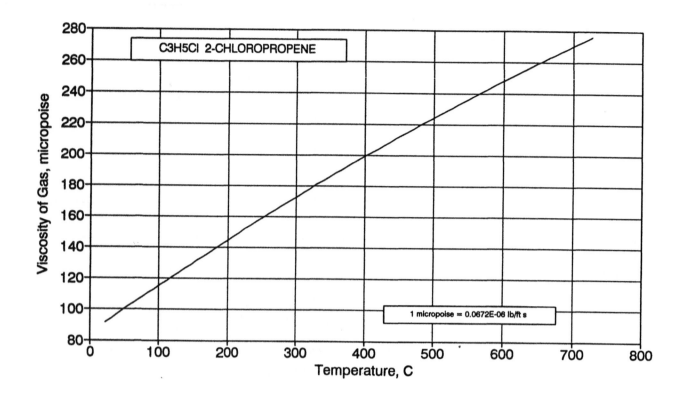

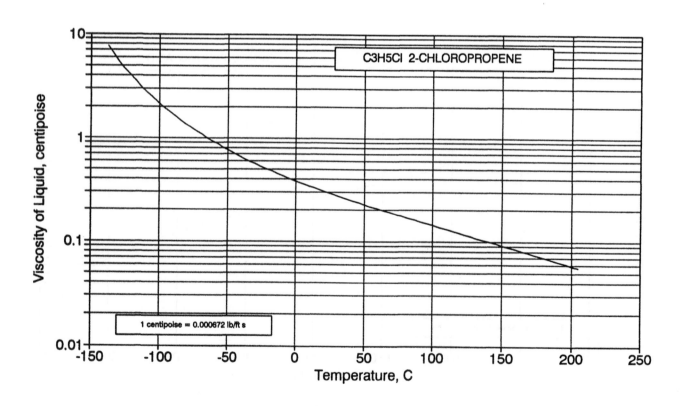

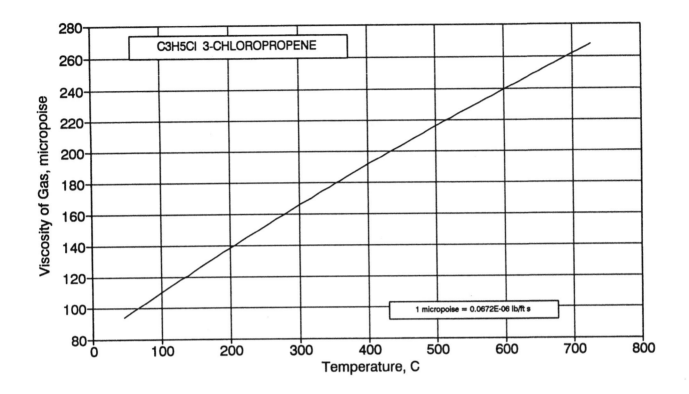

C3H5Cl 3-CHLOROPROPENE

Viscosity of Gas, micropoise

Temperature, C

1 micropoise = 0.0672E-06 lb/ft s

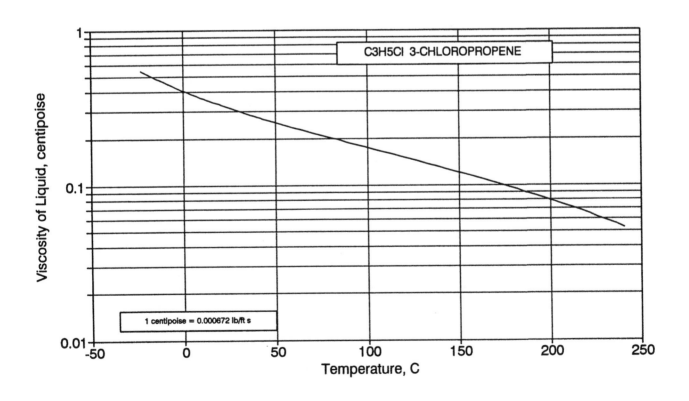

C3H5Cl 3-CHLOROPROPENE

Viscosity of Liquid, centipoise

Temperature, C

1 centipoise = 0.000672 lb/ft s

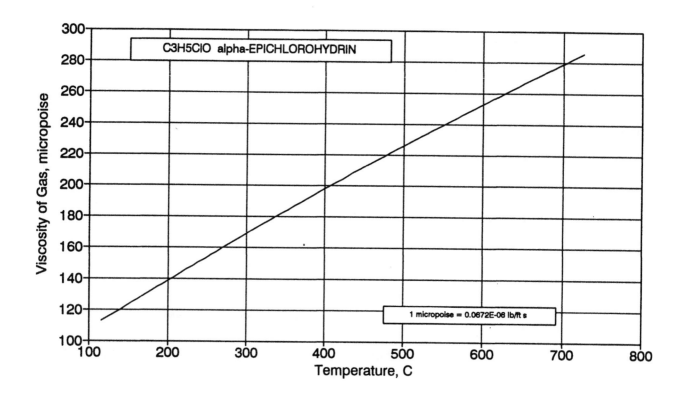

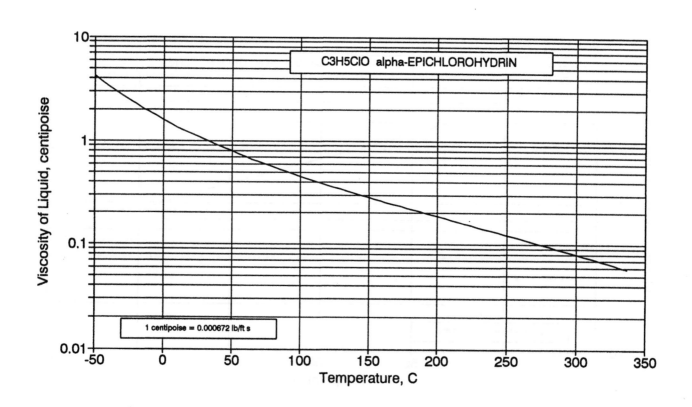

150

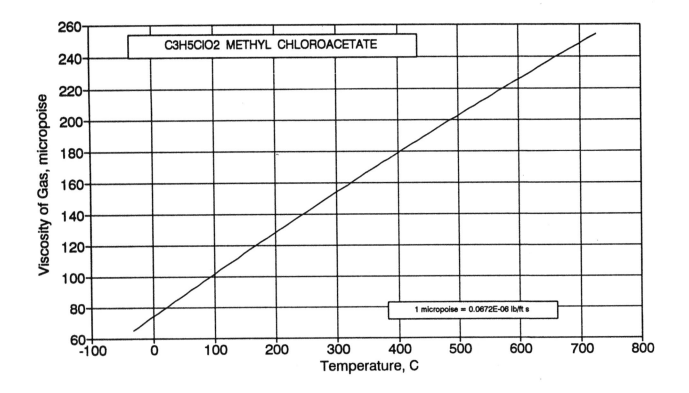

C3H5ClO2 METHYL CHLOROACETATE

1 micropoise = 0.0672E-06 lb/ft s

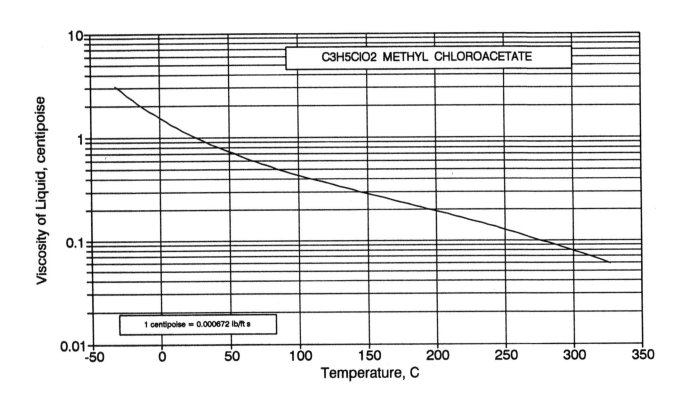

C3H5ClO2 METHYL CHLOROACETATE

1 centipoise = 0.000672 lb/ft s

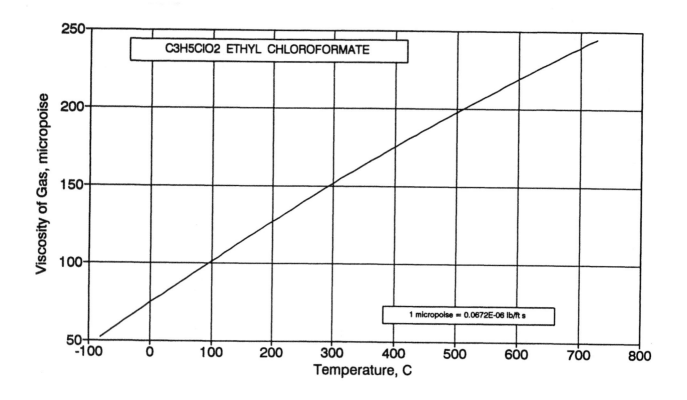

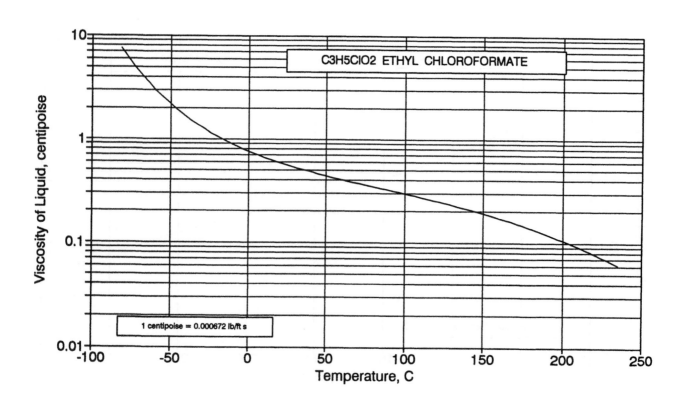

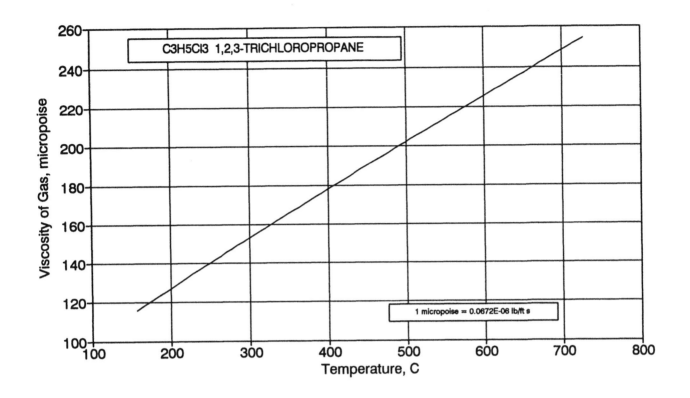

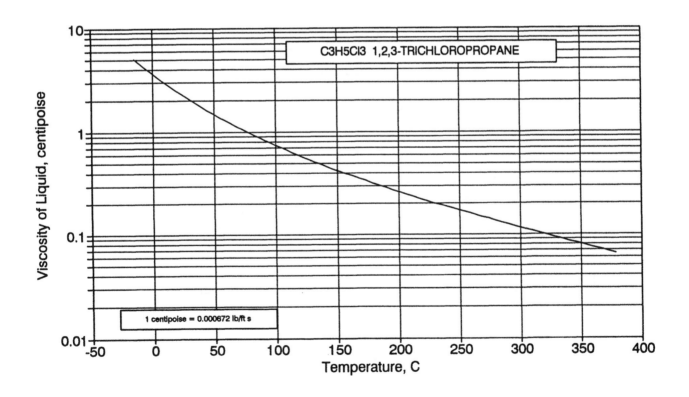

153

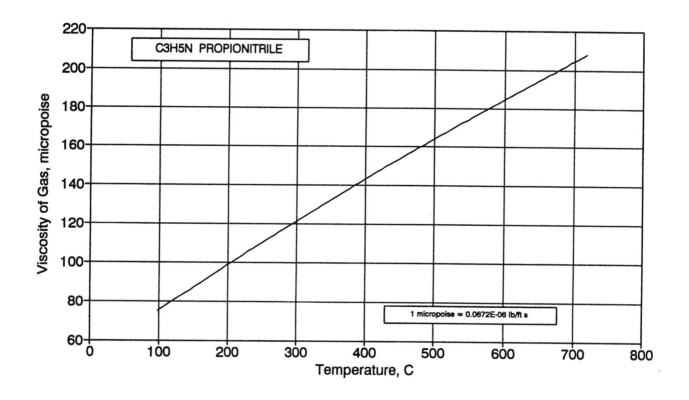

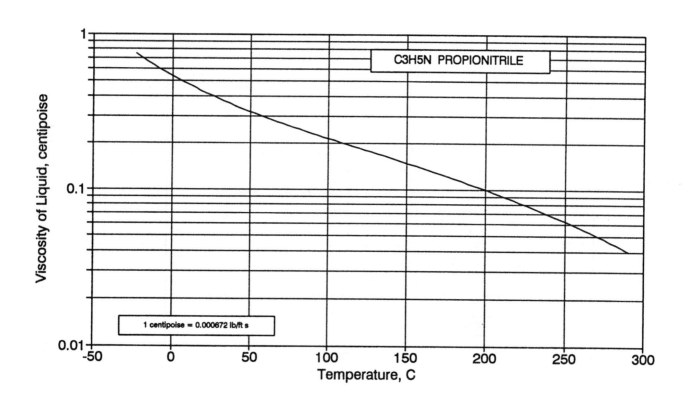

154

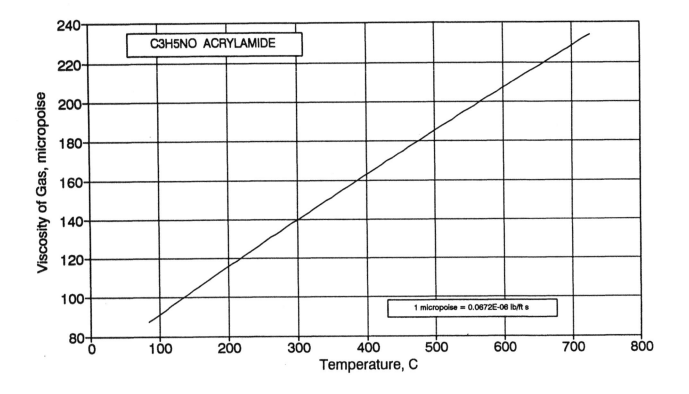

C3H5NO ACRYLAMIDE

1 micropoise = 0.0672E-06 lb/ft s

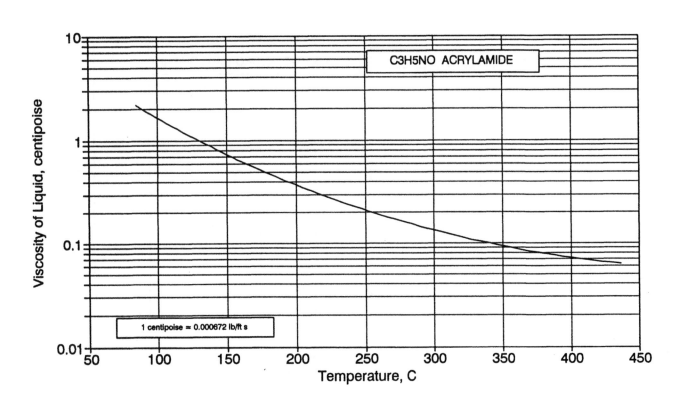

C3H5NO ACRYLAMIDE

1 centipoise = 0.000672 lb/ft s

155

C3H5NO HYDRACRYLONITRILE

Gas viscosity data are not available .

TF , K = 227.15

TB , K = 494.15

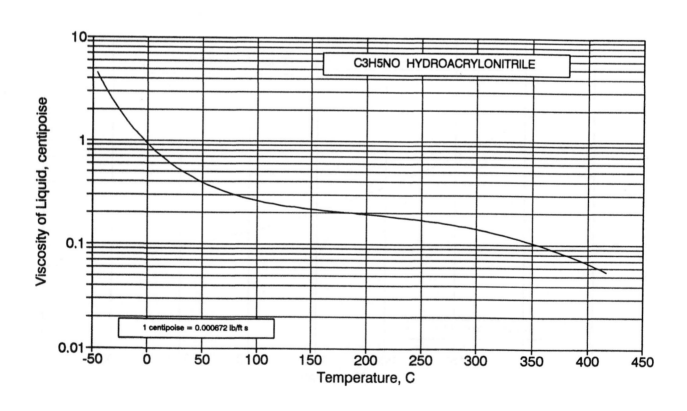

1 centipoise = 0.000672 lb/ft s

C3H5NO HYDROACRYLONITRILE

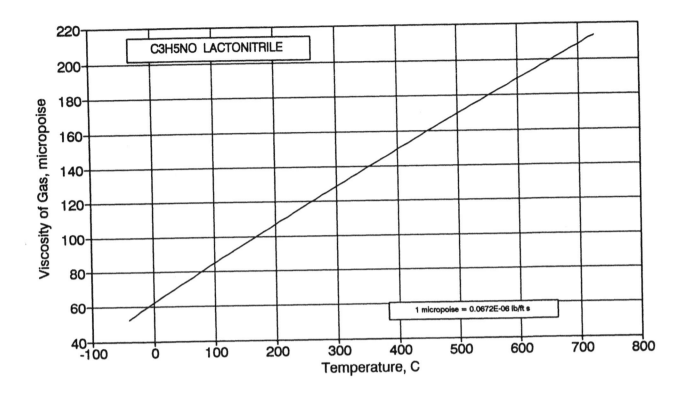

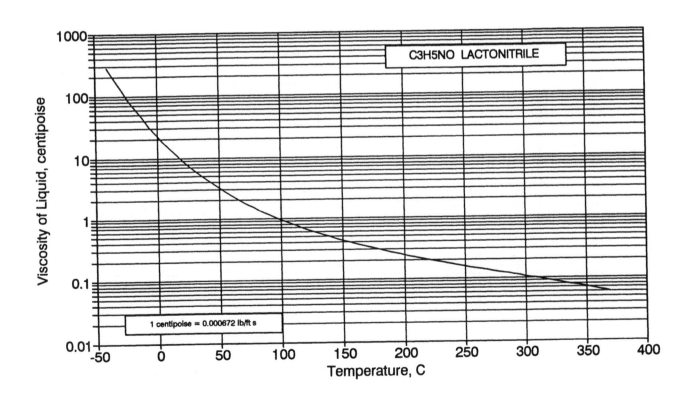

157

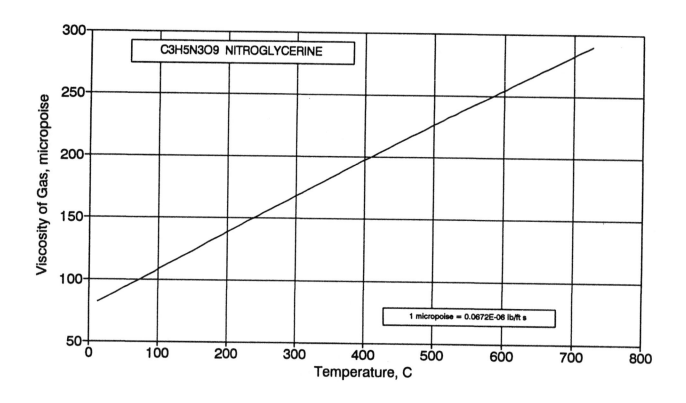

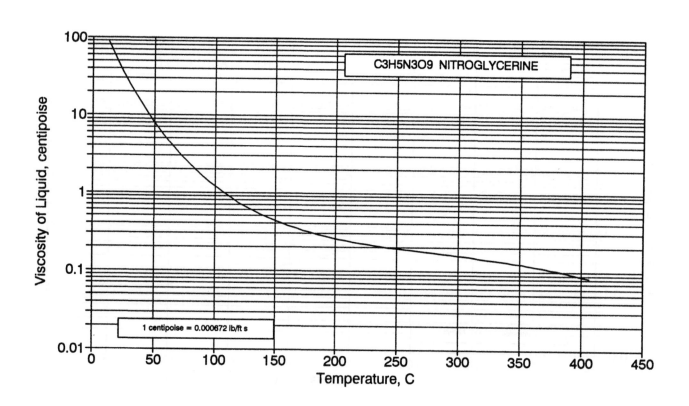

158

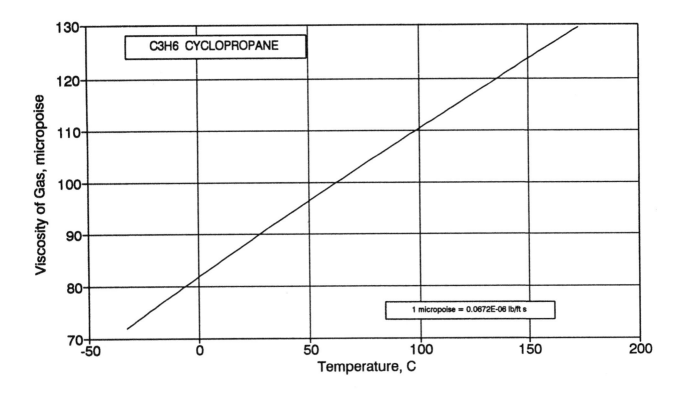

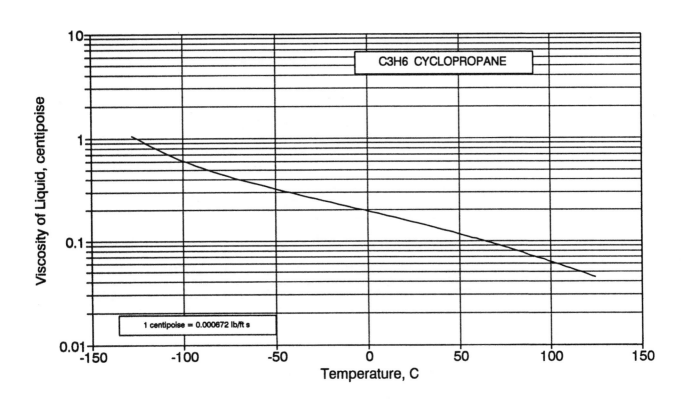

159

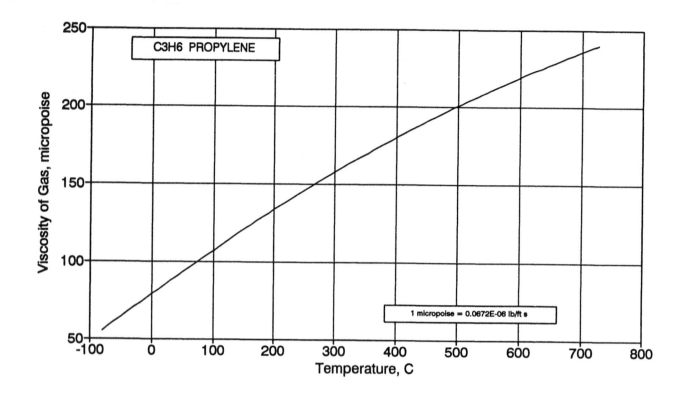

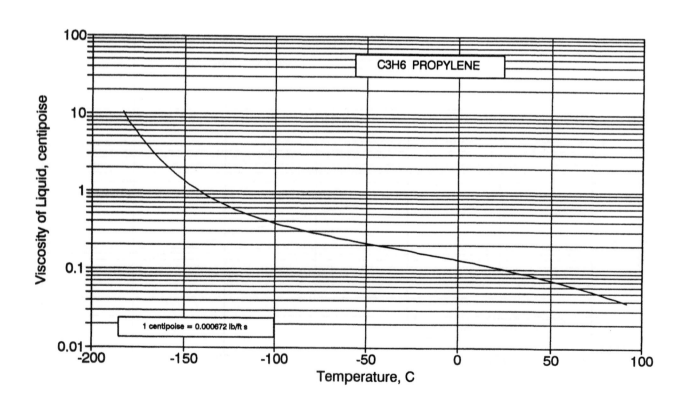

160

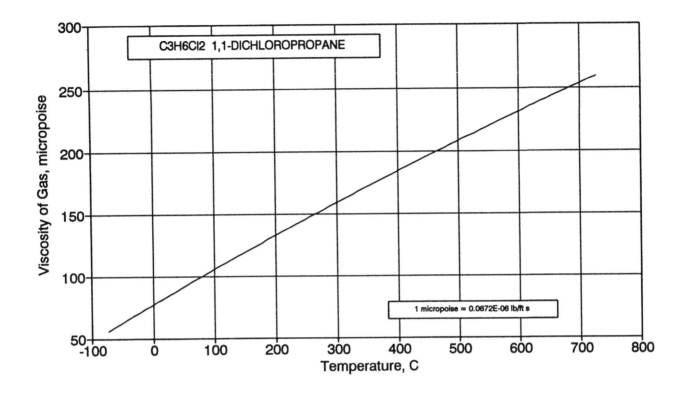

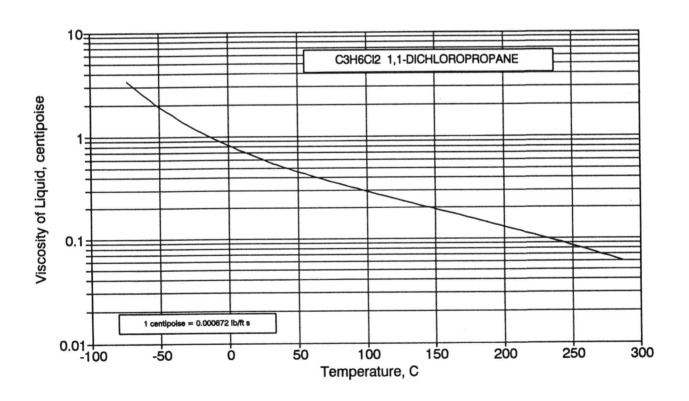

161

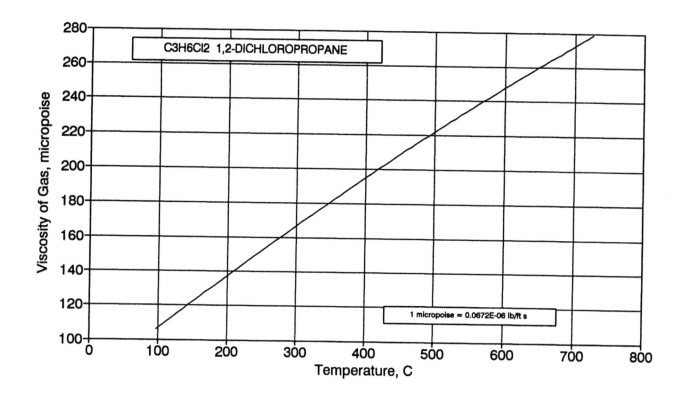

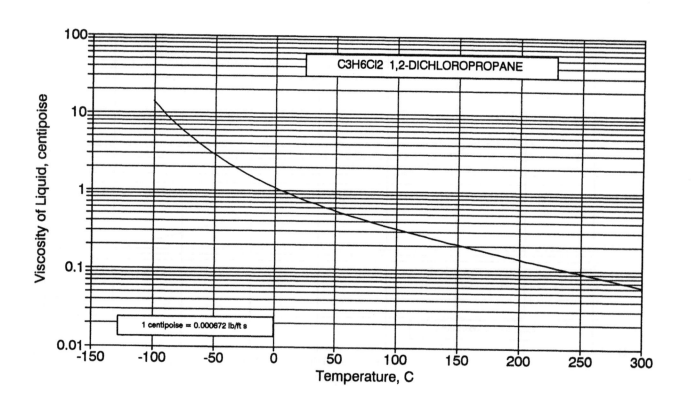

162

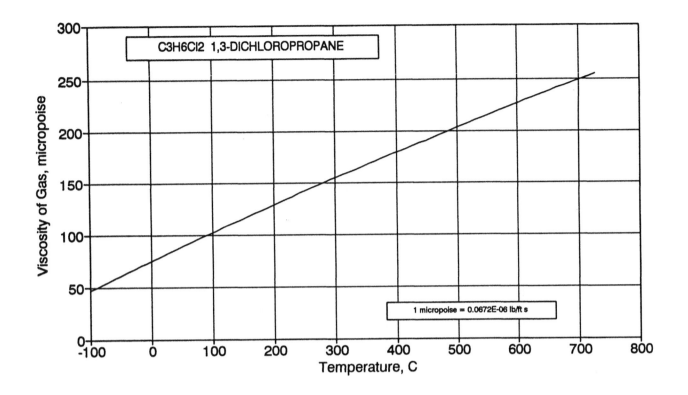

C3H6Cl2 1,3-DICHLOROPROPANE

1 micropoise = 0.0672E-06 lb/ft s

Viscosity of Gas, micropoise

Temperature, C

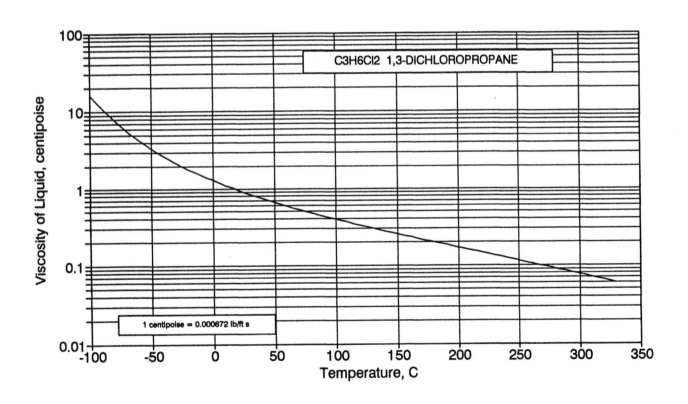

C3H6Cl2 1,3-DICHLOROPROPANE

1 centipoise = 0.000672 lb/ft s

Viscosity of Liquid, centipoise

Temperature, C

163

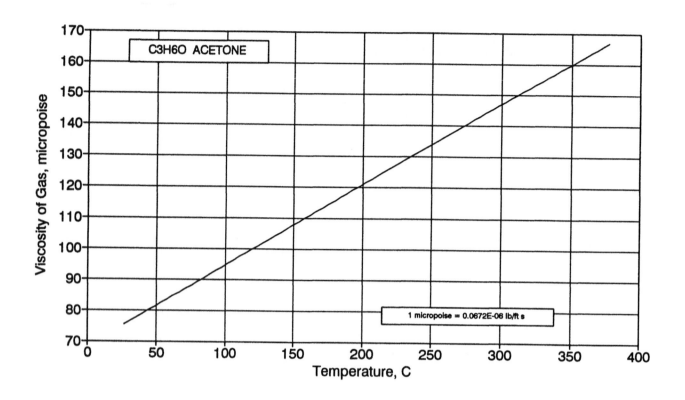

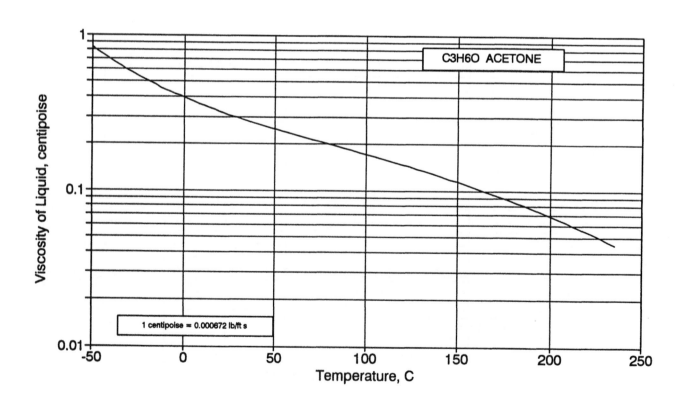

164

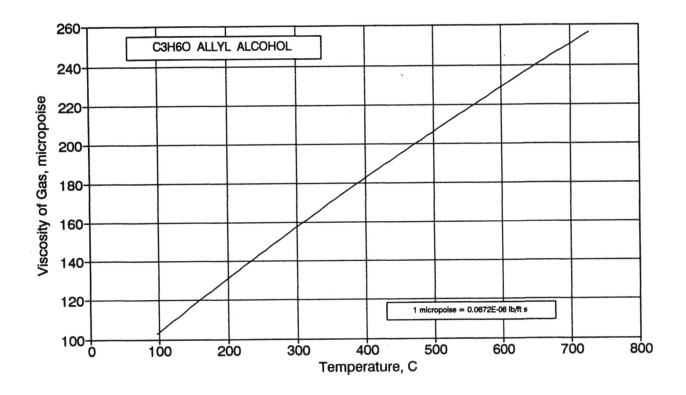

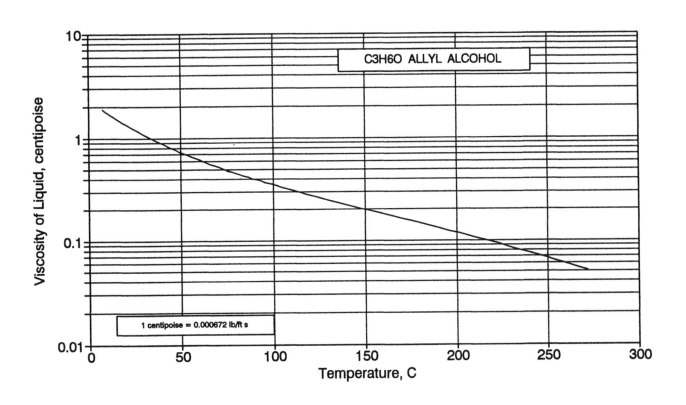

165

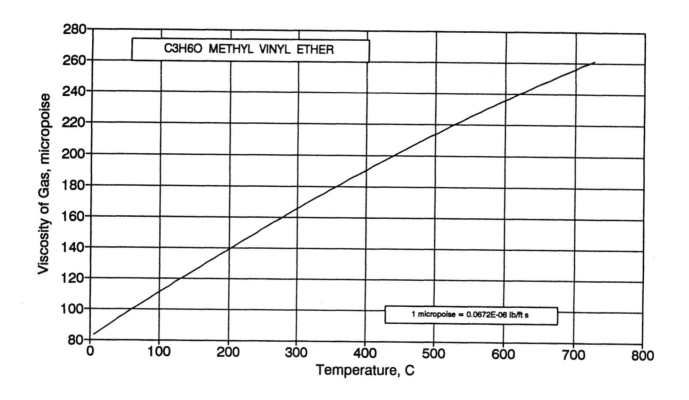

C3H6O METHYL VINYL ETHER

1 micropoise = 0.0672E-06 lb/ft s

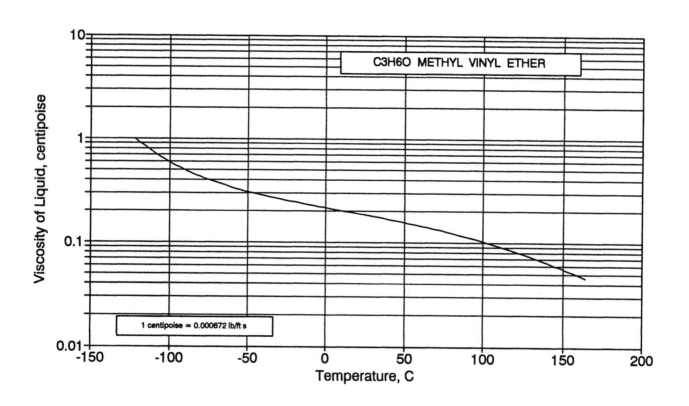

C3H6O METHYL VINYL ETHER

1 centipoise = 0.000672 lb/ft s

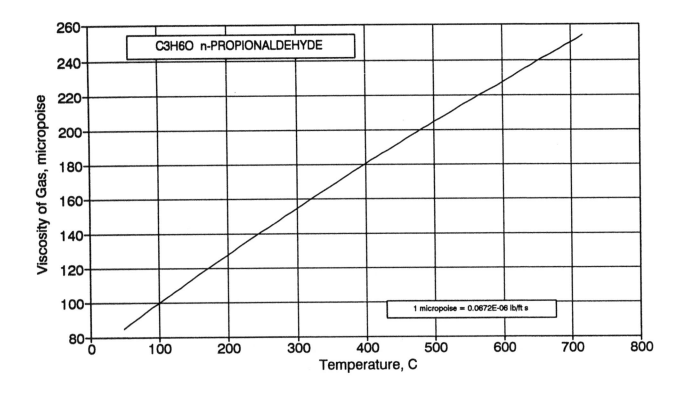

C3H6O n-PROPIONALDEHYDE

Viscosity of Gas, micropoise

1 micropoise = 0.0672E-06 lb/ft s

Temperature, C

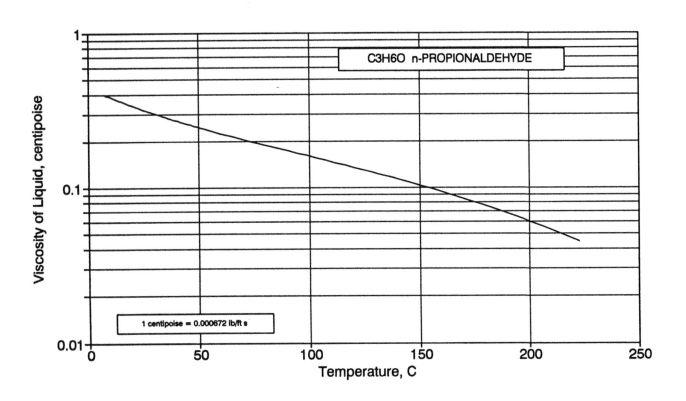

C3H6O n-PROPIONALDEHYDE

Viscosity of Liquid, centipoise

1 centipoise = 0.000672 lb/ft s

Temperature, C

167

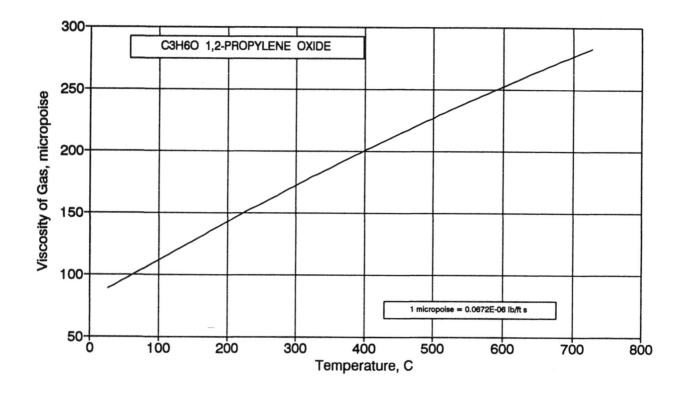

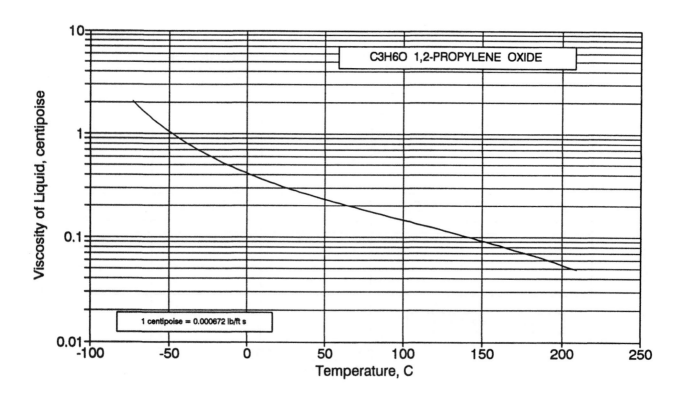

168

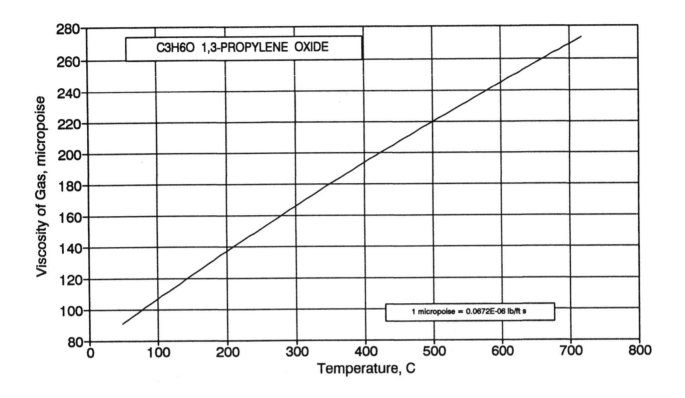

C3H6O 1,3-PROPYLENE OXIDE

1 micropoise = 0.0672E-06 lb/ft s

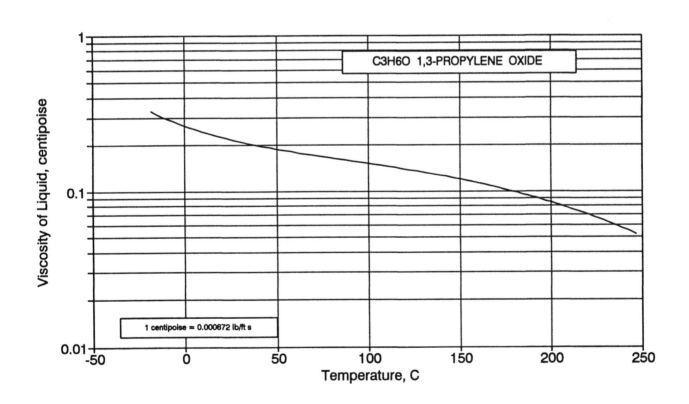

C3H6O 1,3-PROPYLENE OXIDE

1 centipoise = 0.000672 lb/ft s

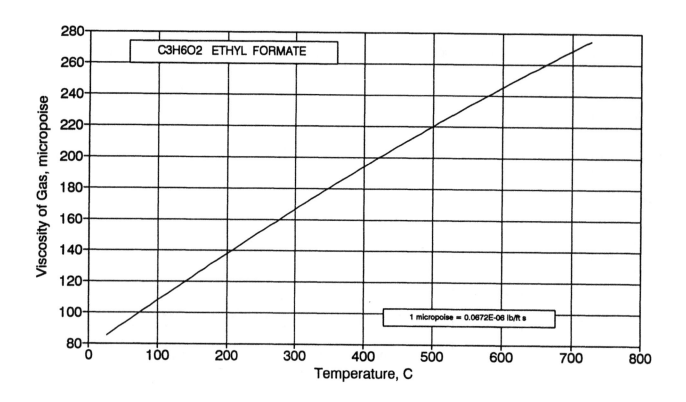

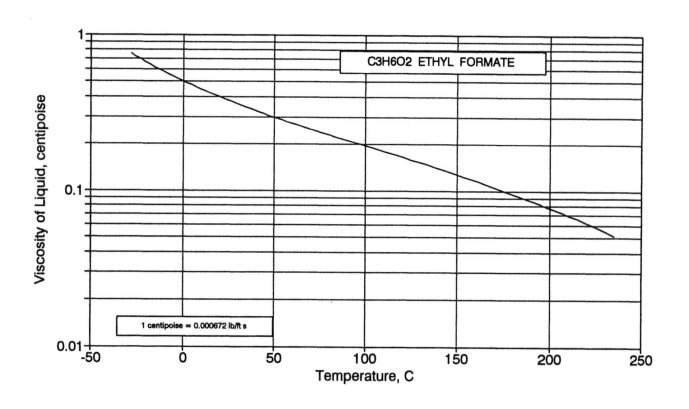

170

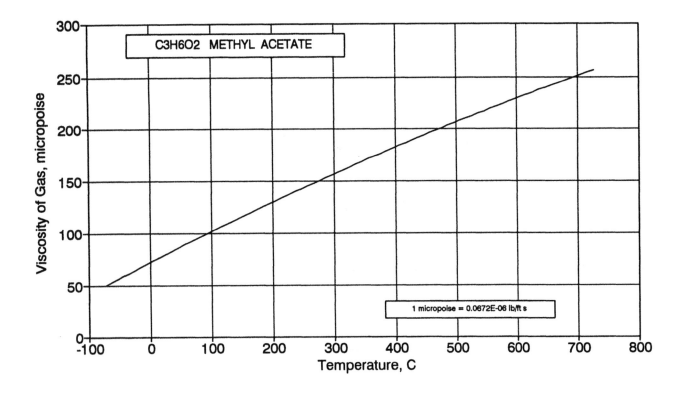

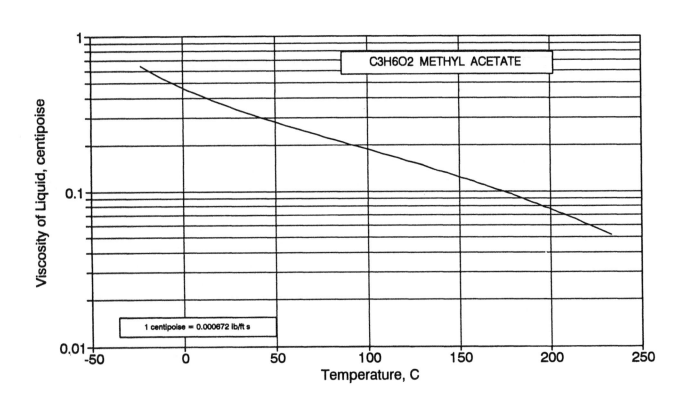

171

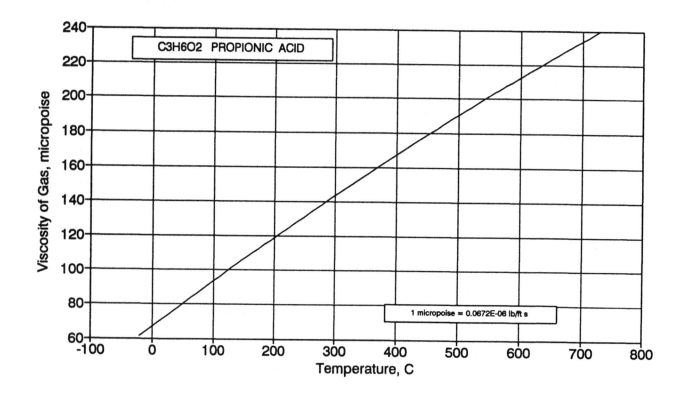

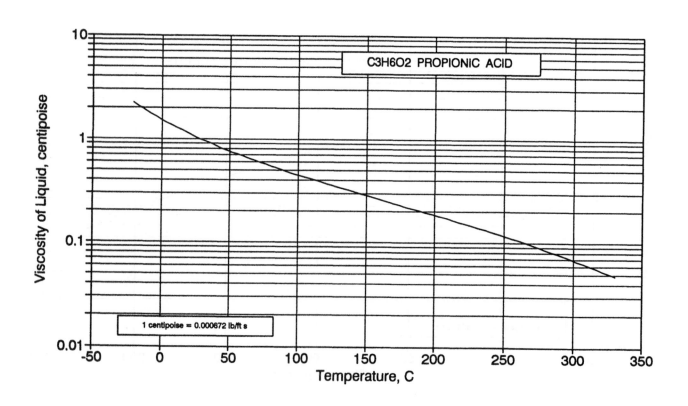

172

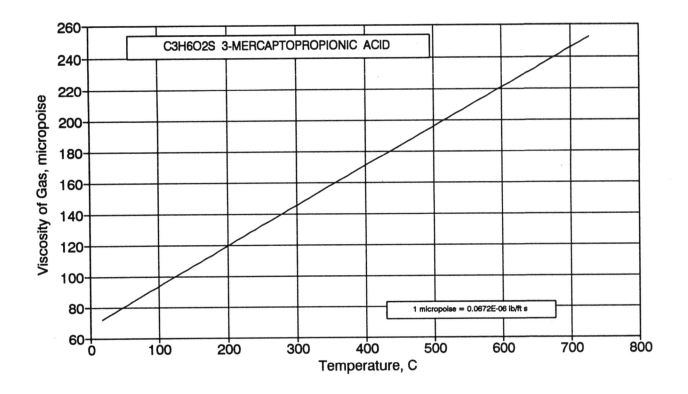

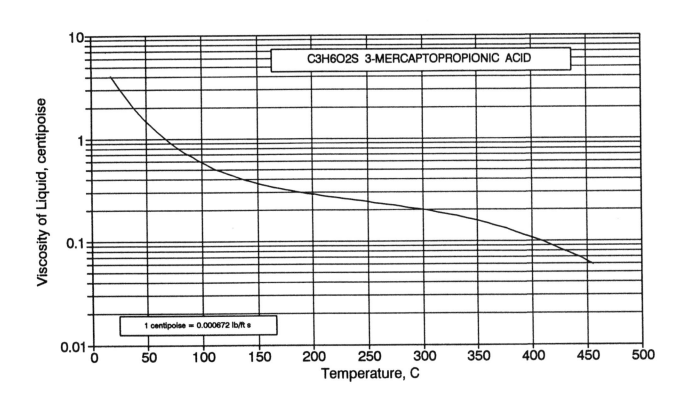

173

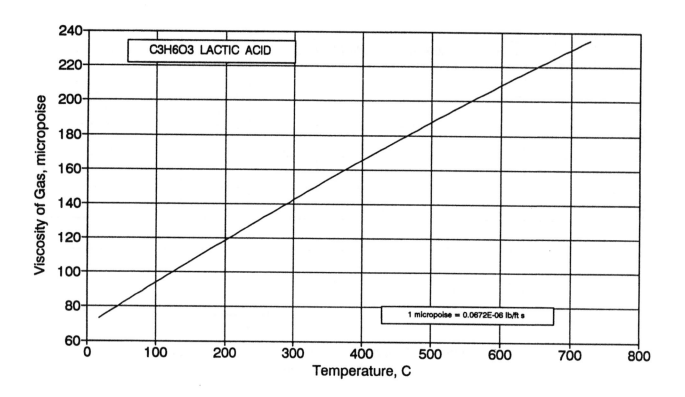

C3H6O3 LACTIC ACID

1 micropoise = 0.0672E-06 lb/ft s

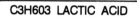

C3H603 LACTIC ACID

Liquid viscosity data are not available .

TF , K = 291.15

TB , K = 447 (estimated)

Decomposes before the boiiling point.

174

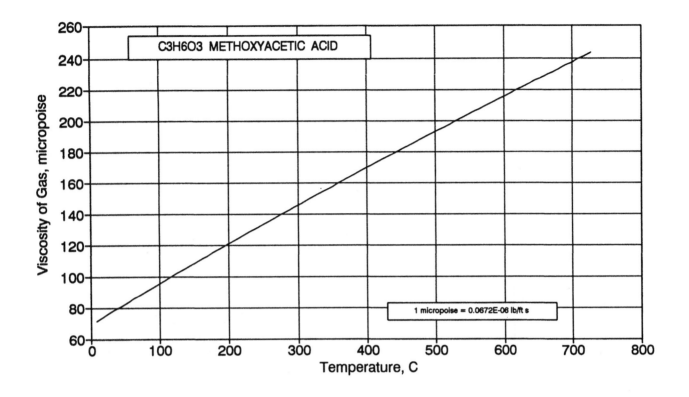

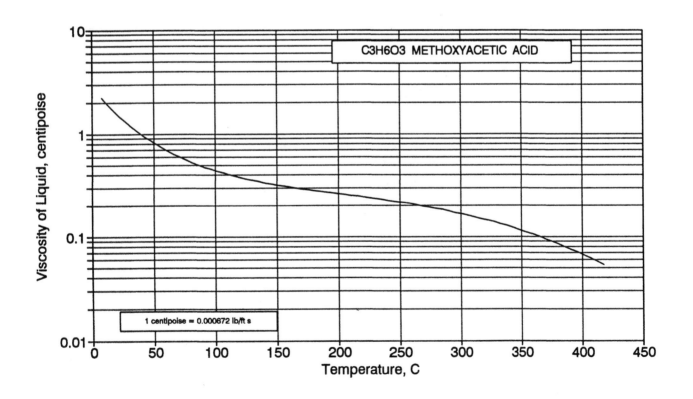

175

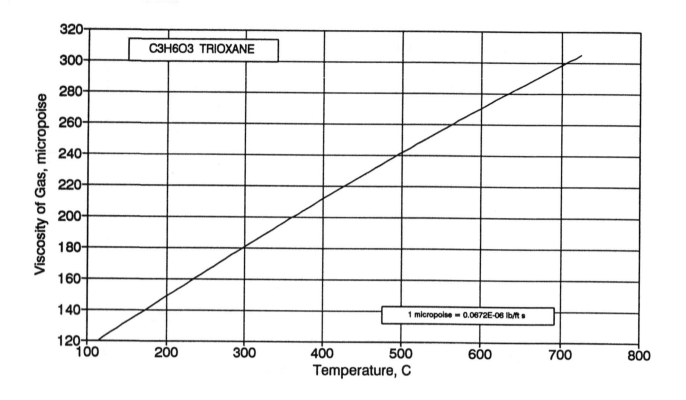

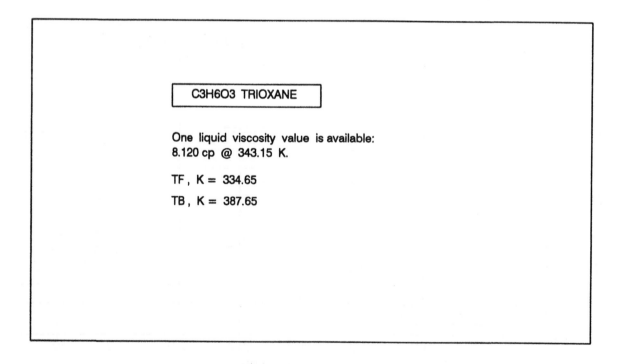

C3H6O3 TRIOXANE

One liquid viscosity value is available:
8.120 cp @ 343.15 K.

TF , K = 334.65

TB , K = 387.65

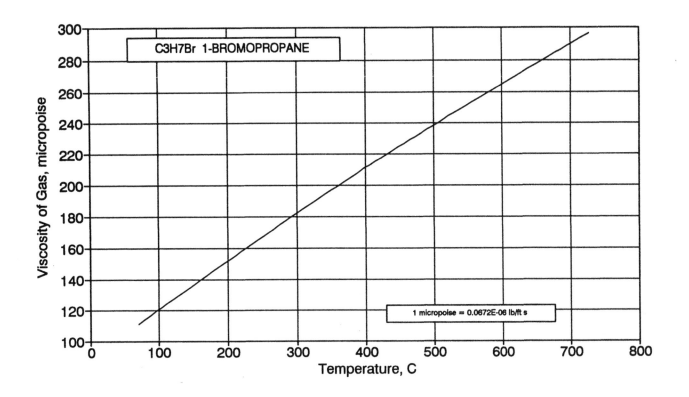

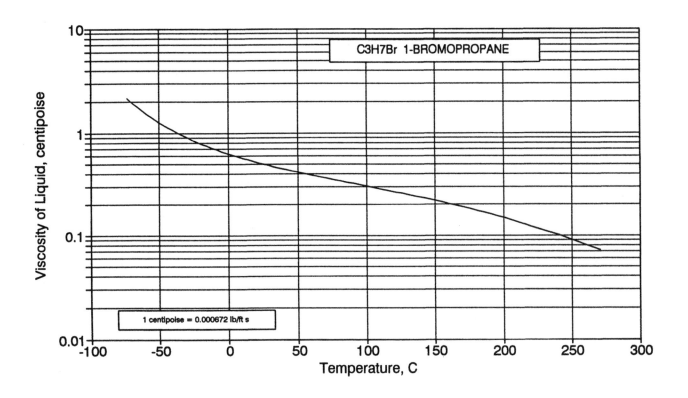

177

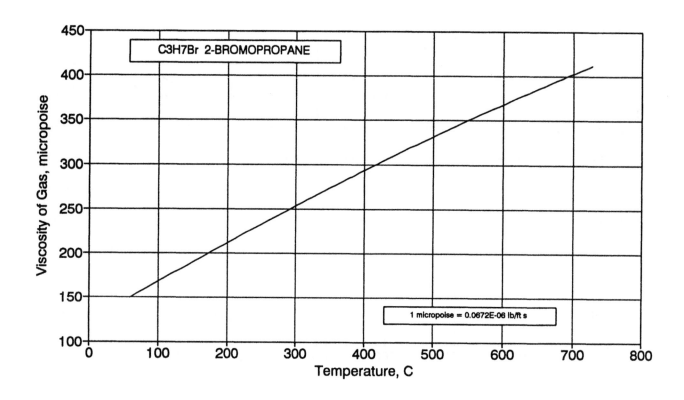

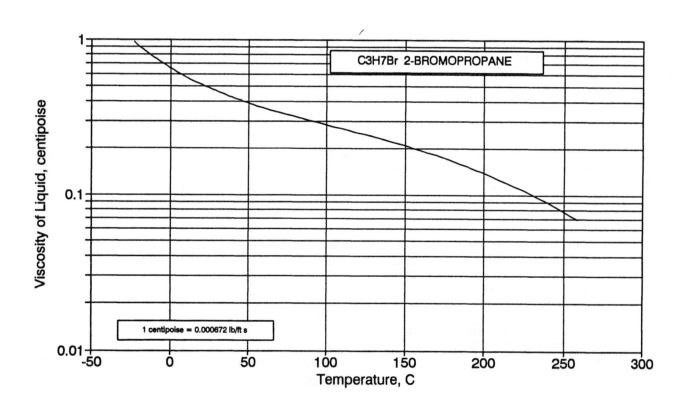

178

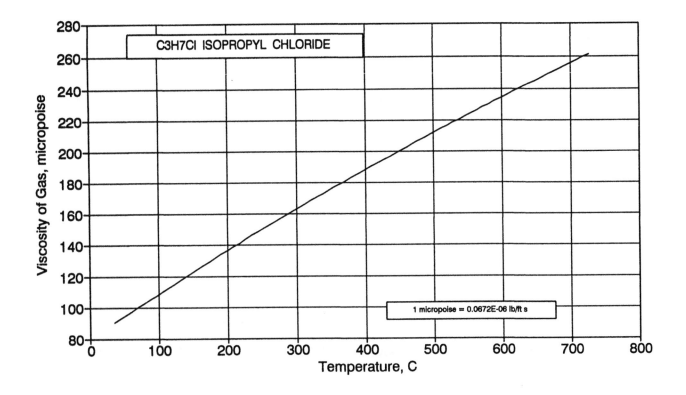

C3H7Cl ISOPROPYL CHLORIDE

1 micropoise = 0.0672E-06 lb/ft s

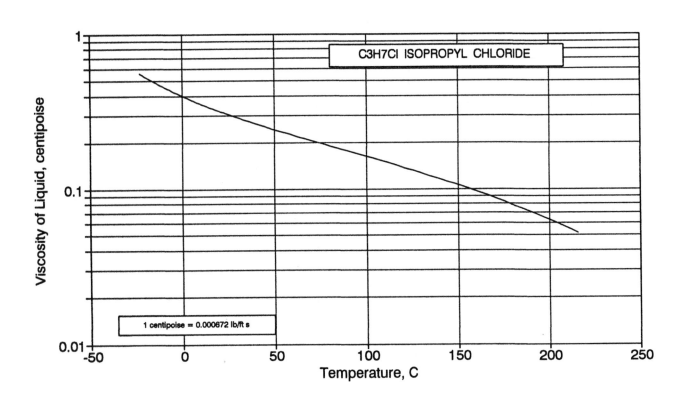

C3H7Cl ISOPROPYL CHLORIDE

1 centipoise = 0.000672 lb/ft s

179

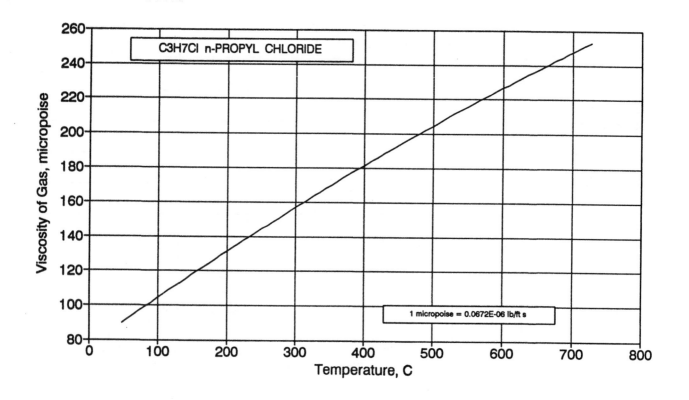

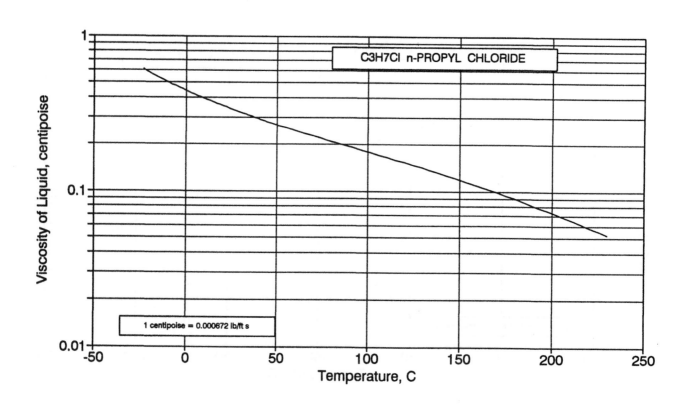

180

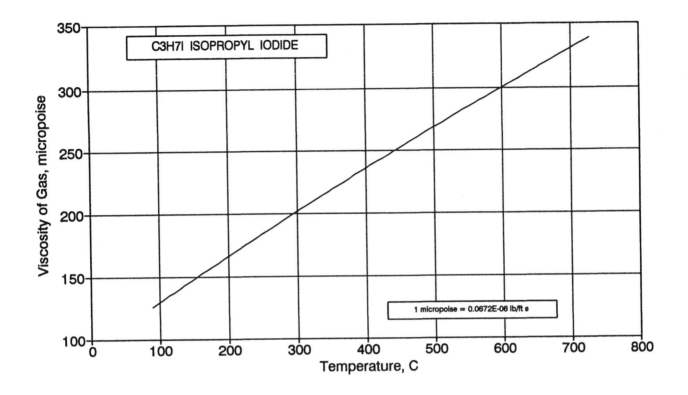

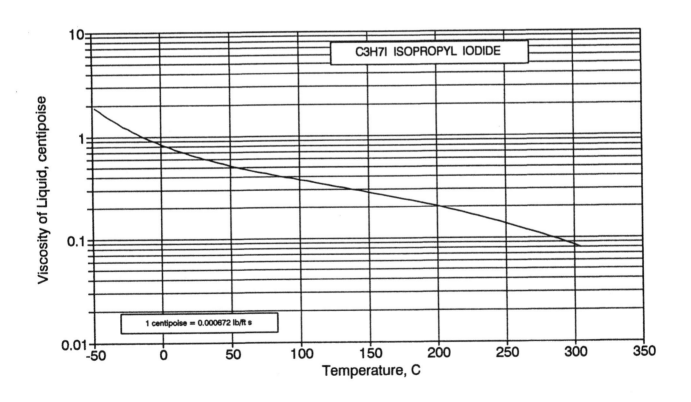

181

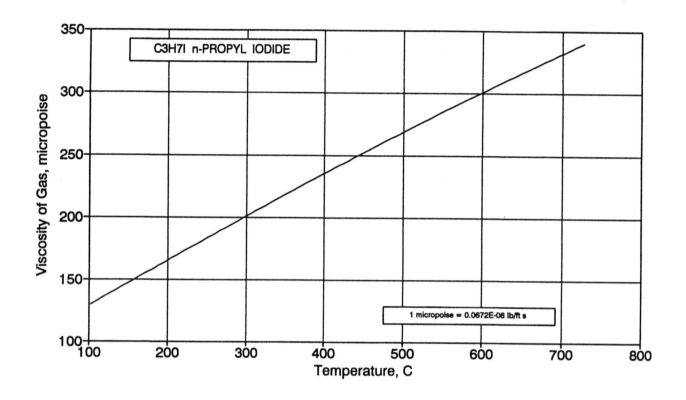

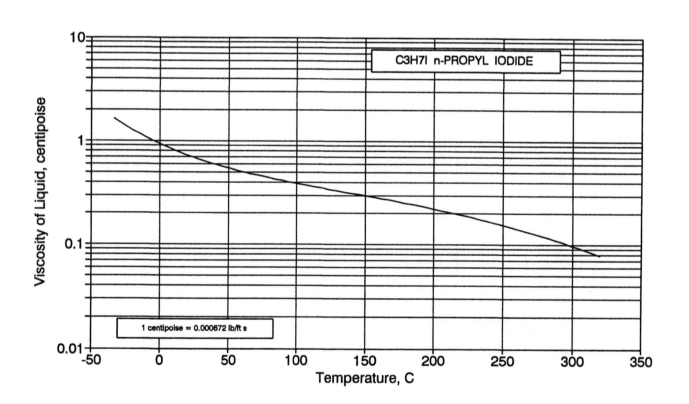

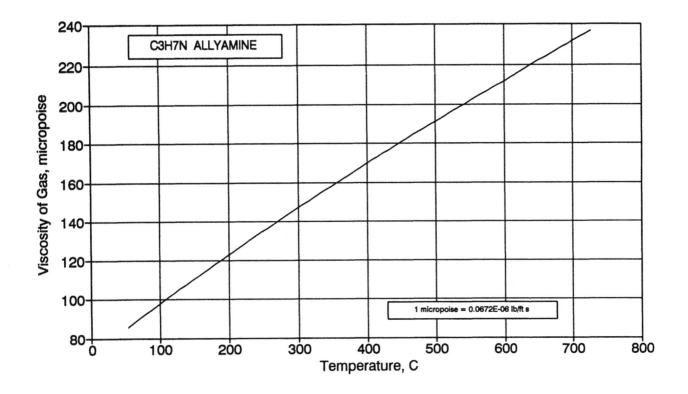

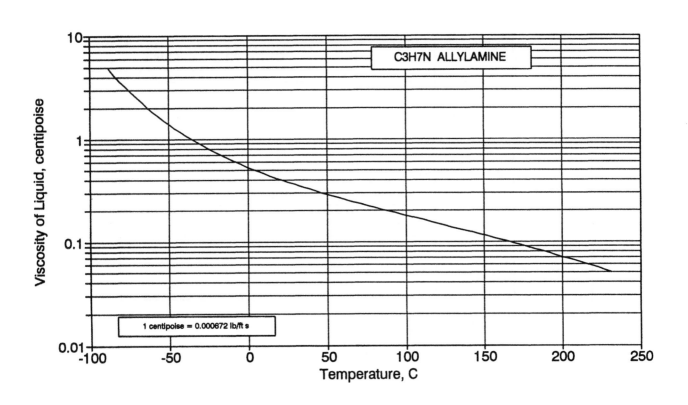

183

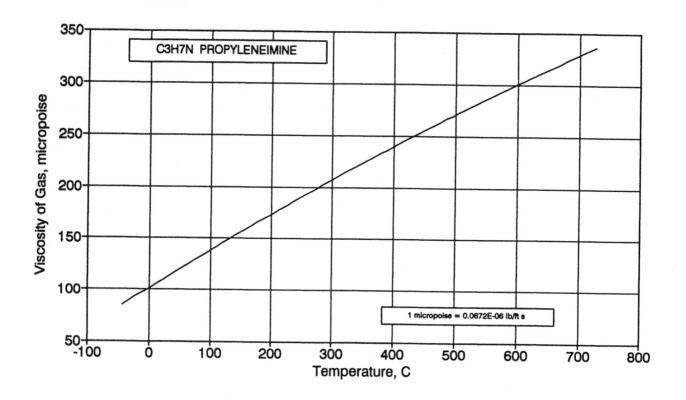

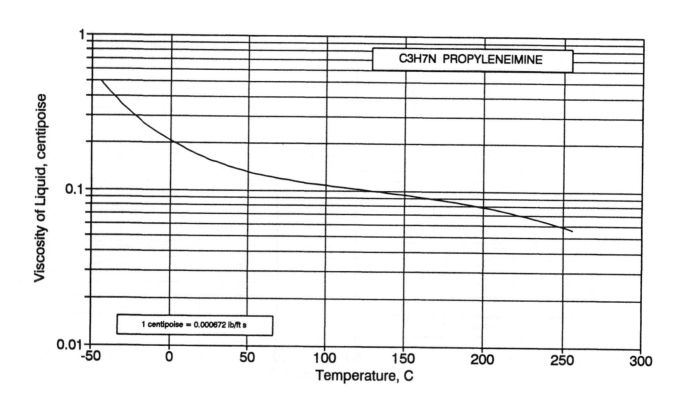

184

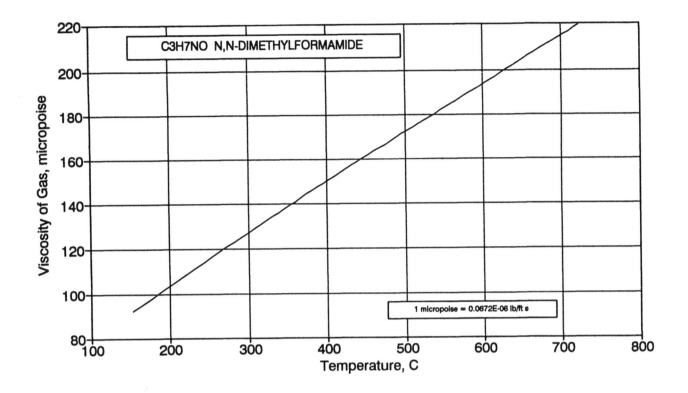

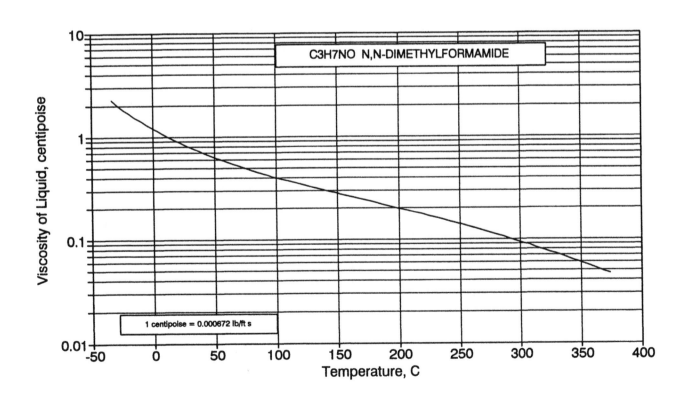

185

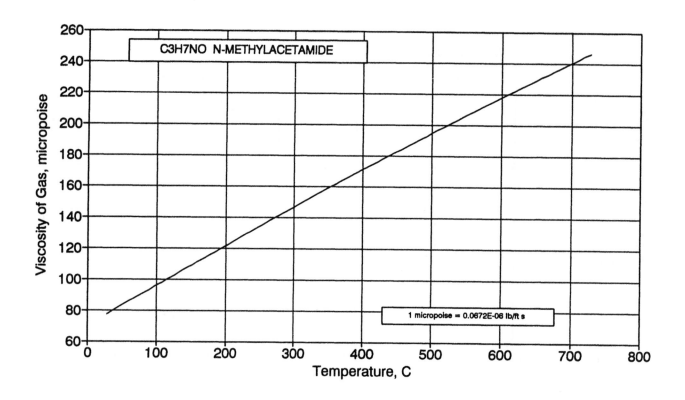

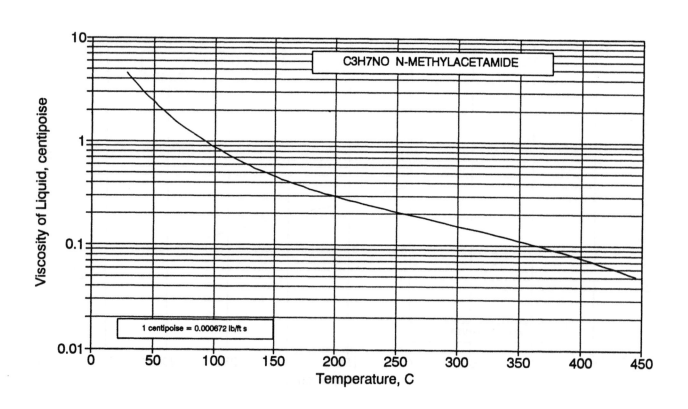

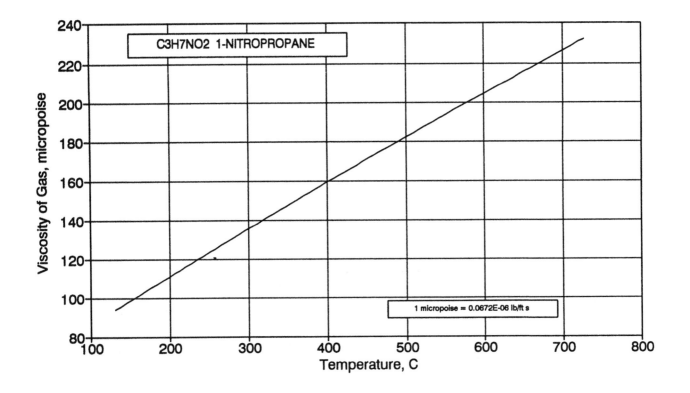

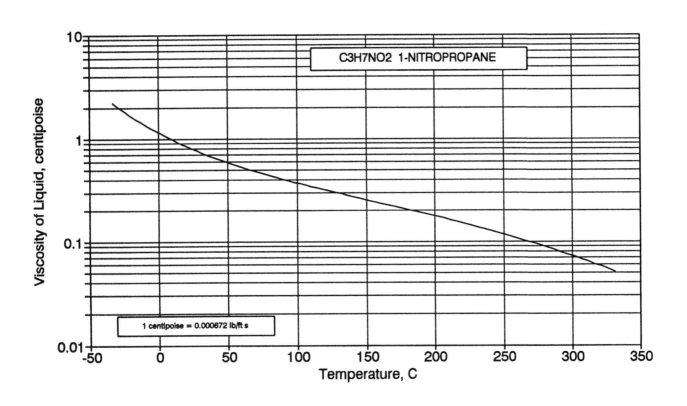

187

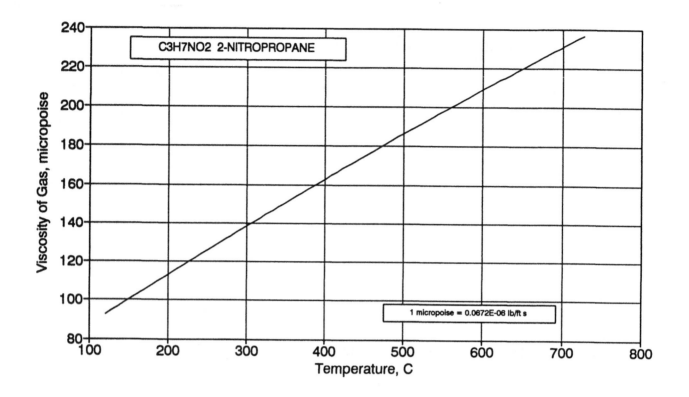

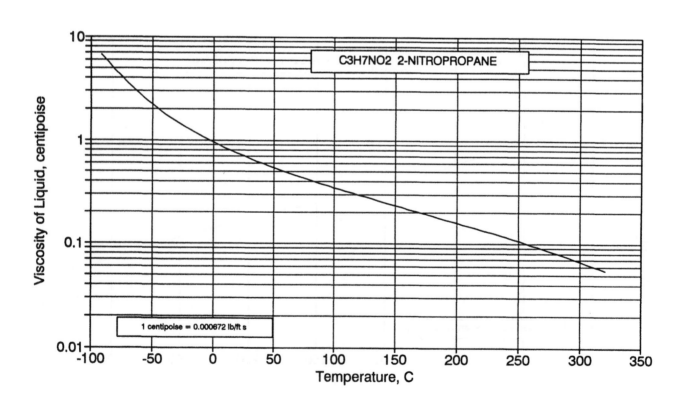

188

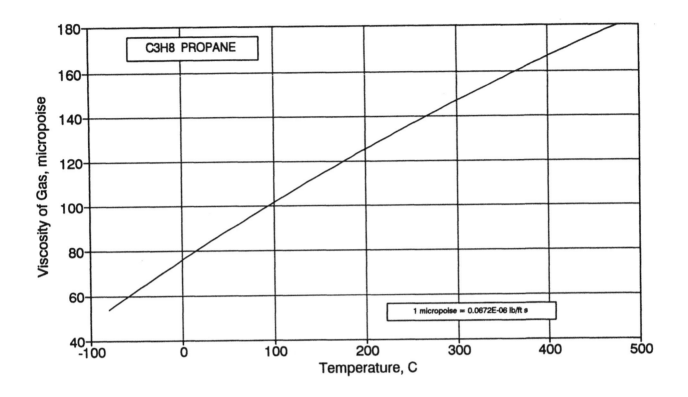

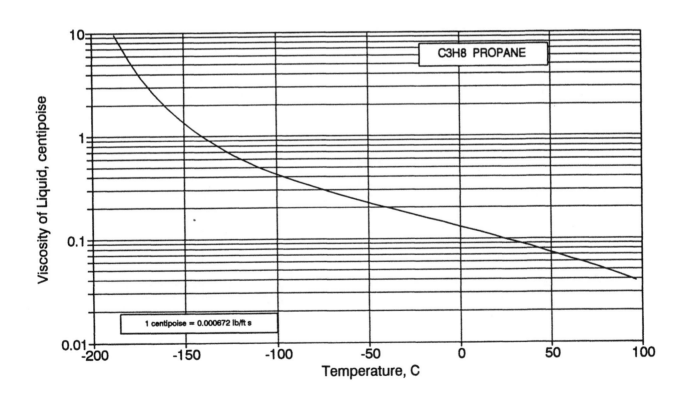

189

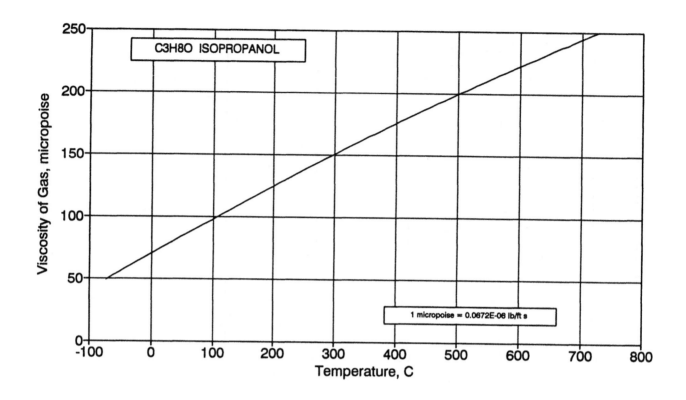

1 micropoise = 0.0672E-06 lb/ft s

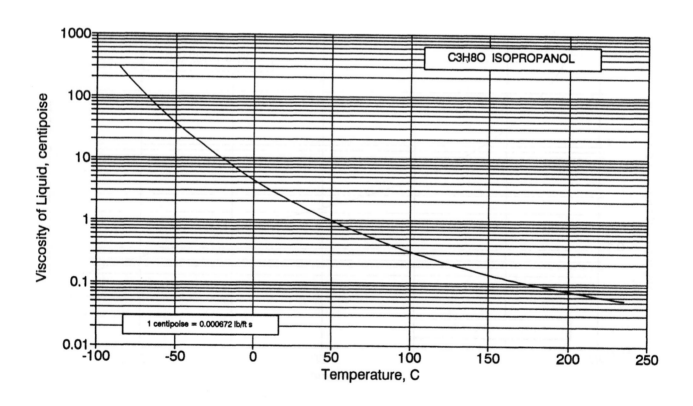

1 centipoise = 0.000672 lb/ft s

190

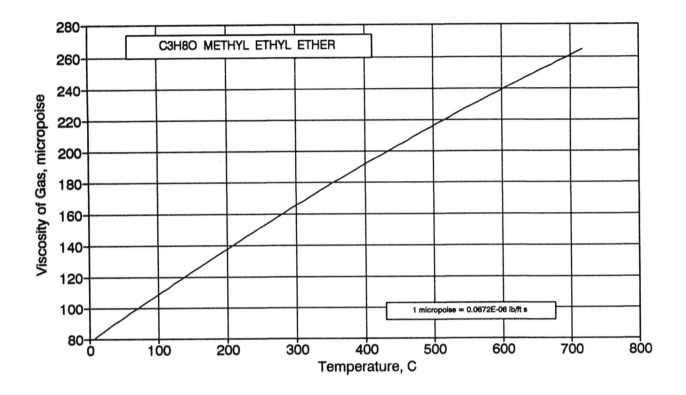

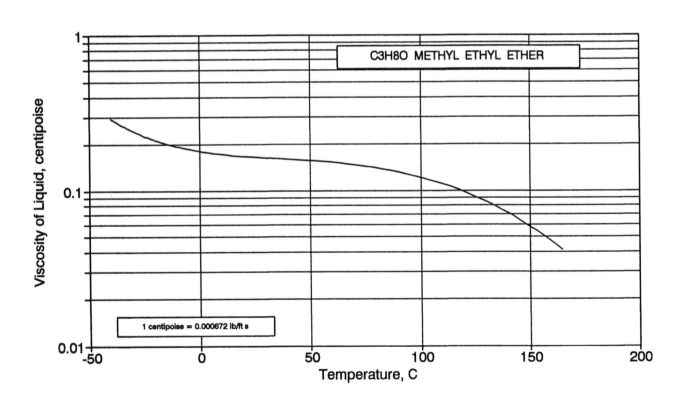

191

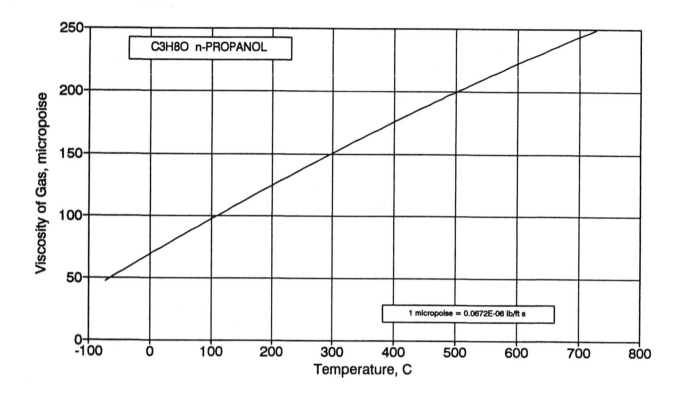

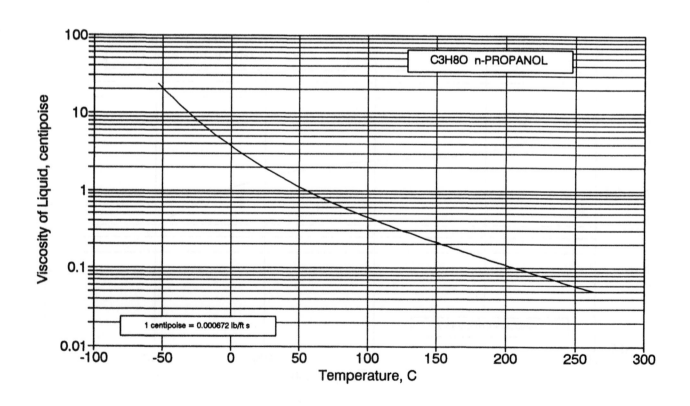

192

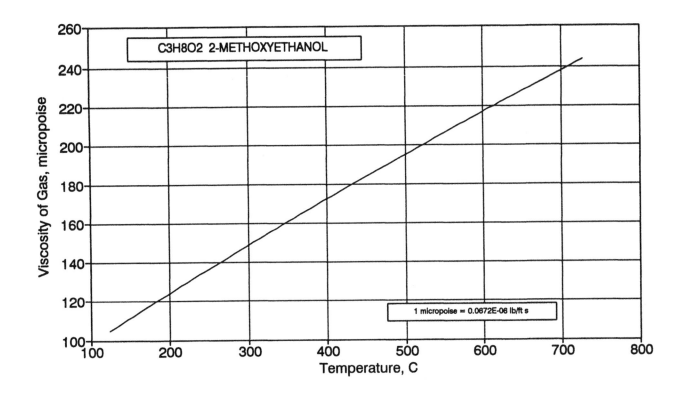

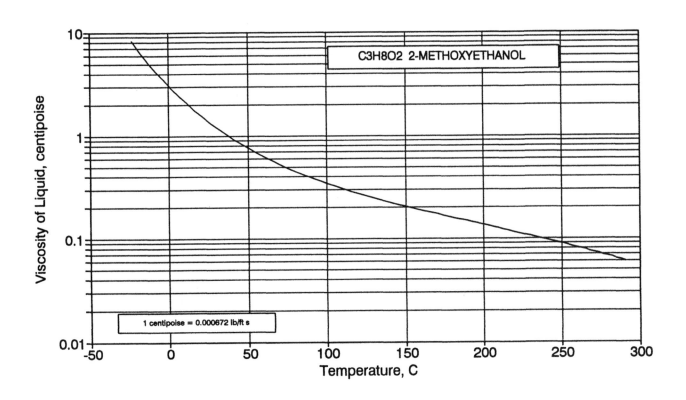

193

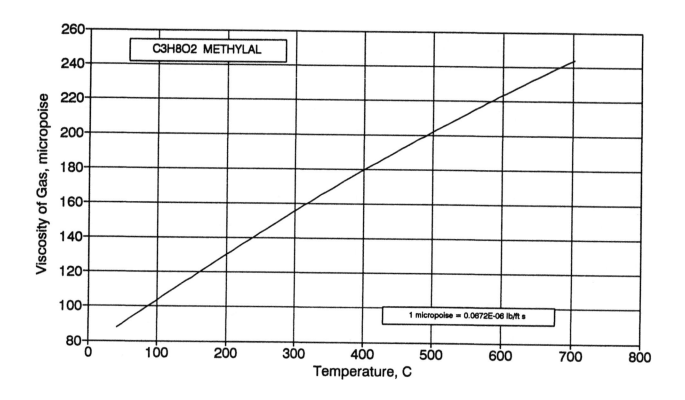

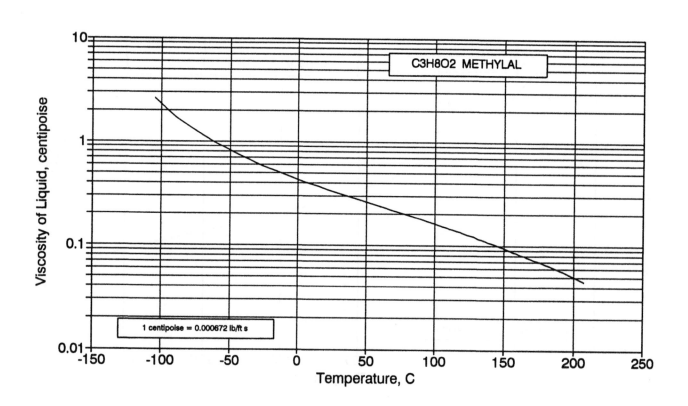

194

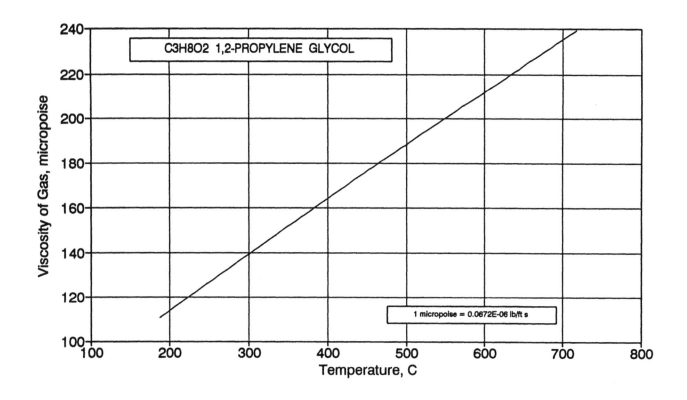

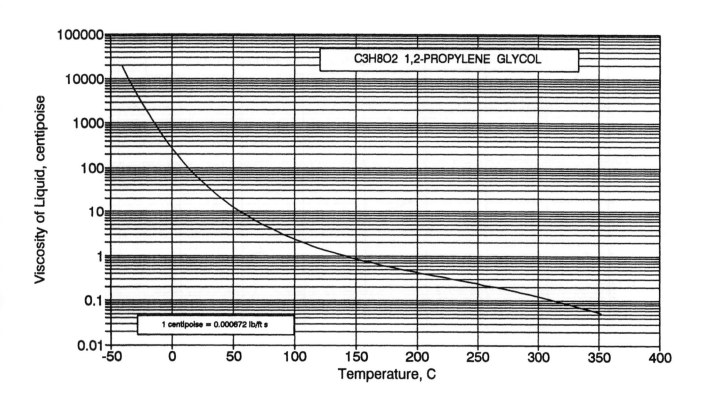

195

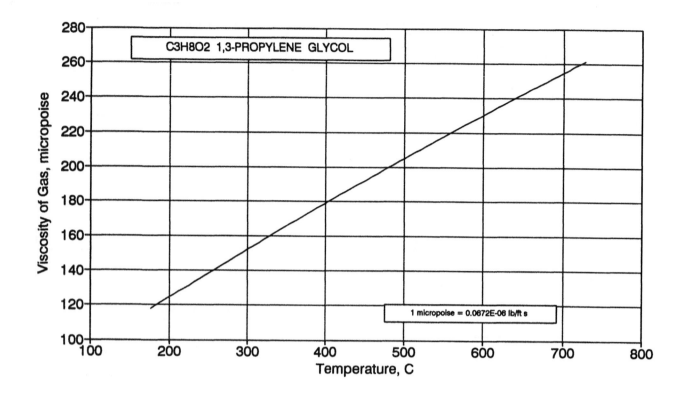

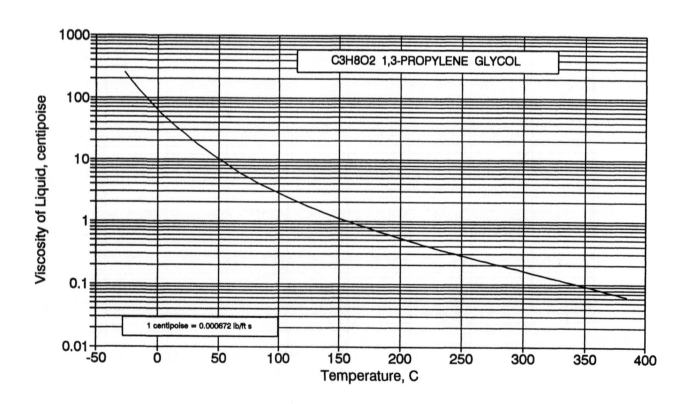

196

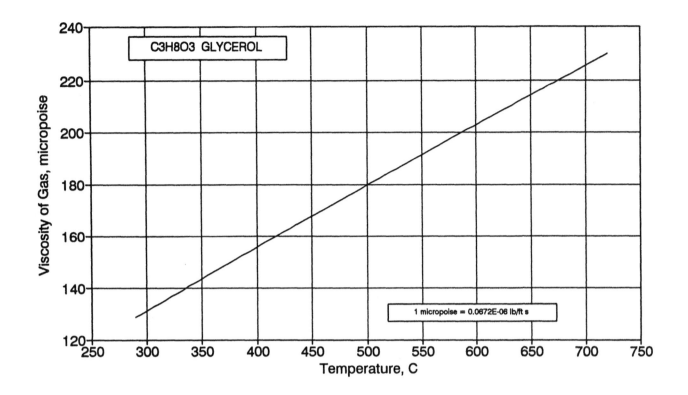

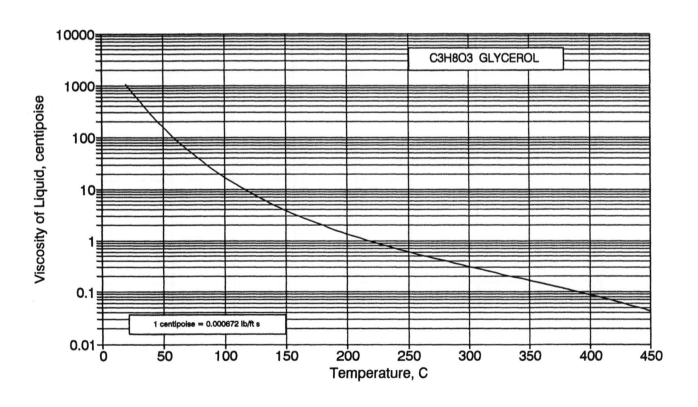

197

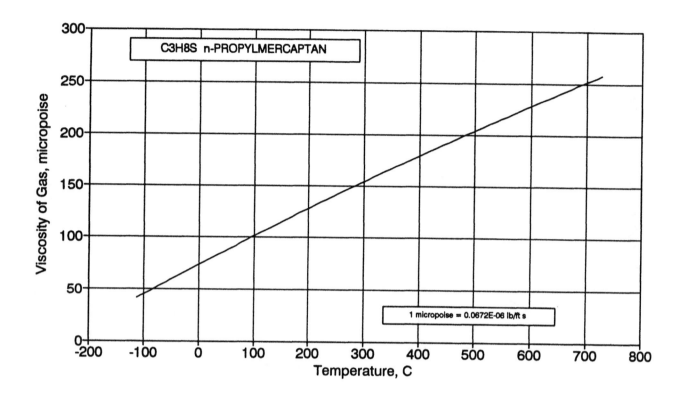

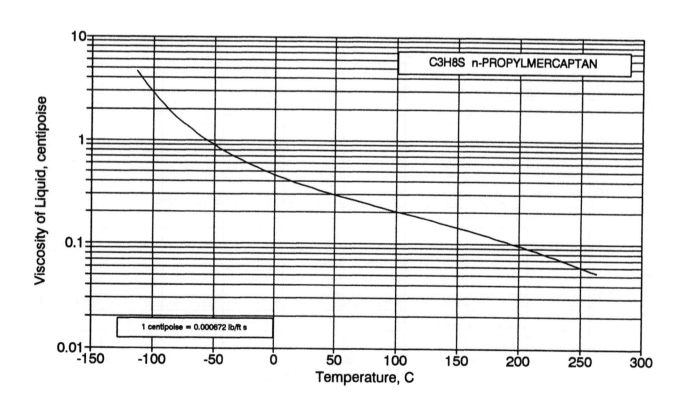

198

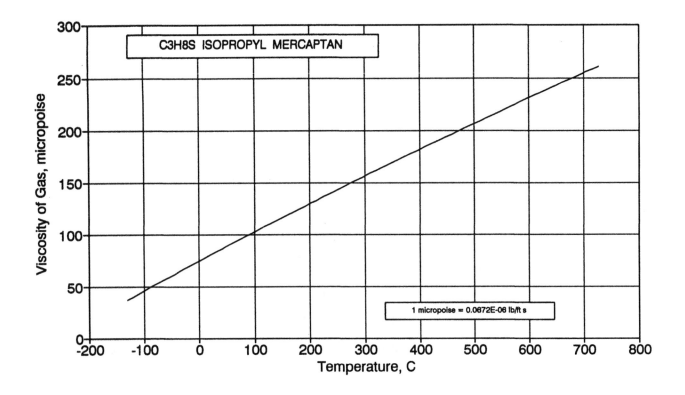

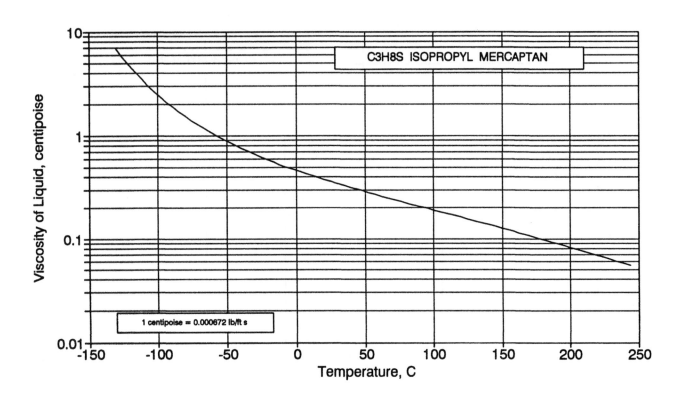

199

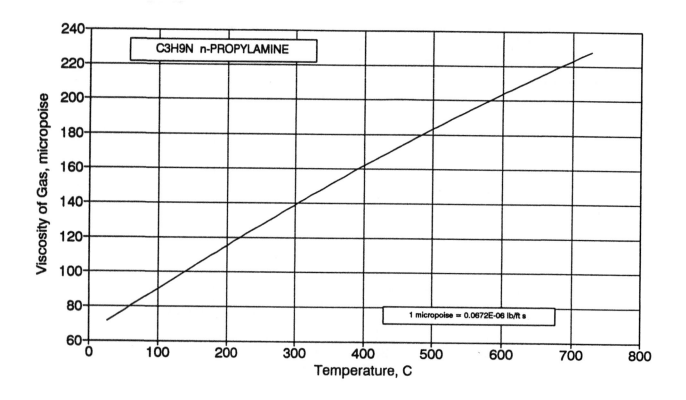

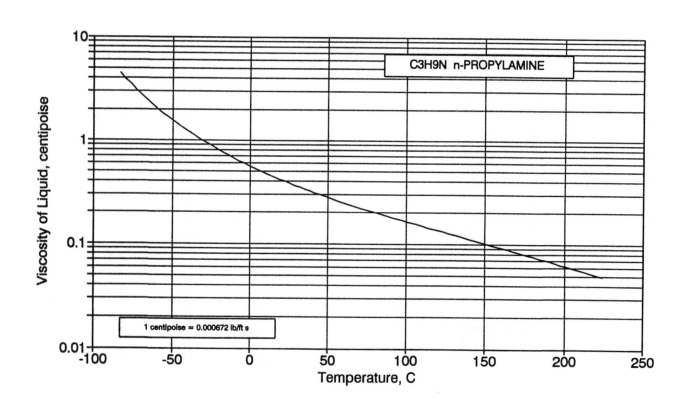

200

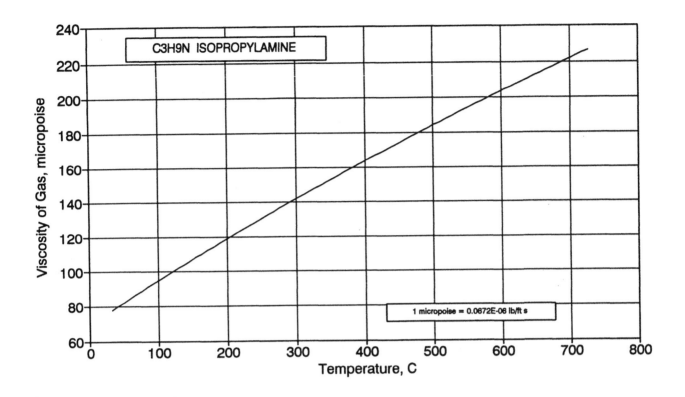

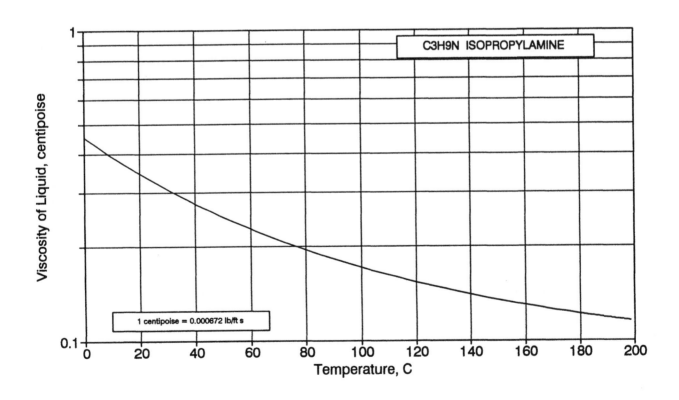

201

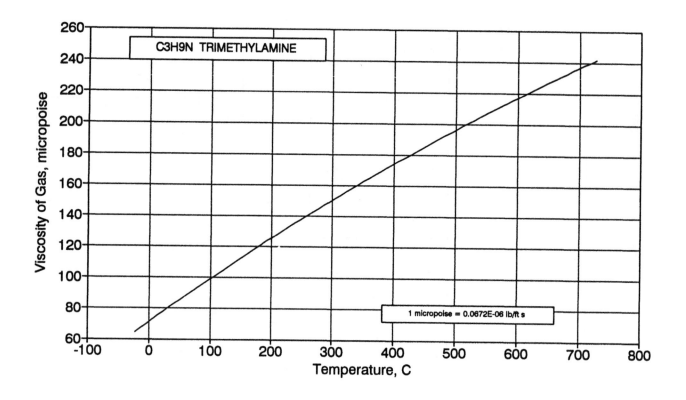

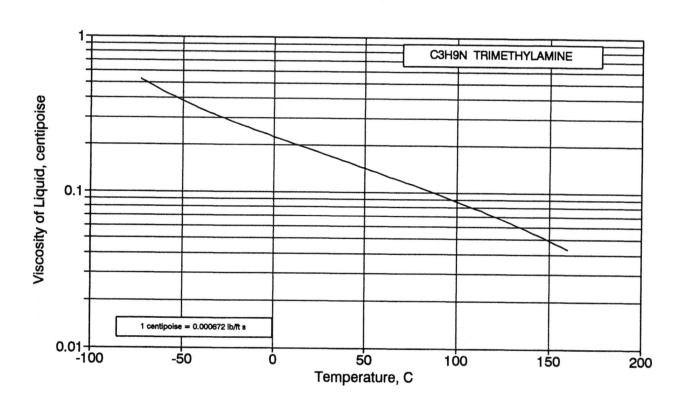

202

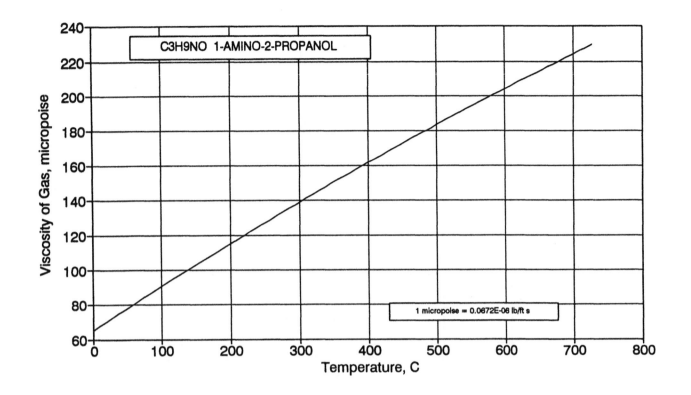

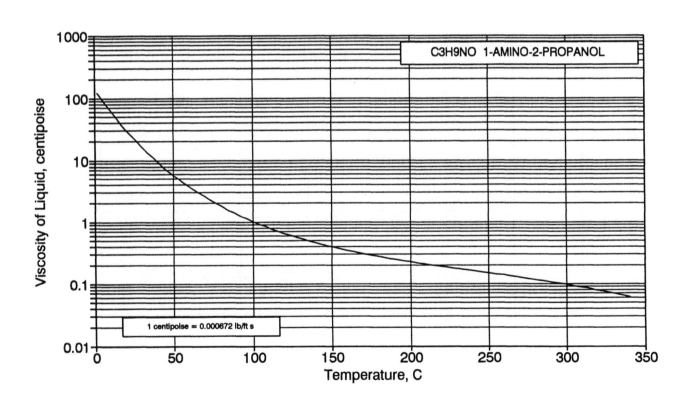

203

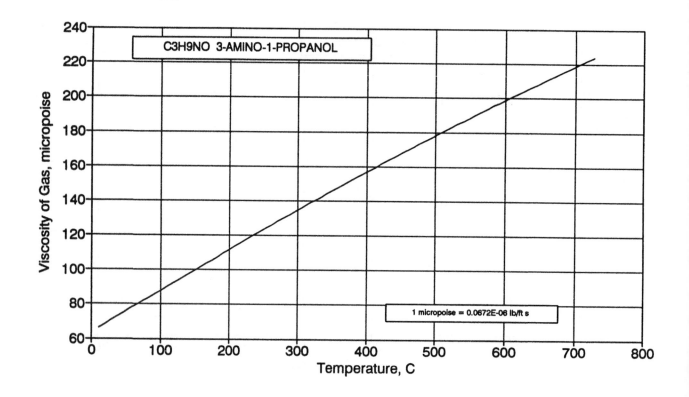

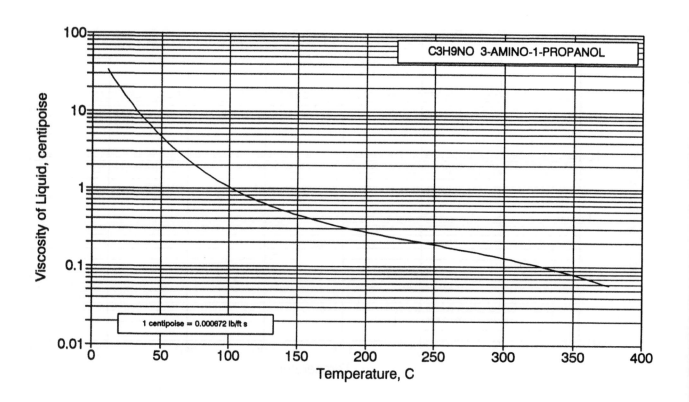

204

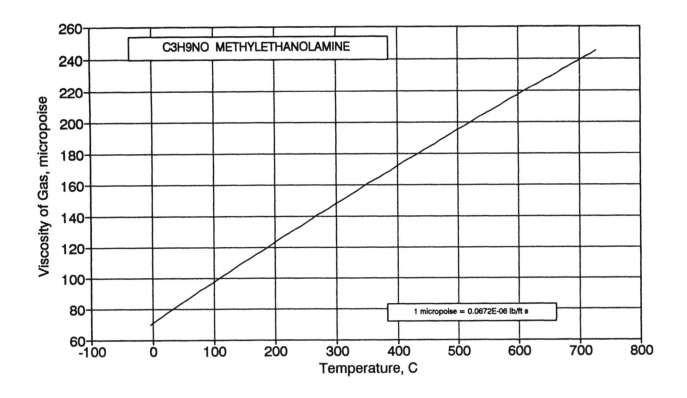

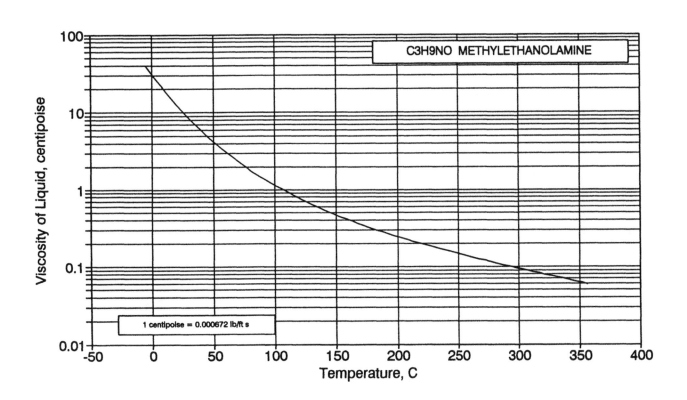

205

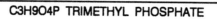

C3H9O4P TRIMETHYL PHOSPHATE

Gas viscosity data are not available .

TF , K = 227.00

TB , K = 465.85

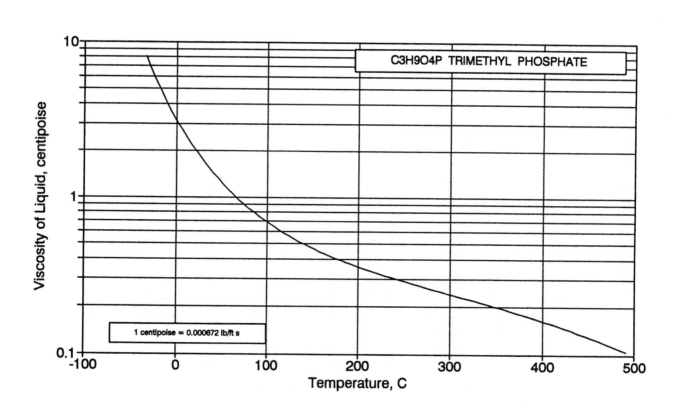

206

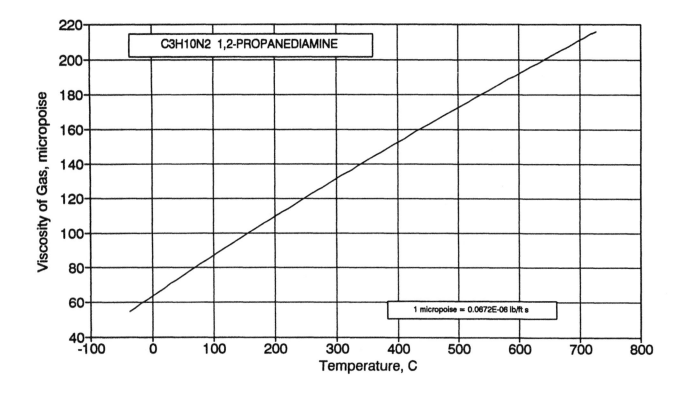

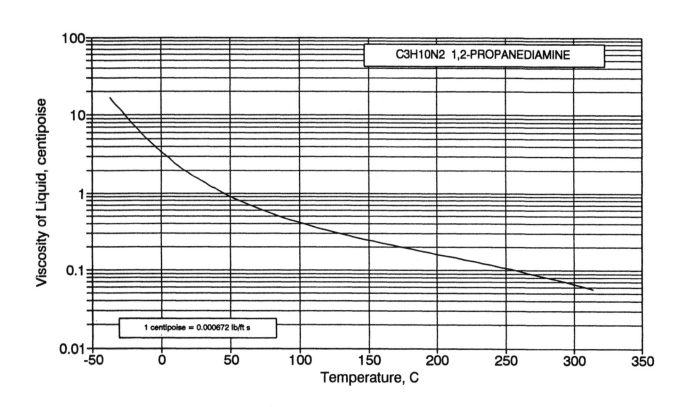

207

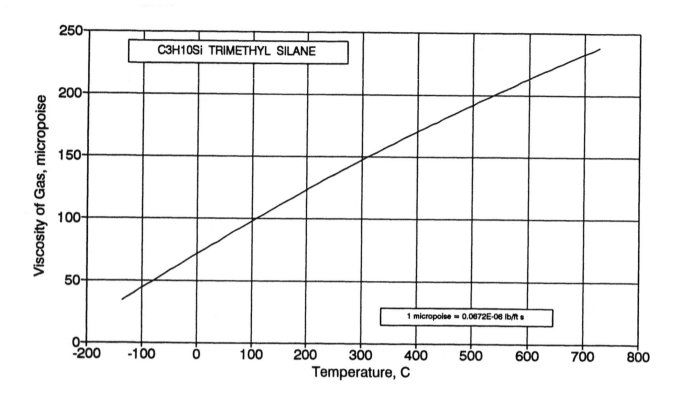

1 micropoise = 0.0672E-06 lb/ft s

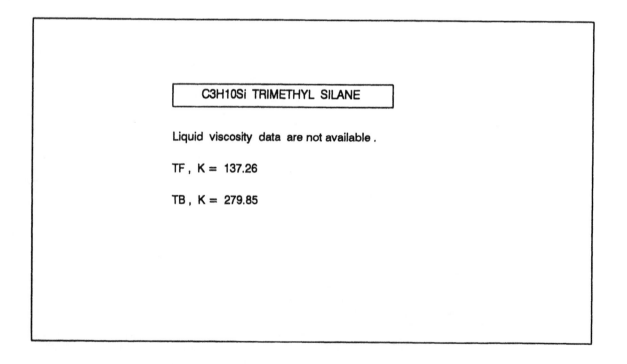

C3H10Si TRIMETHYL SILANE

Liquid viscosity data are not available.

TF, K = 137.26

TB, K = 279.85

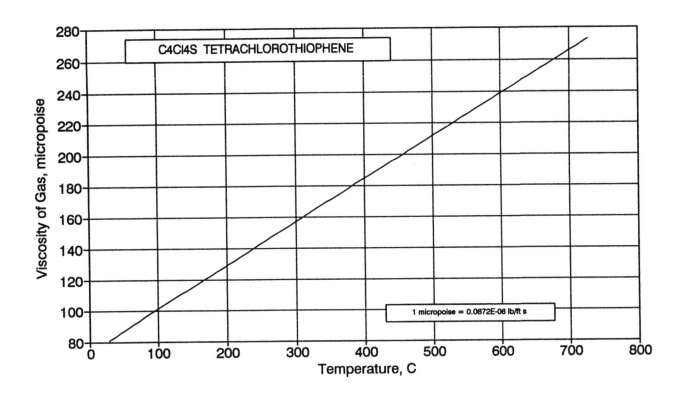

C4Cl4S TETRACHLOROTHIOPHENE

One liquid viscosity value is available:
3.318 cp @ 303.15 K.

TF , K = 301.97

TB , K = 506.54

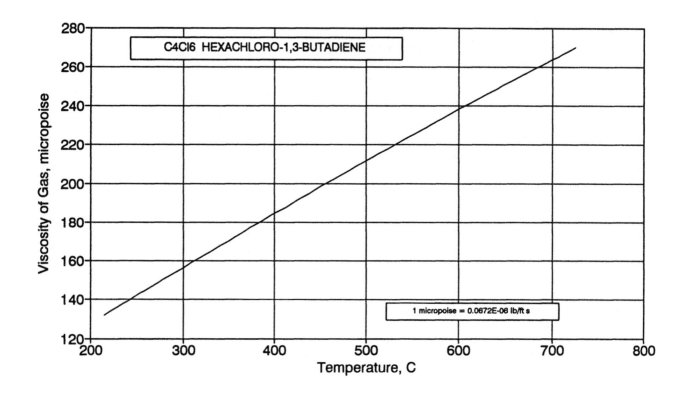

C4Cl6 HEXACHLORO-1,3-BUTADIENE

1 micropoise = 0.0672E-06 lb/ft s

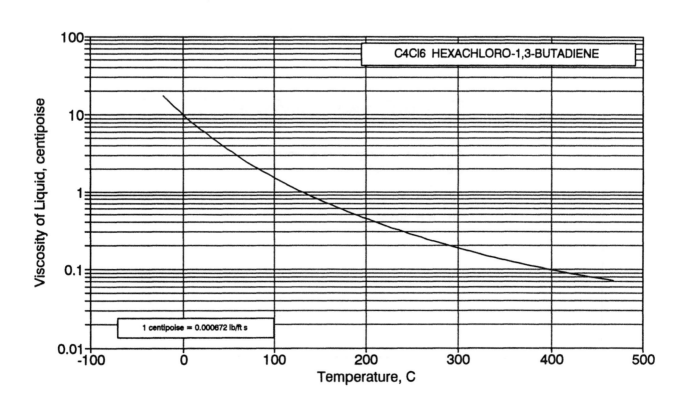

C4Cl6 HEXACHLORO-1,3-BUTADIENE

1 centipoise = 0.000672 lb/ft s

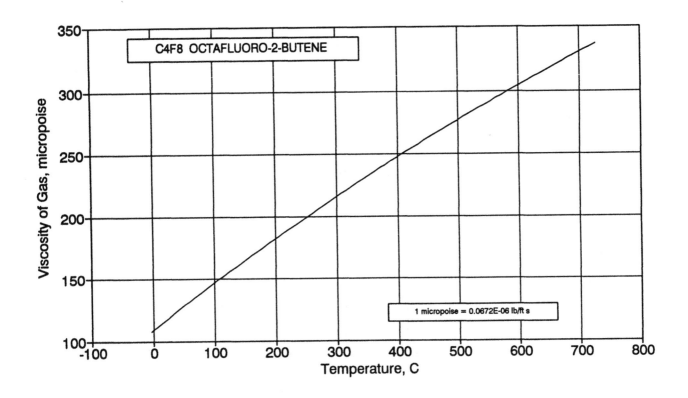

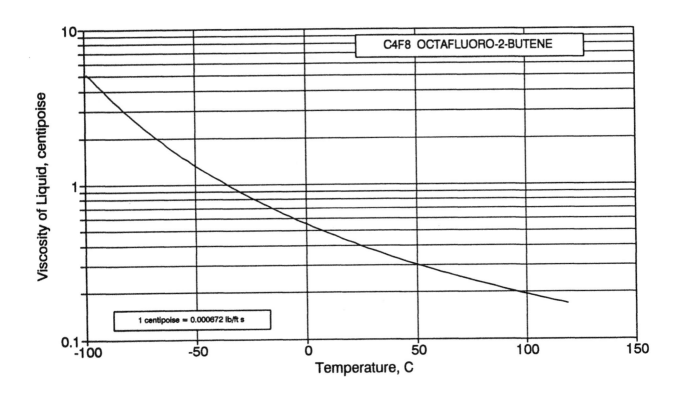

211

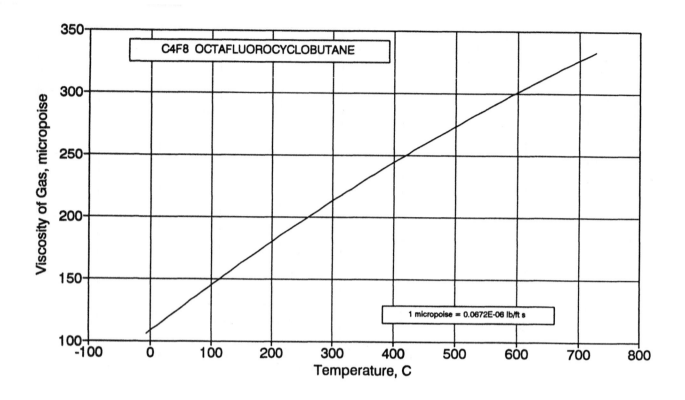

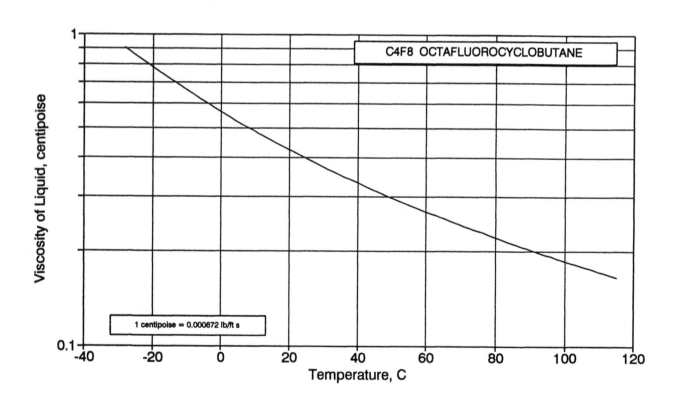

212

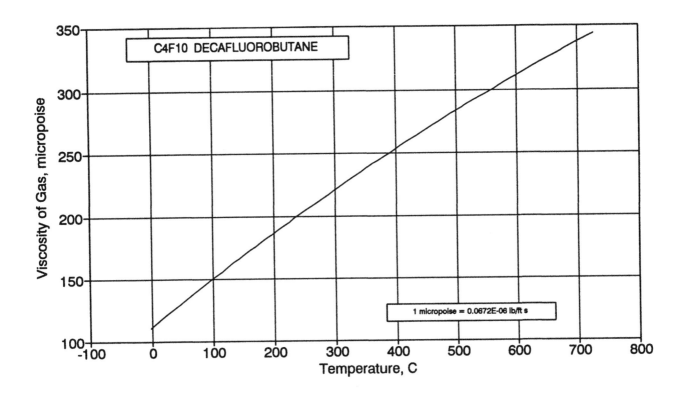

C4F10 DECAFLUOROBUTANE

1 micropoise = 0.0672E-06 lb/ft s

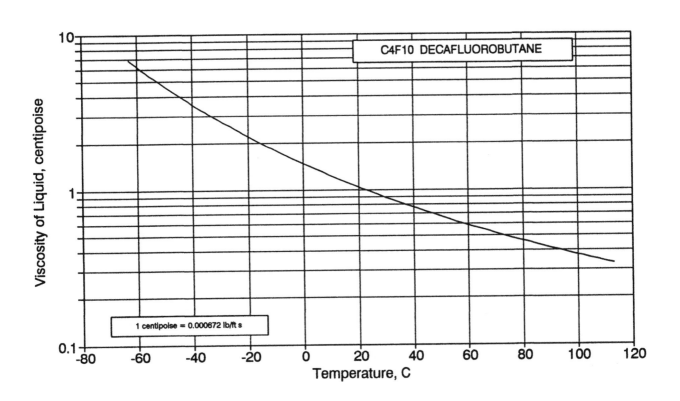

C4F10 DECAFLUOROBUTANE

1 centipoise = 0.000672 lb/ft s

213

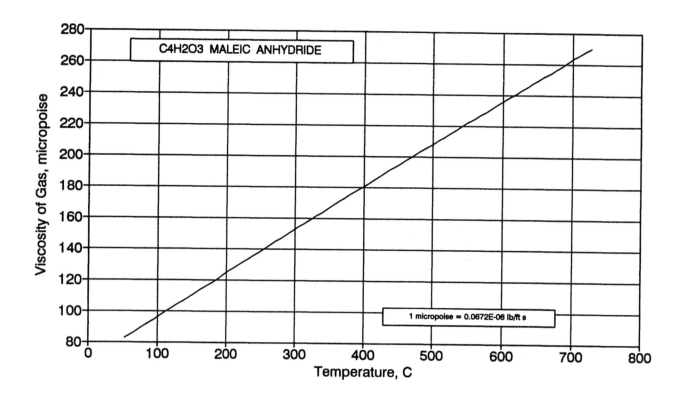

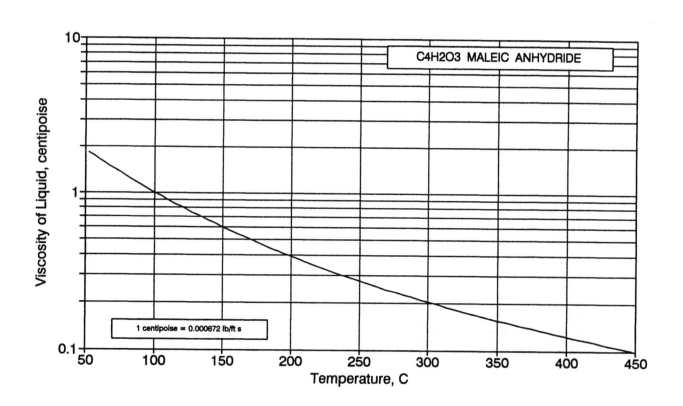

214

C4H4 VINYLACETYLENE

Gas viscosity data are not available.

TF , K = not available

TB , K = 278.25

C4H4 VINYLACETYLENE

Liquid viscosity data are not available.

TF , K = not available

TB , K = 278.25

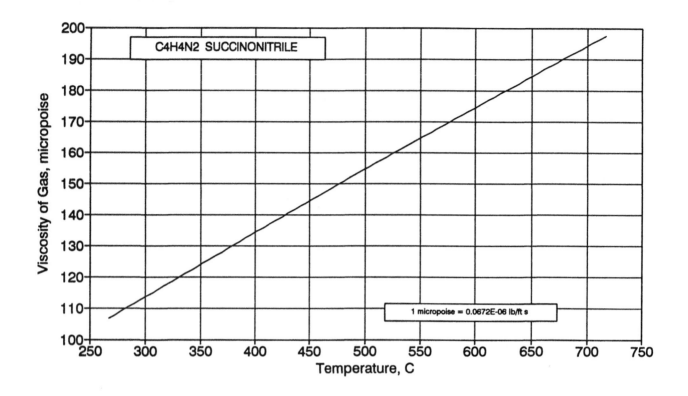

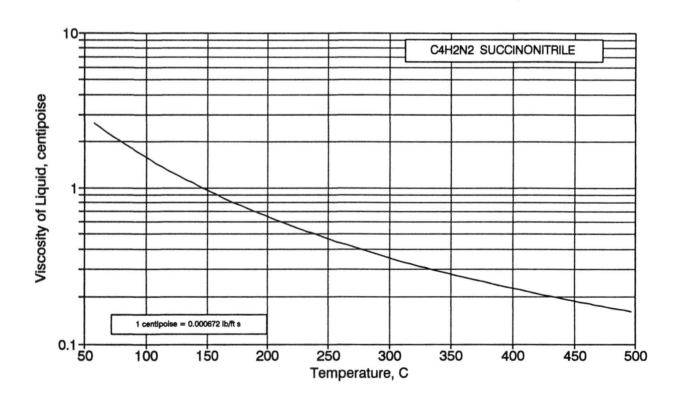

216

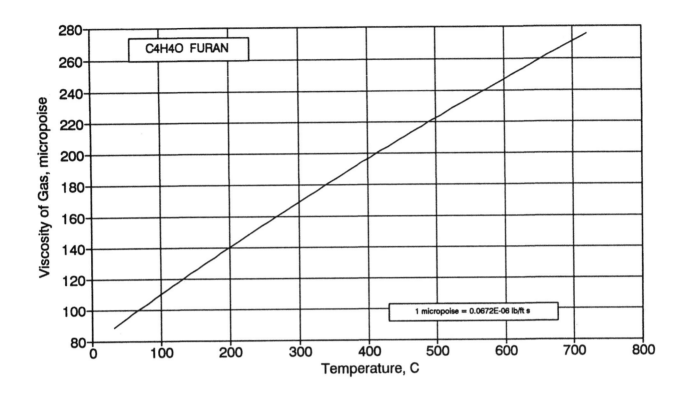

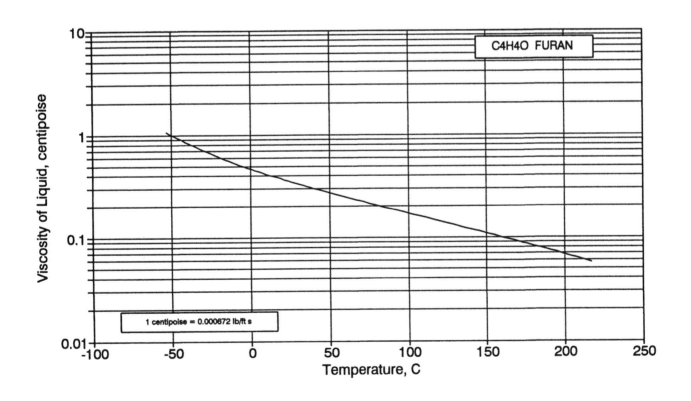

217

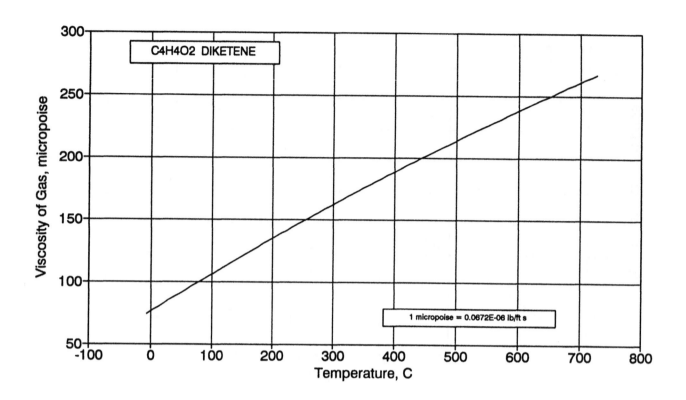

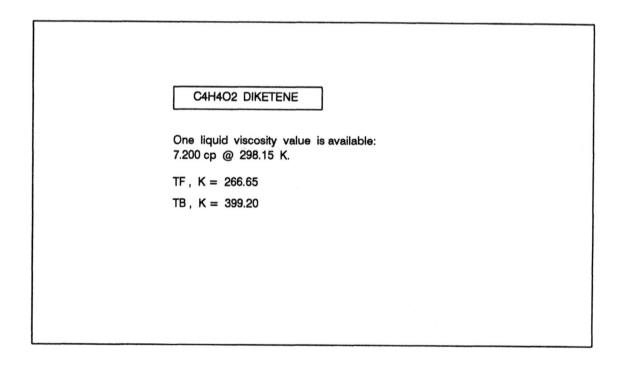

C4H4O2 DIKETENE

One liquid viscosity value is available:
7.200 cp @ 298.15 K.

TF , K = 266.65
TB , K = 399.20

218

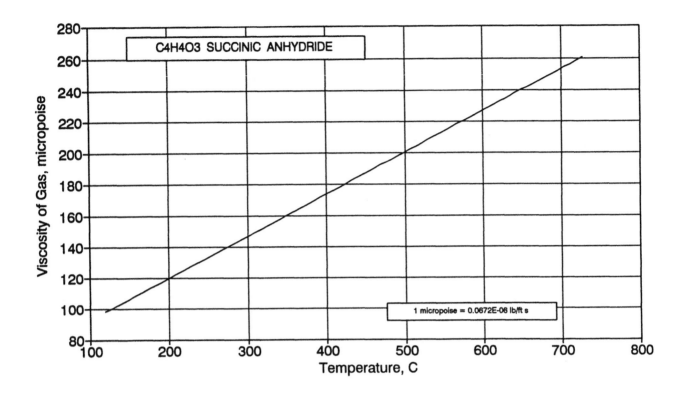

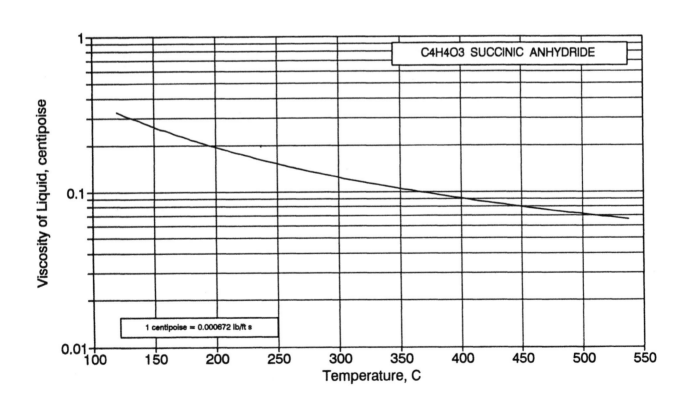

219

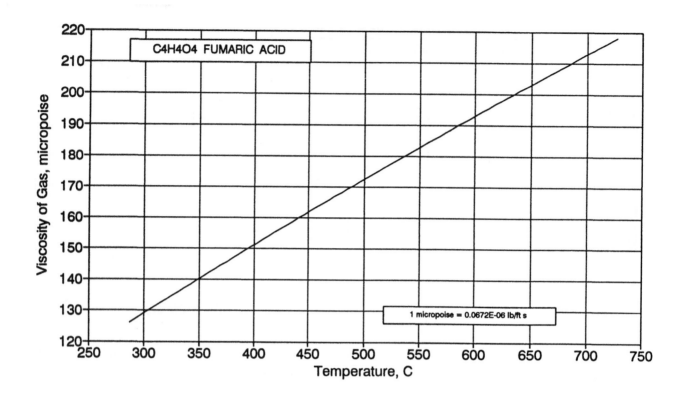

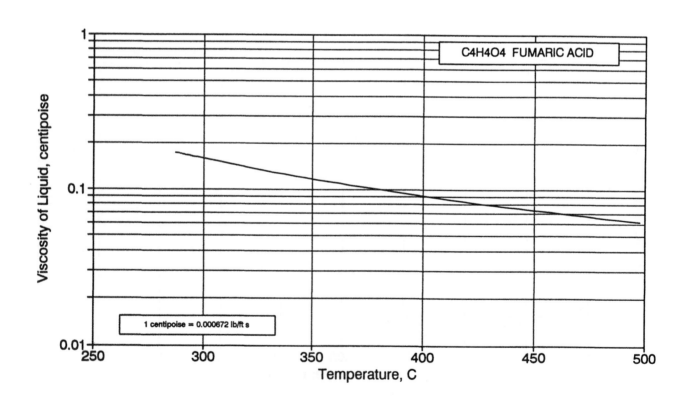

220

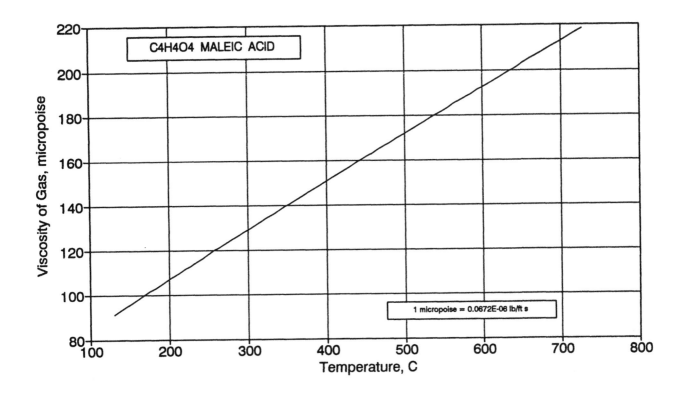

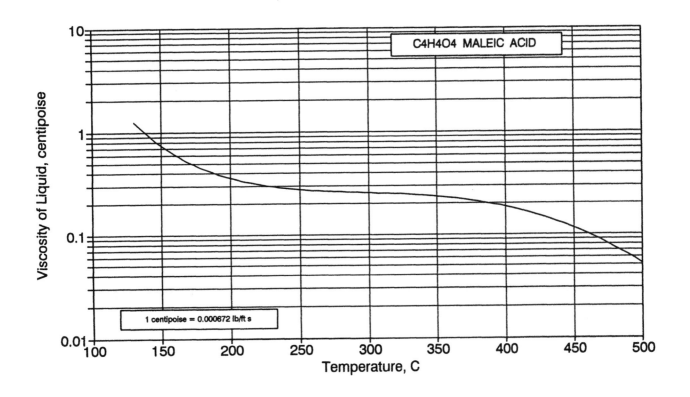

221

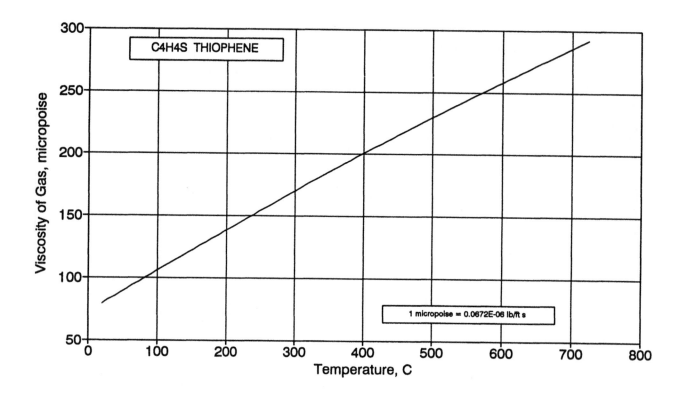

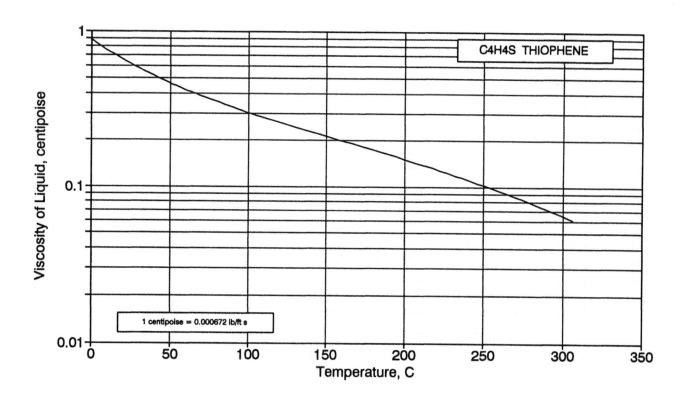

222

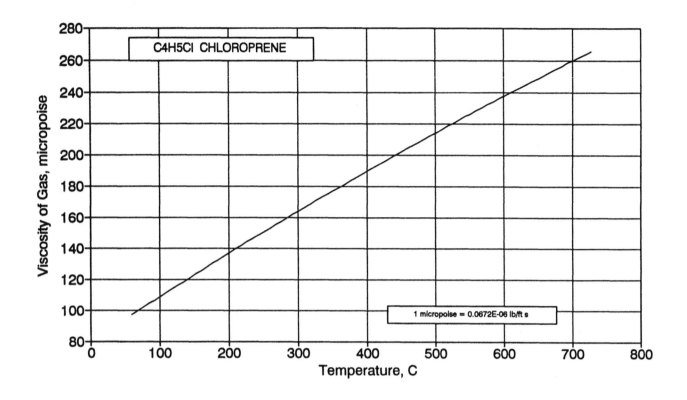

Viscosity of Gas, micropoise

C4H5Cl CHLOROPRENE

1 micropoise = 0.0672E-06 lb/ft s

Temperature, C

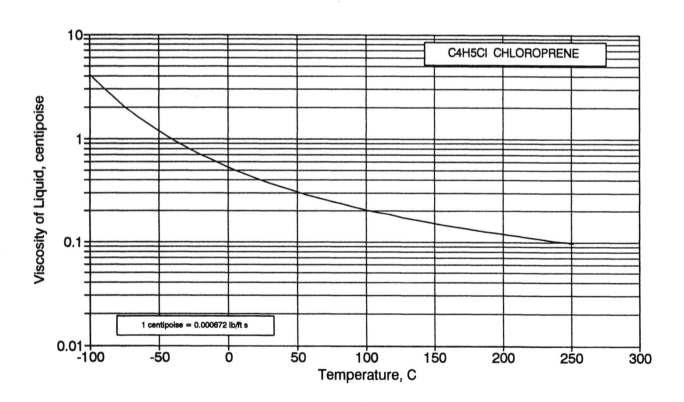

Viscosity of Liquid, centipoise

C4H5Cl CHLOROPRENE

1 centipoise = 0.000672 lb/ft s

Temperature, C

223

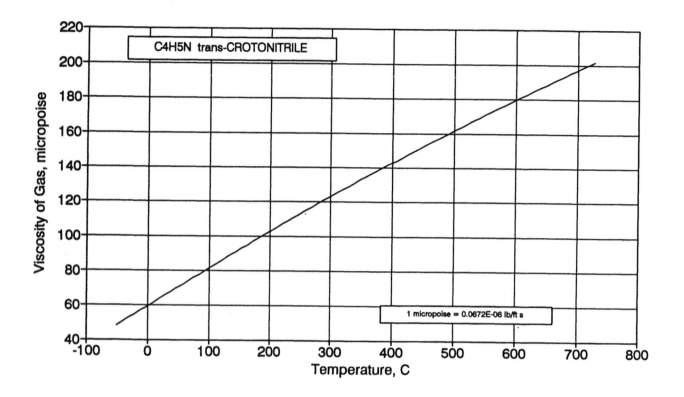

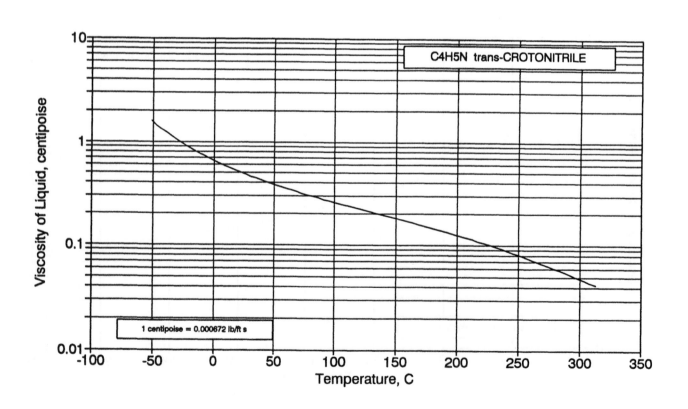

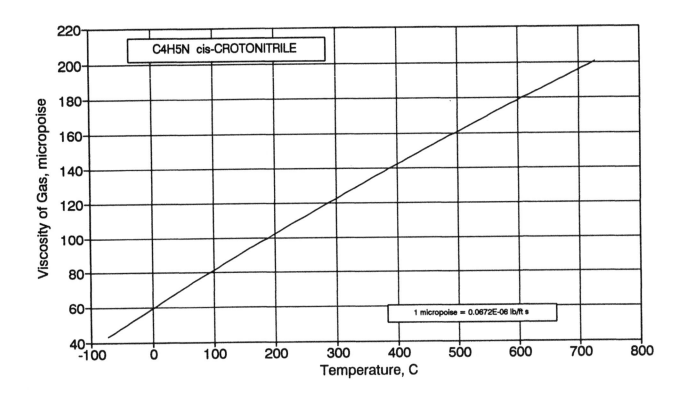

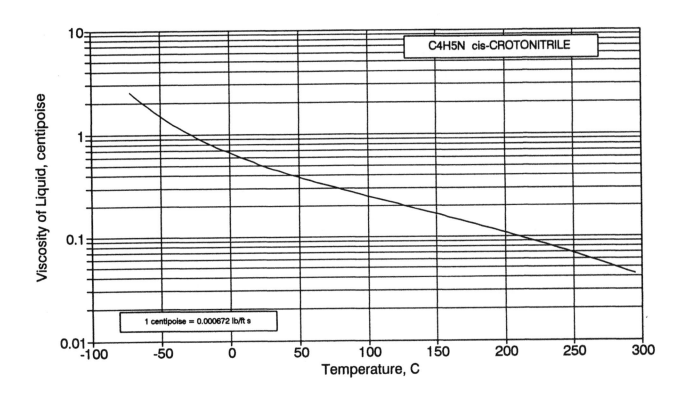

225

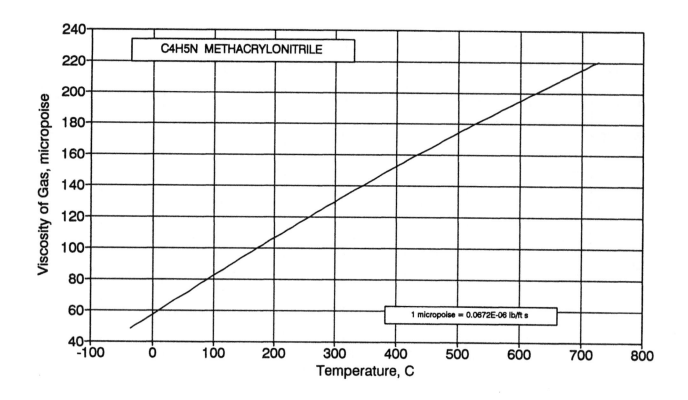

C4H5N METHACRYLONITRILE

Viscosity of Gas, micropoise vs Temperature, C

1 micropoise = 0.0672E-06 lb/ft s

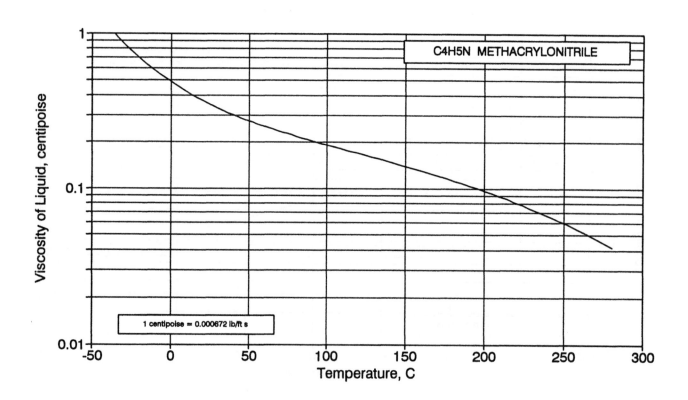

C4H5N METHACRYLONITRILE

Viscosity of Liquid, centipoise vs Temperature, C

1 centipoise = 0.000672 lb/ft s

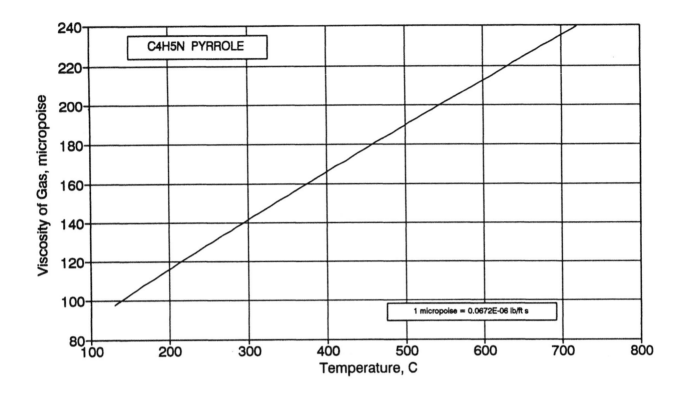

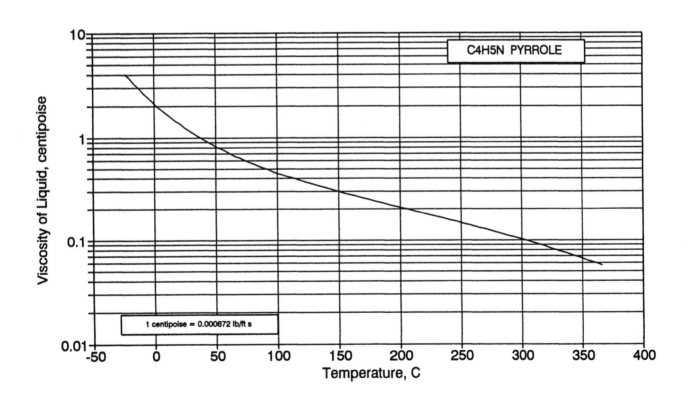

227

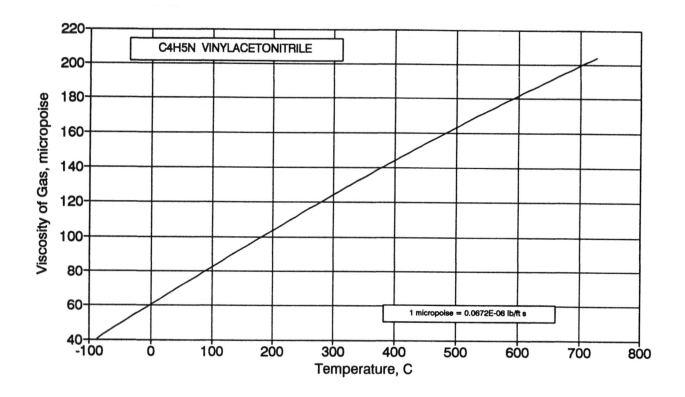

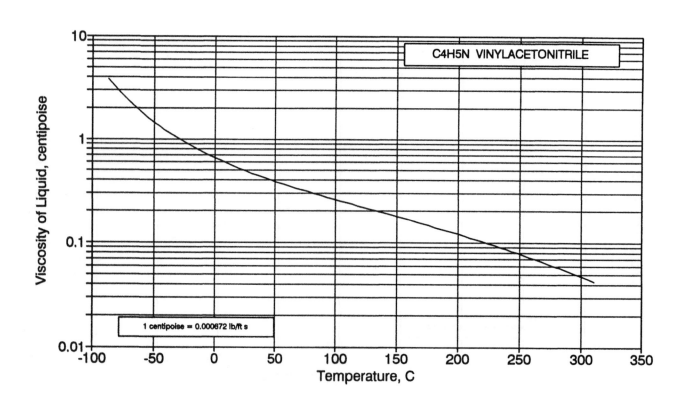

228

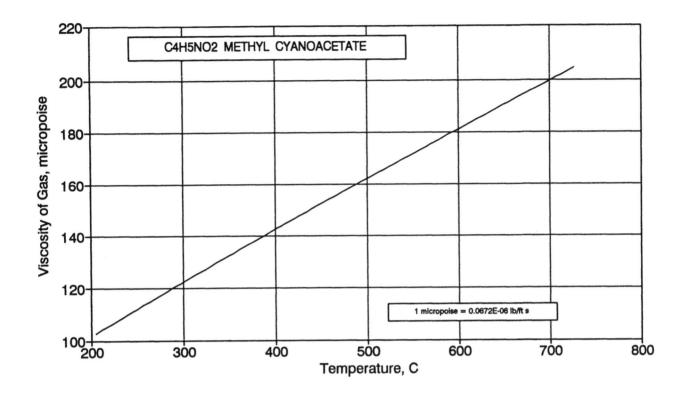

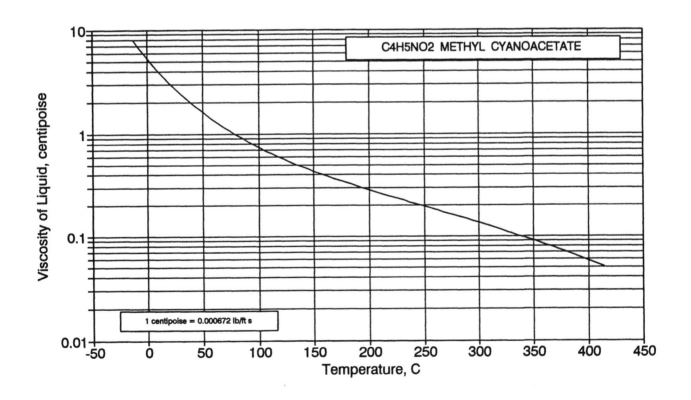

229

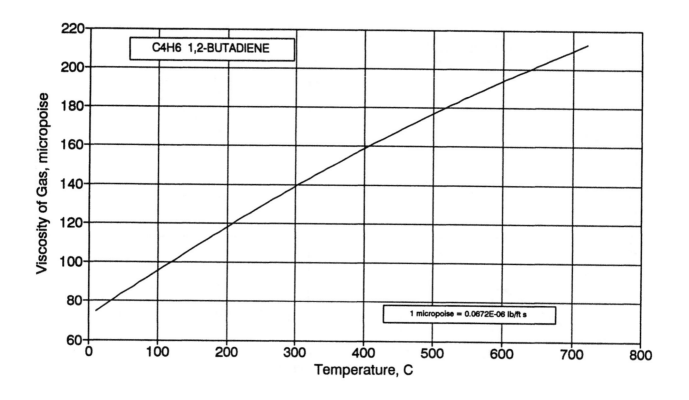

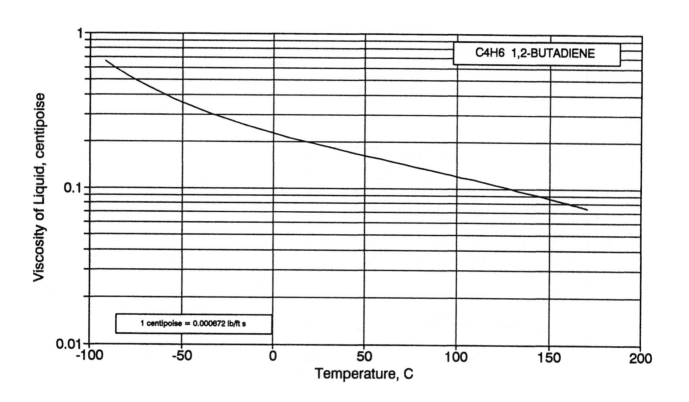

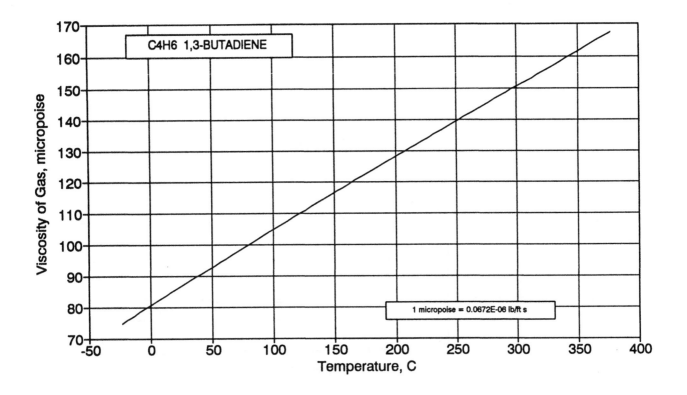

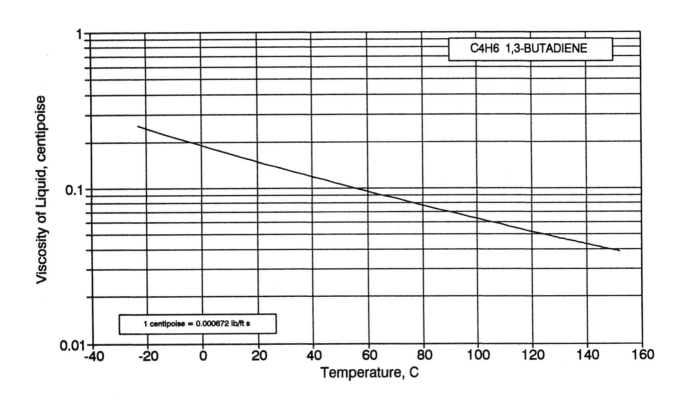

231

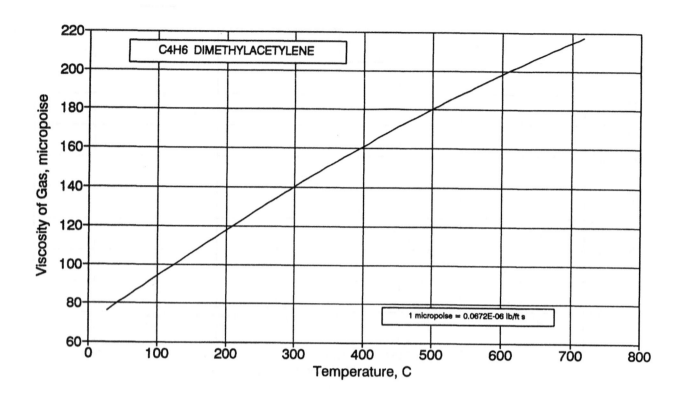

C4H6 DIMETHYLACETYLENE

1 micropoise = 0.0672E-06 lb/ft s

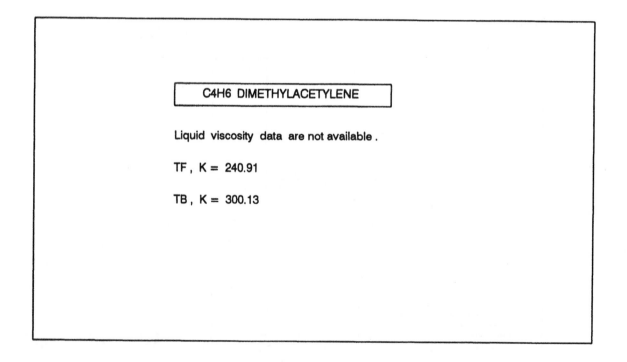

C4H6 DIMETHYLACETYLENE

Liquid viscosity data are not available.

TF , K = 240.91

TB , K = 300.13

232

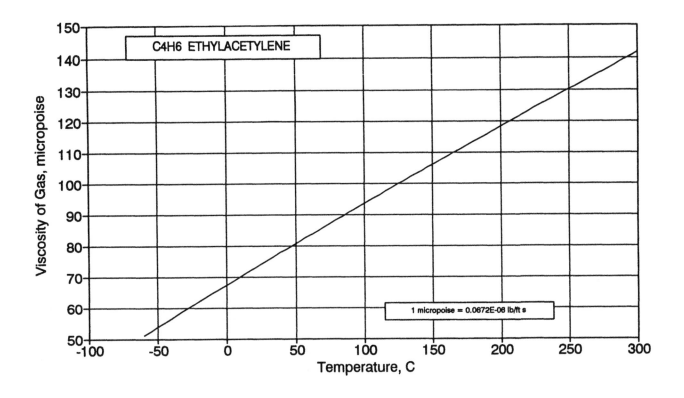

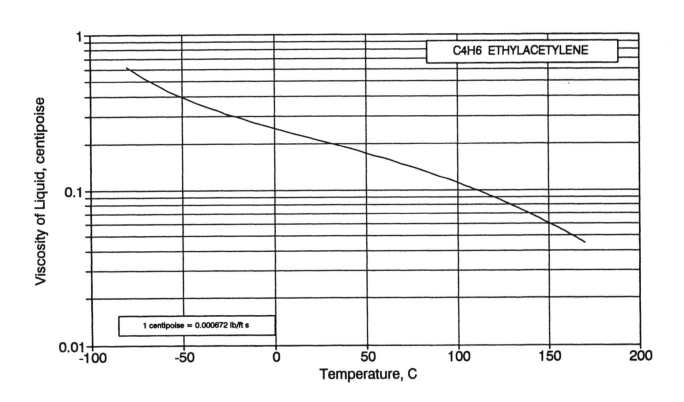

233

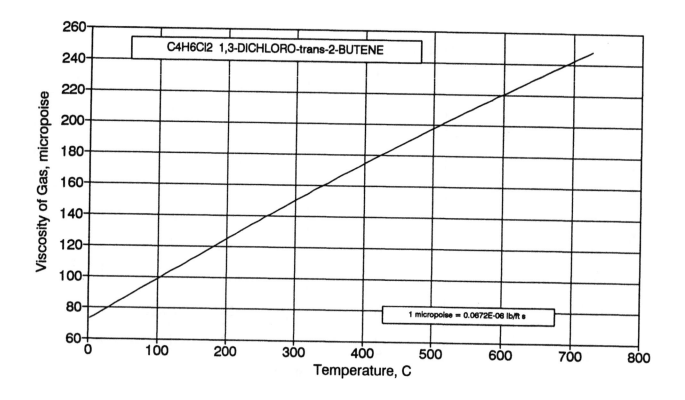

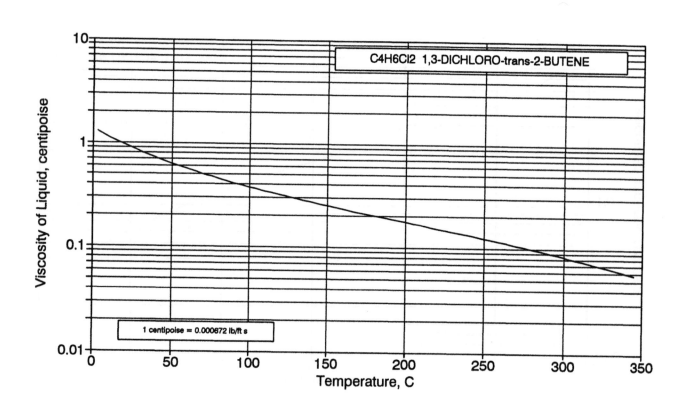

234

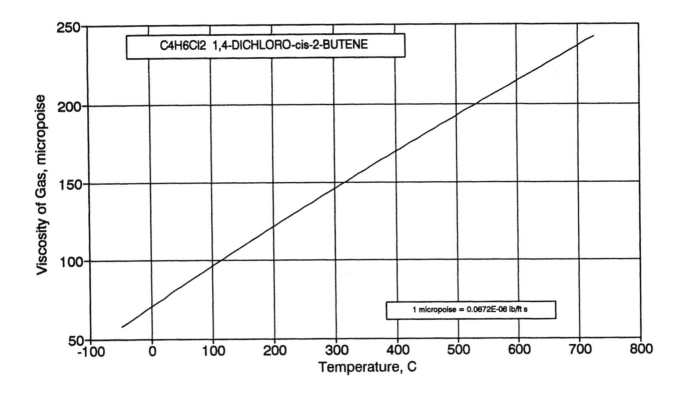

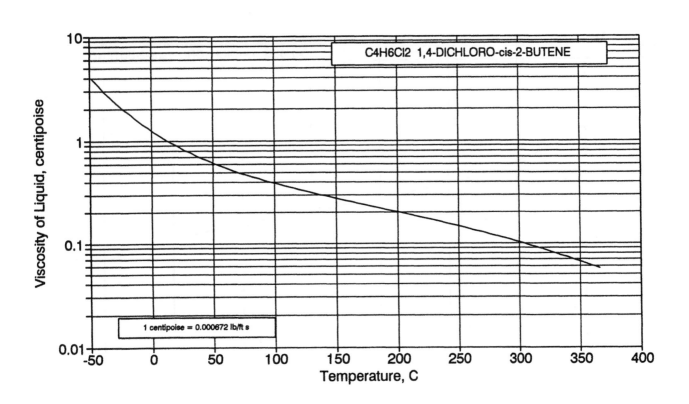

235

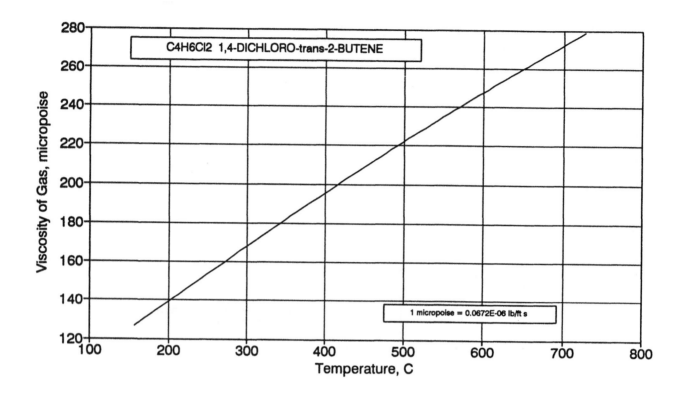

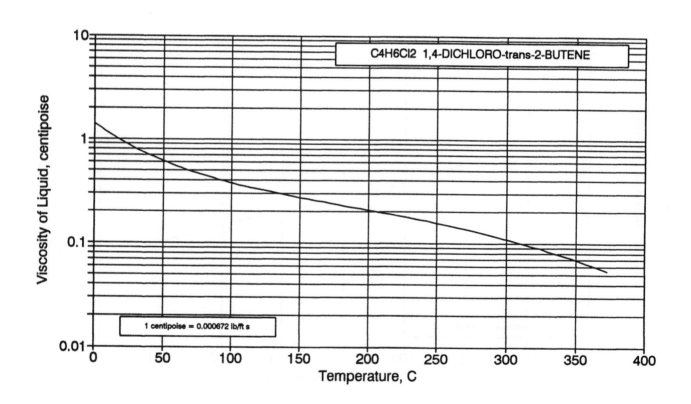

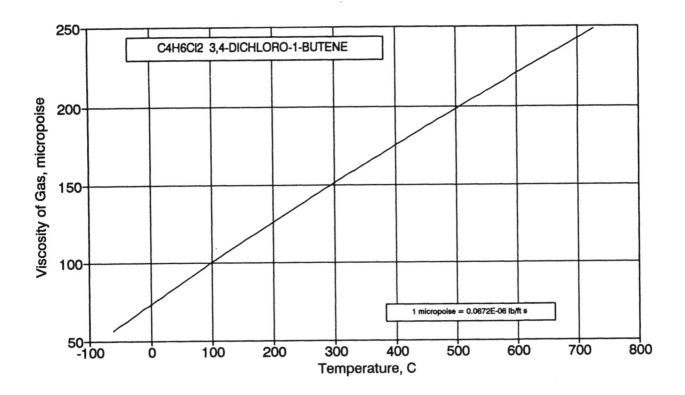

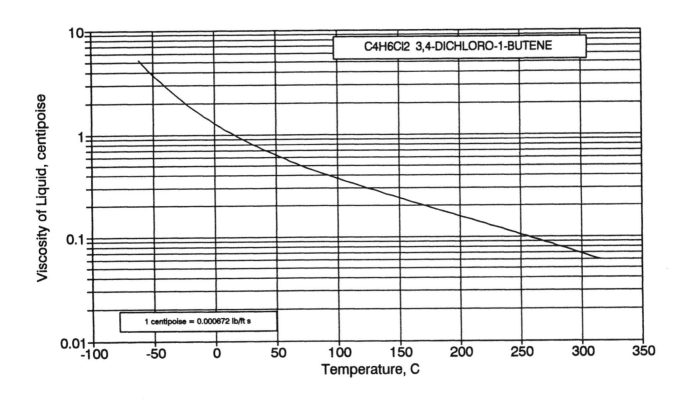

237

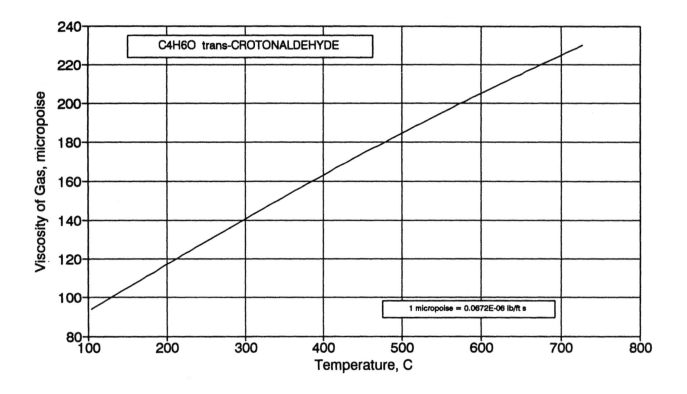

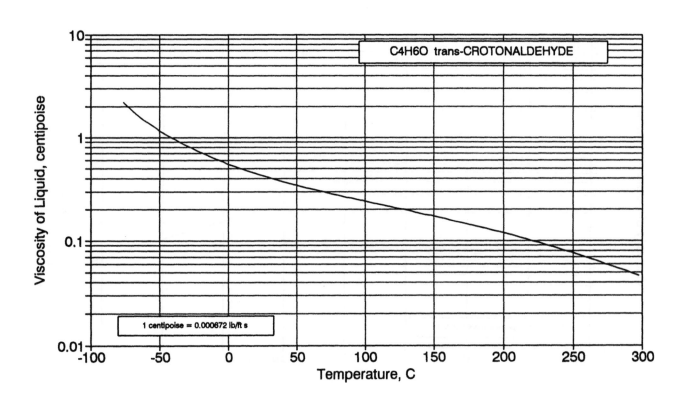

238

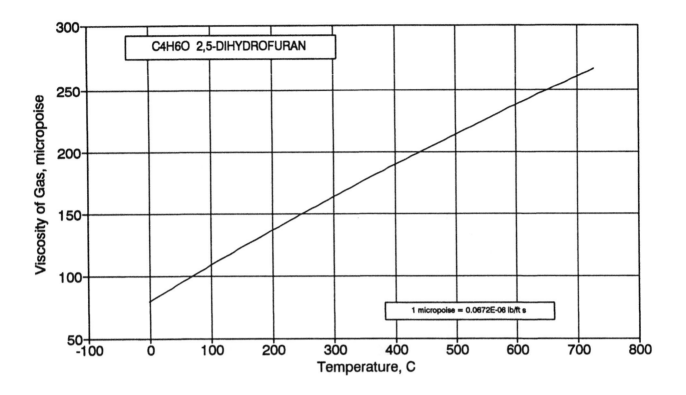

C4H6O 2,5-DIHYDROFURAN

1 micropoise = 0.0672E-06 lb/ft s

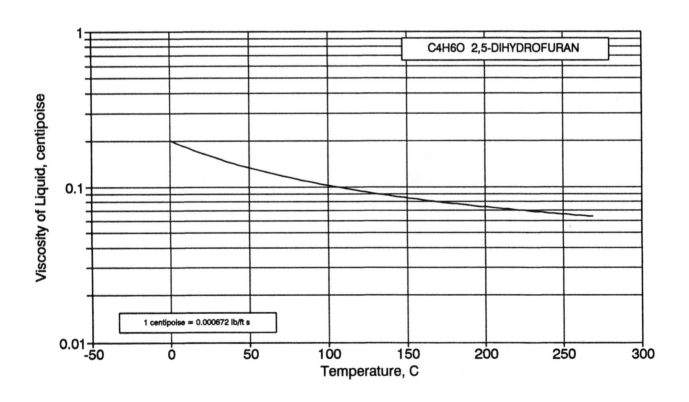

C4H6O 2,5-DIHYDROFURAN

1 centipoise = 0.000672 lb/ft s

239

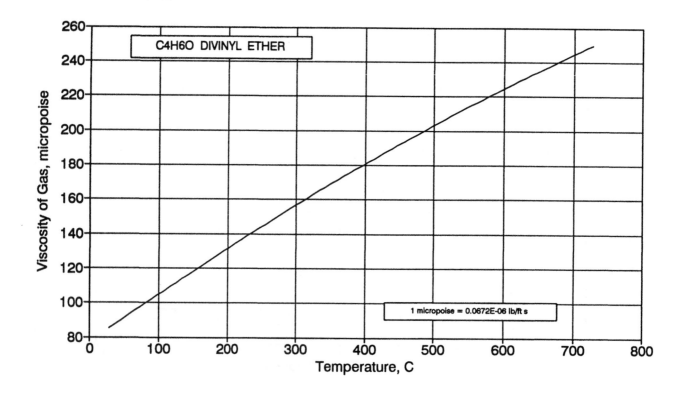

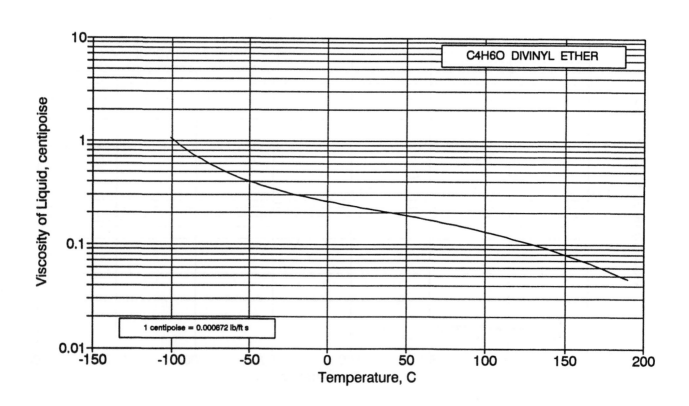

240

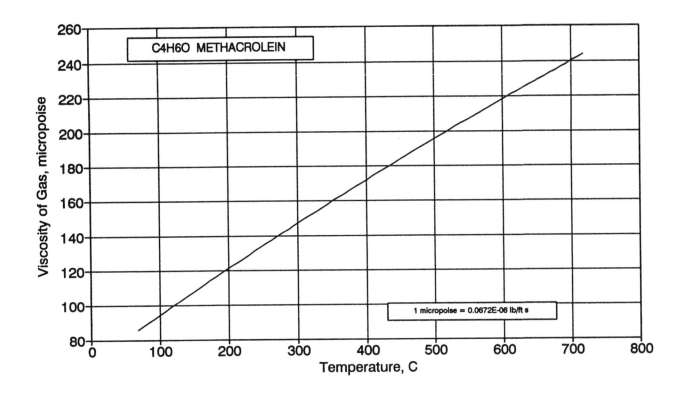

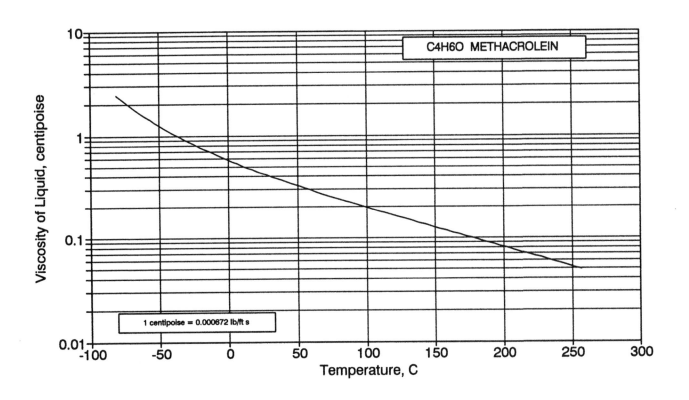

241

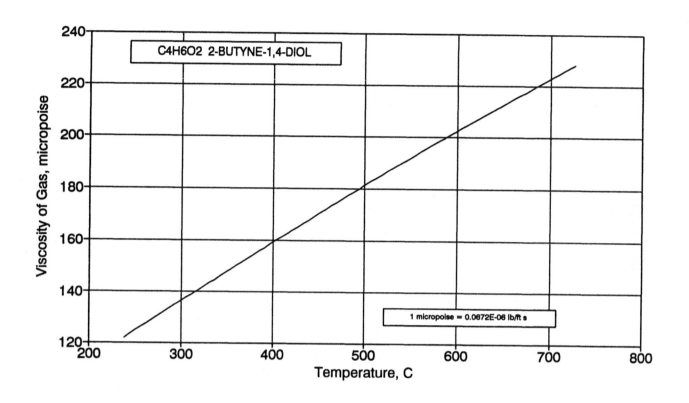

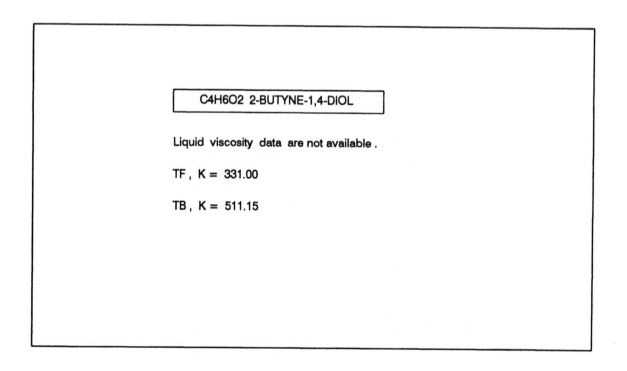

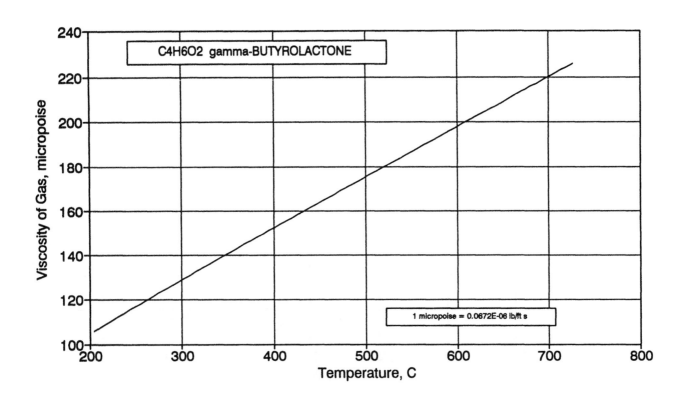

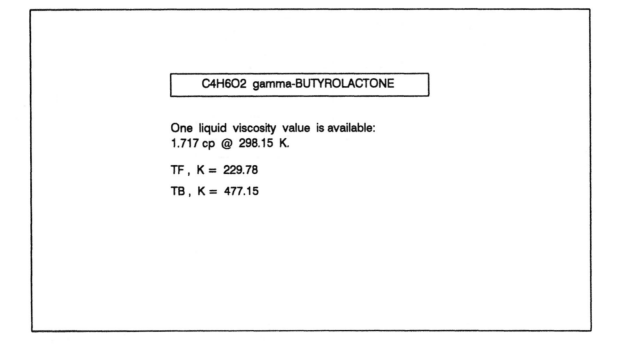

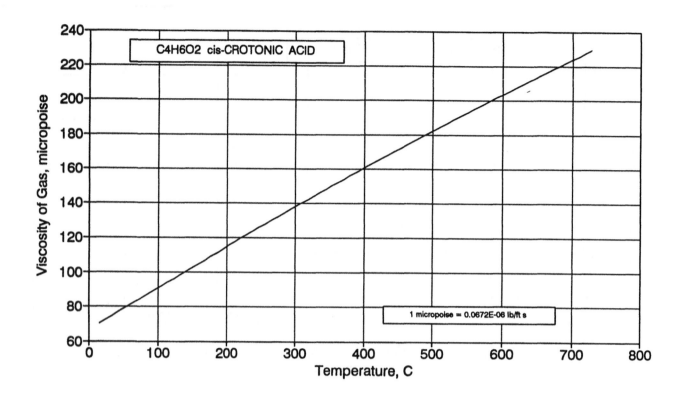

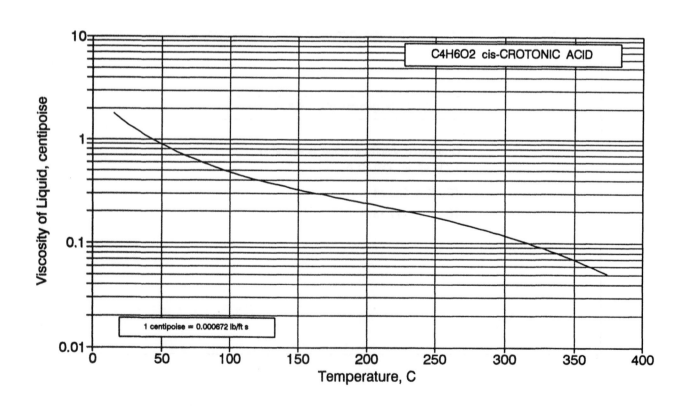

244

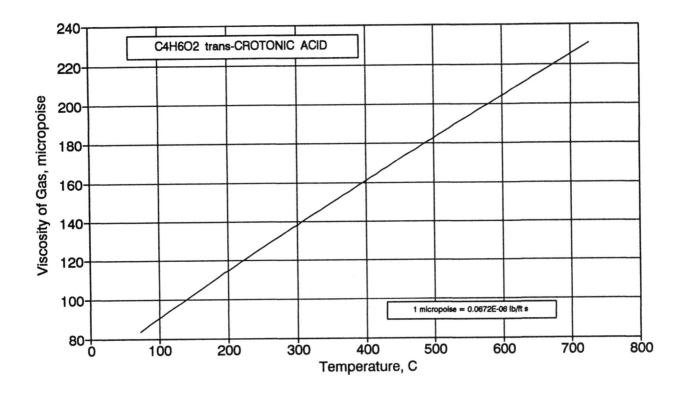

Viscosity of Gas, micropoise vs Temperature, C

C4H6O2 trans-CROTONIC ACID

1 micropoise = 0.0672E-06 lb/ft s

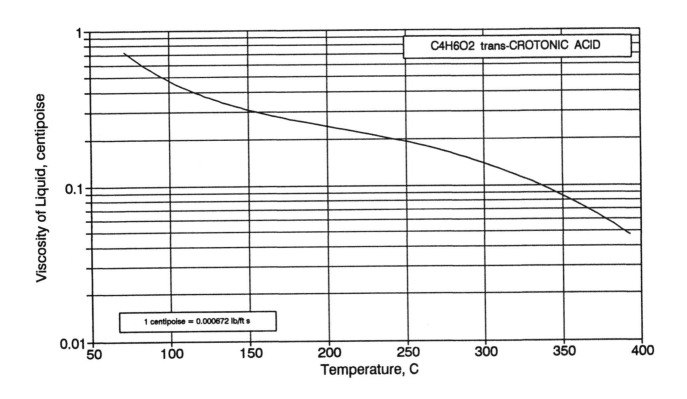

Viscosity of Liquid, centipoise vs Temperature, C

C4H6O2 trans-CROTONIC ACID

1 centipoise = 0.000672 lb/ft s

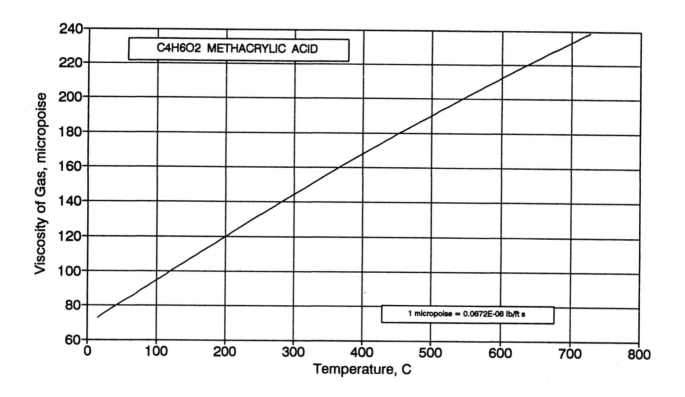

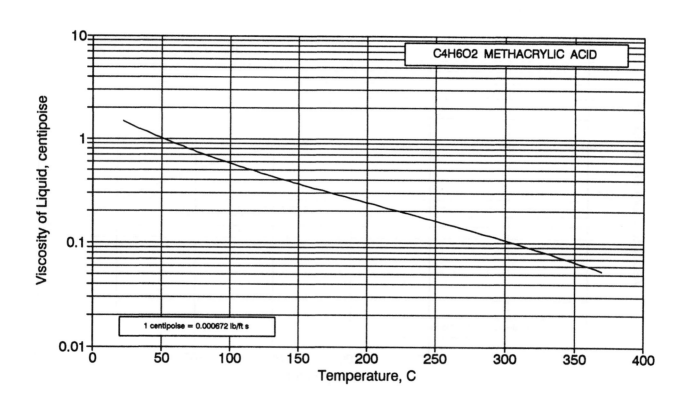

246

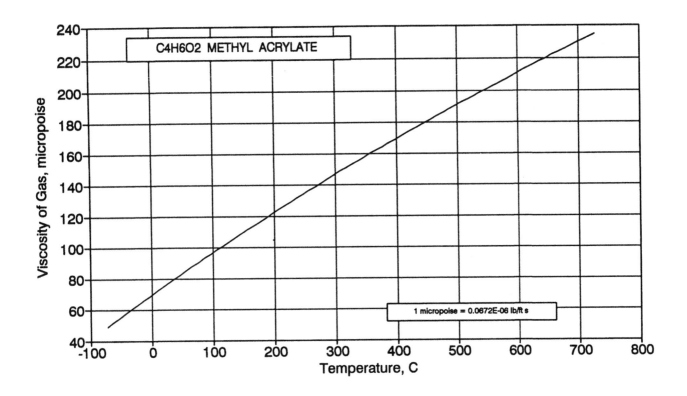

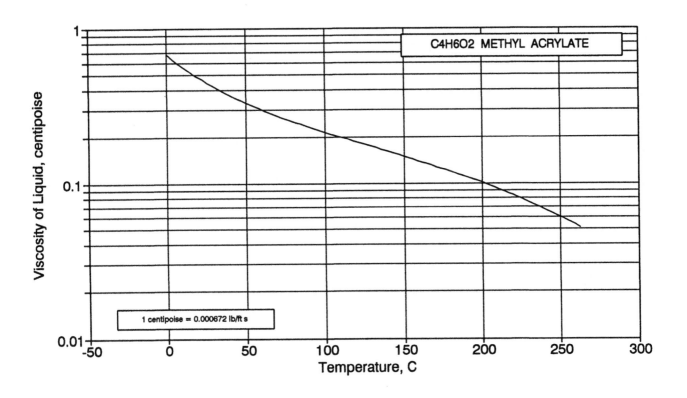

247

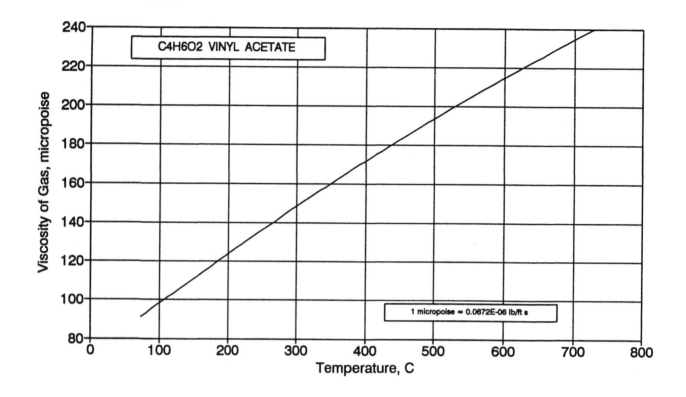

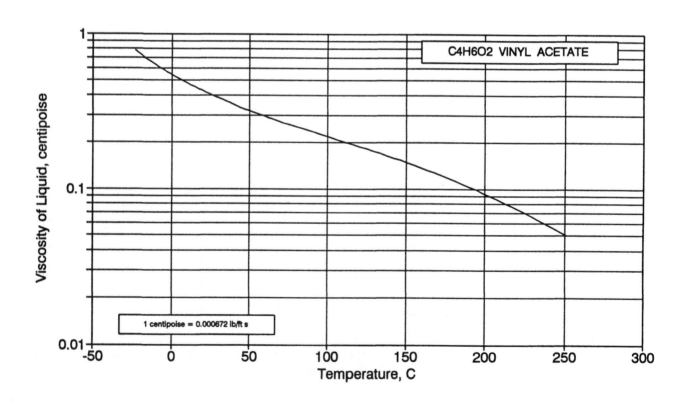

248

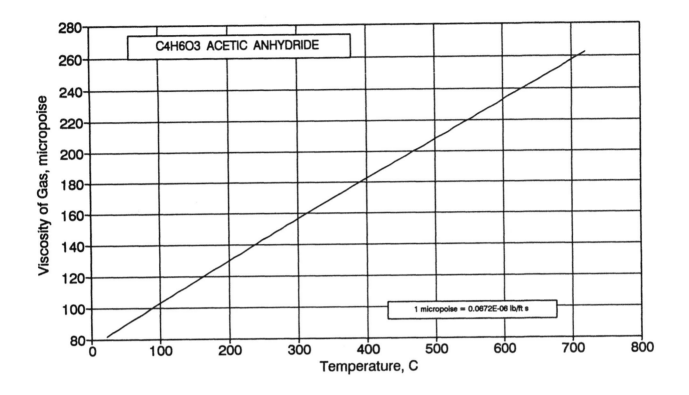

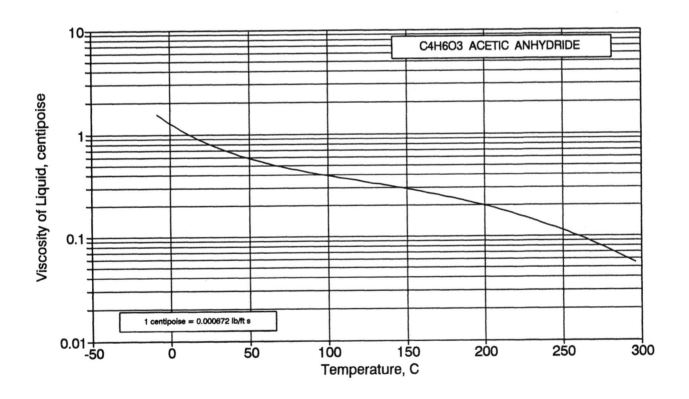

249

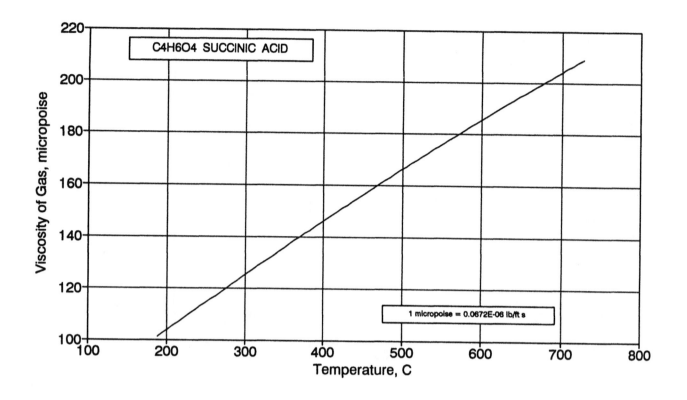

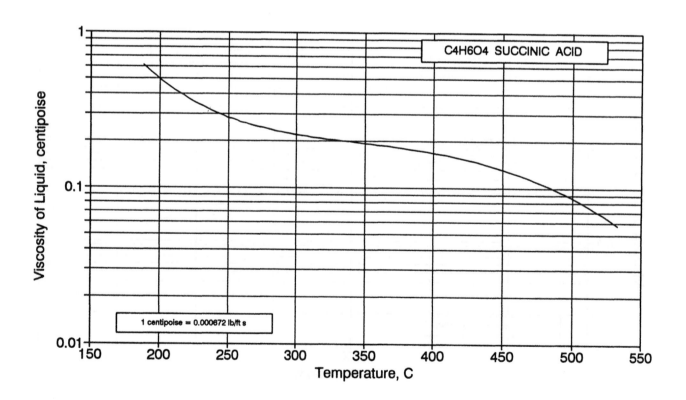

250

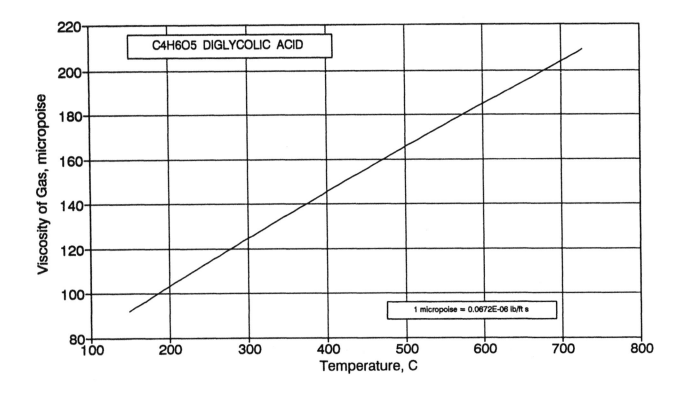

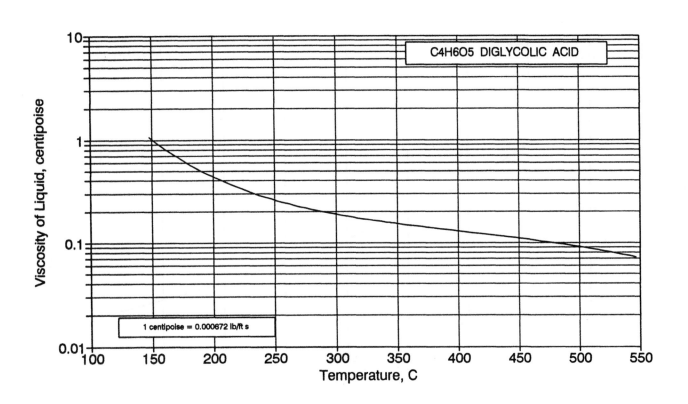

251

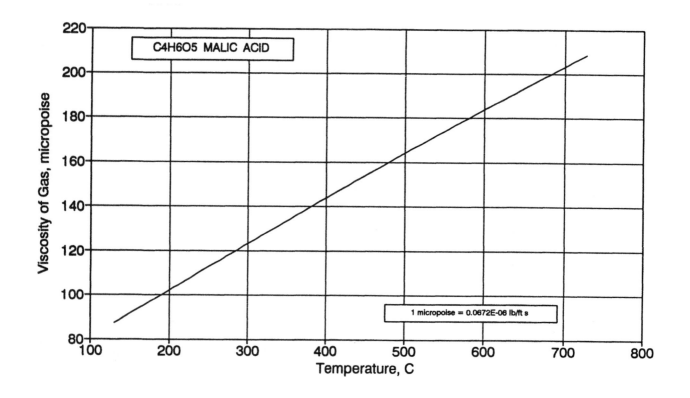

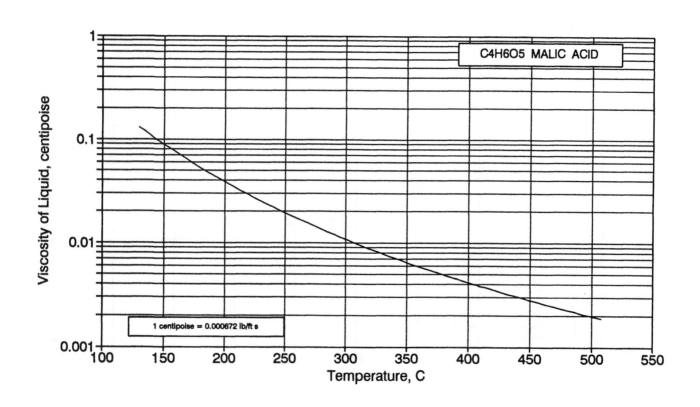

252

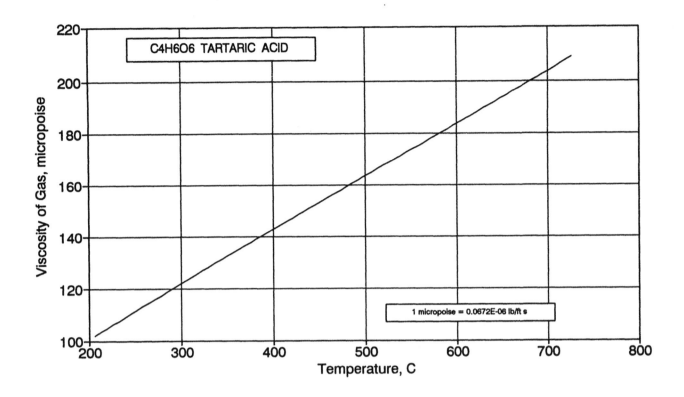

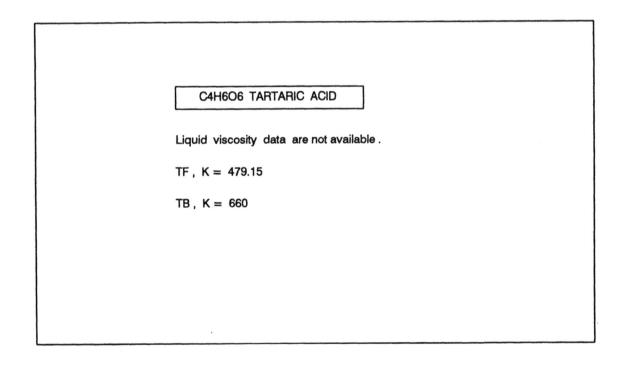

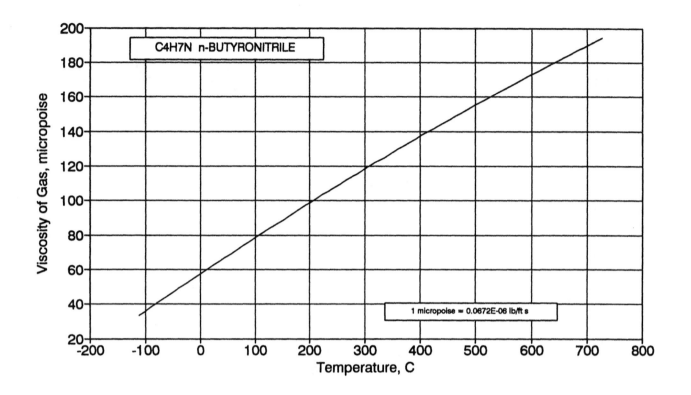

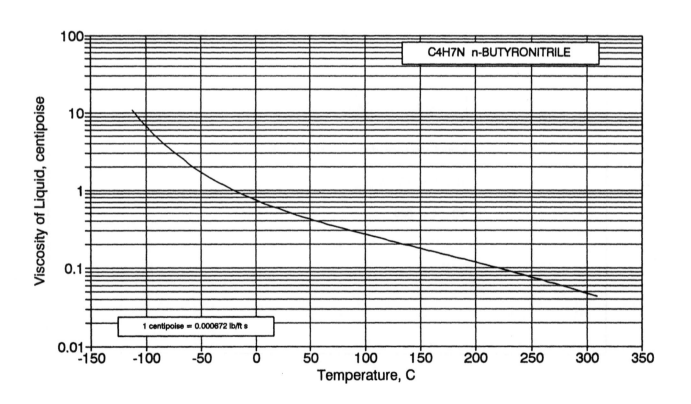

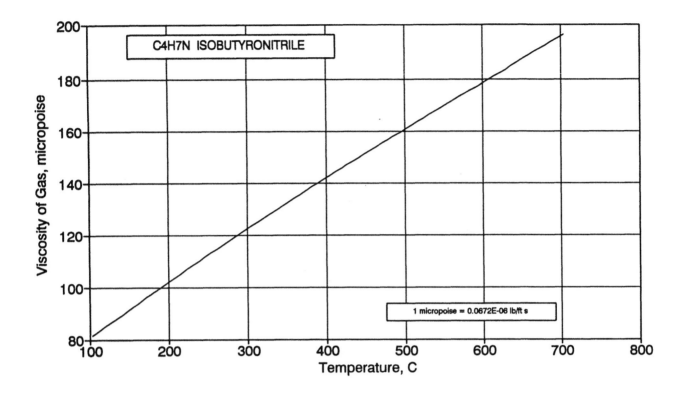

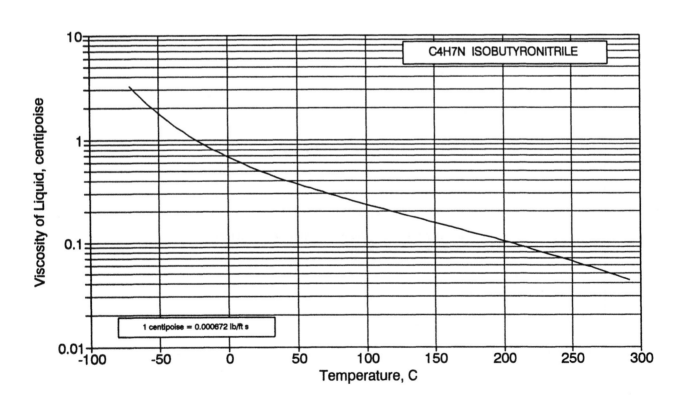

255

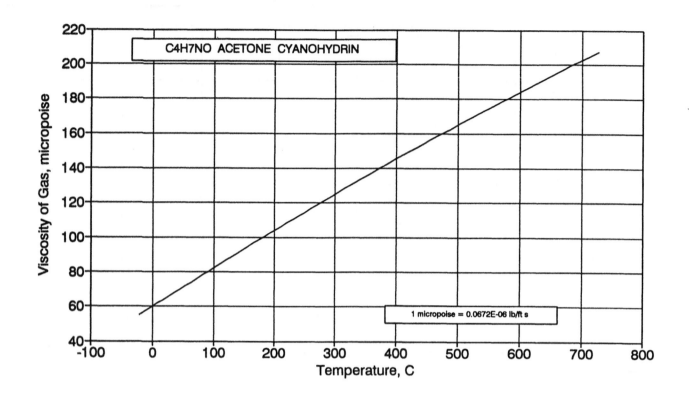

C4H7NO ACETONE CYANOHYDRIN

Liquid viscosity data are not available .

TF , K = 253.15

TB , K = 463 (estimated)

Decomposes to hydrogen cyanide and acetone under basic conditions.

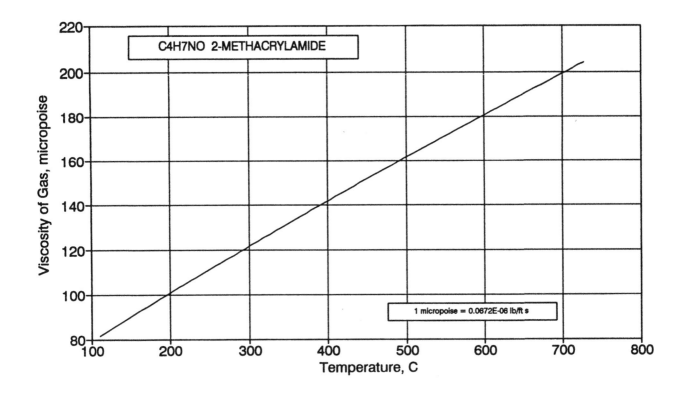

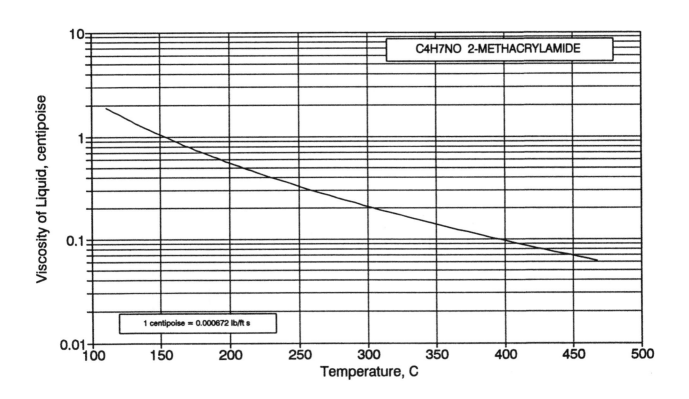

257

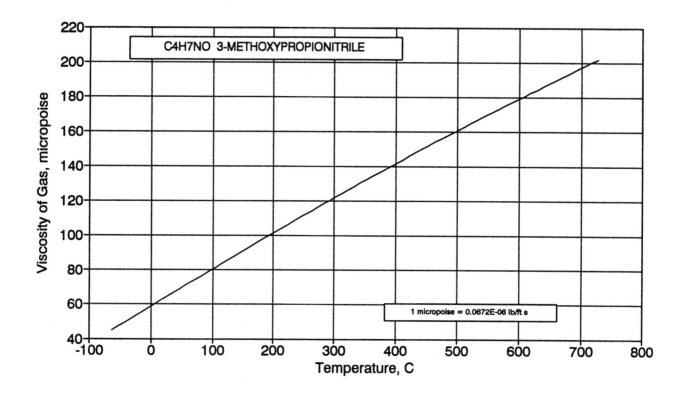

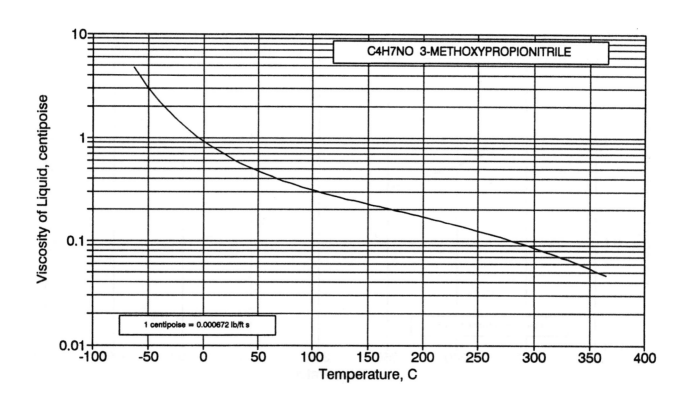

258

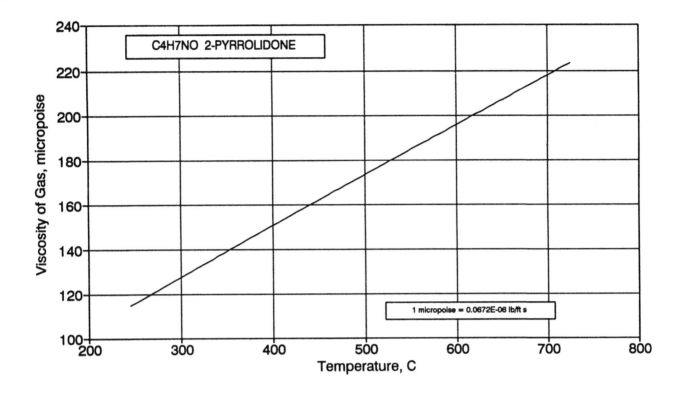

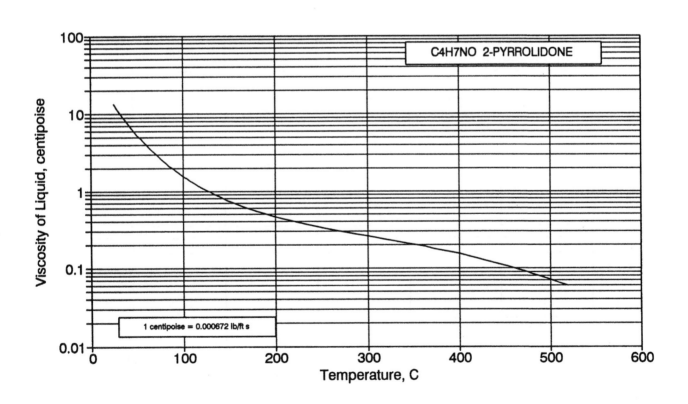

259

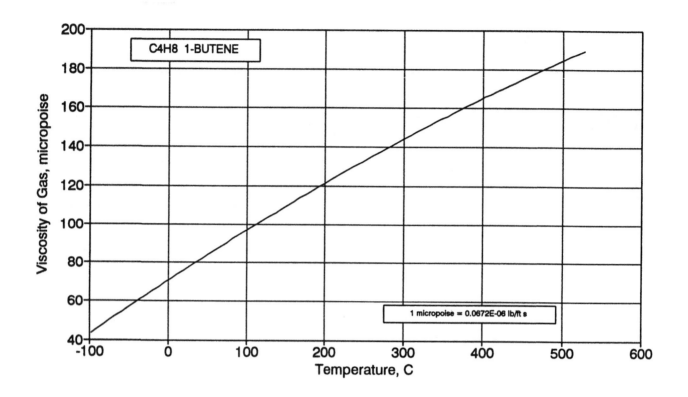

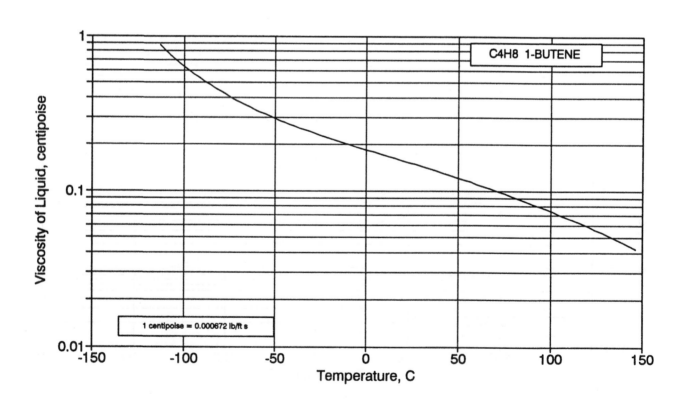

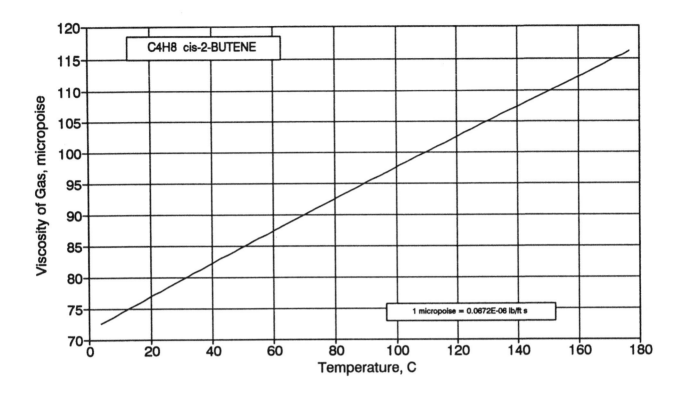

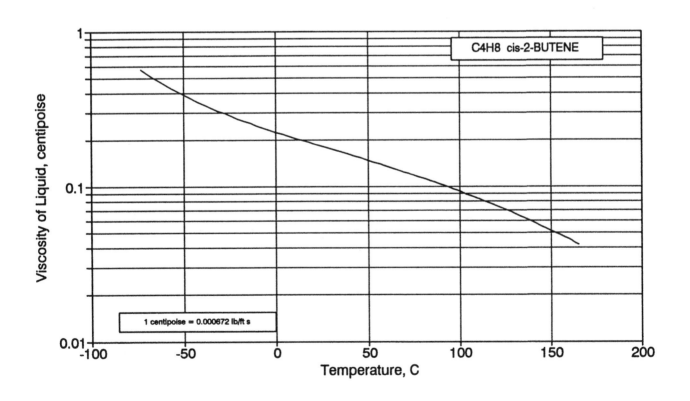

261

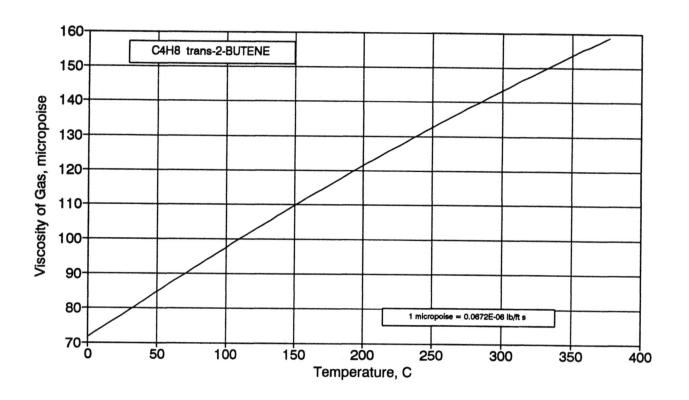

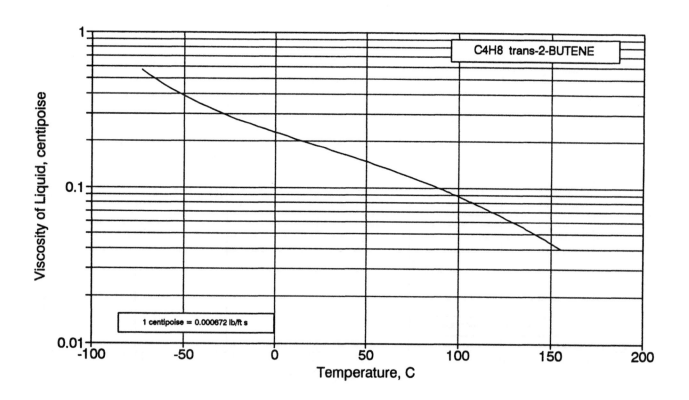

262

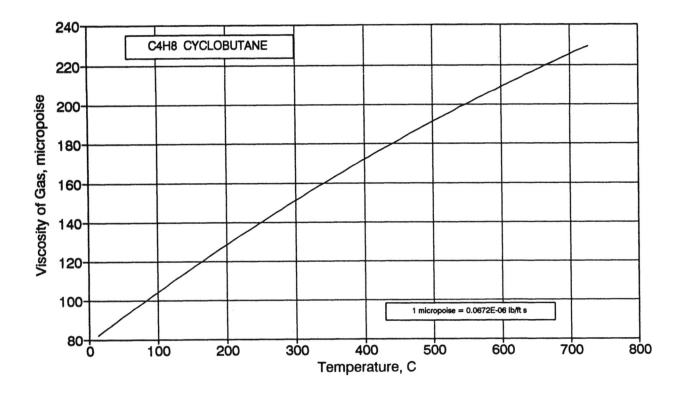

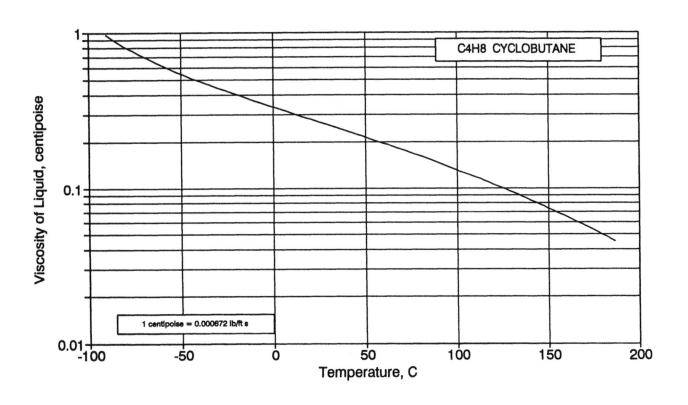

263

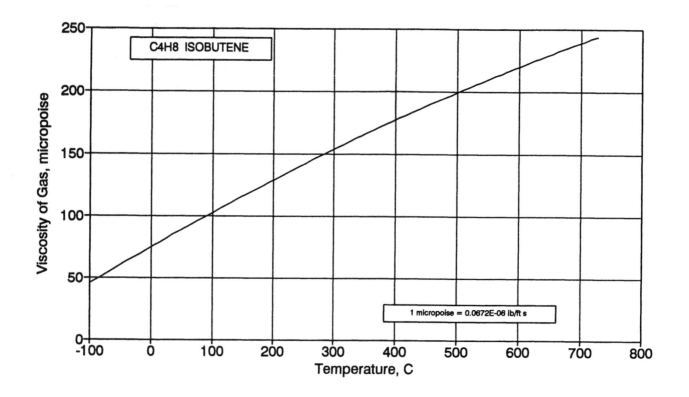

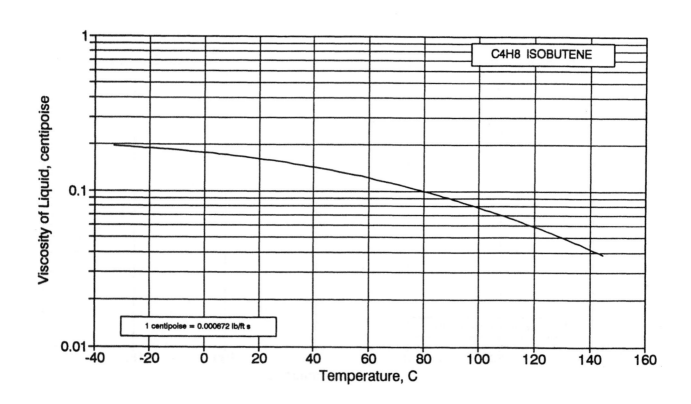

264

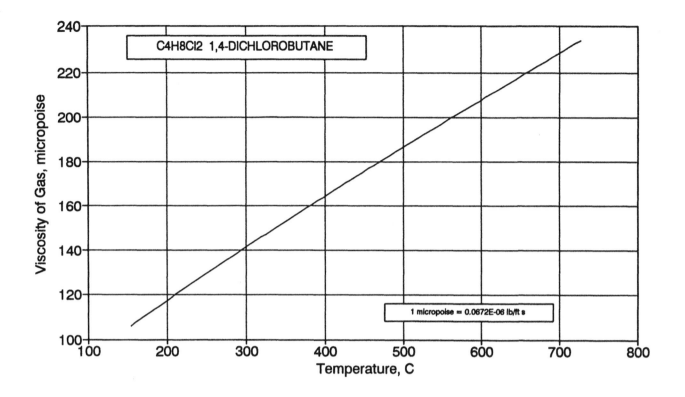

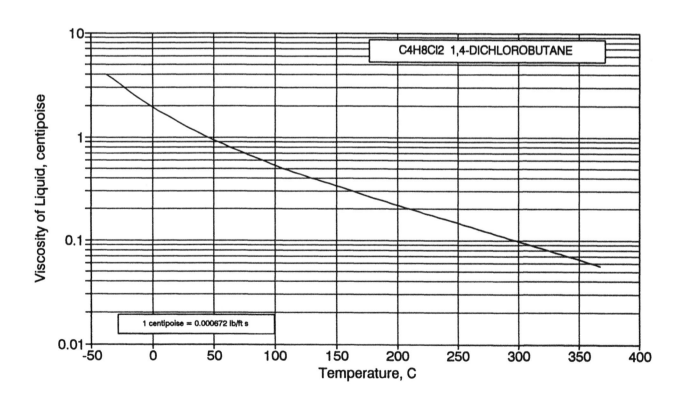

265

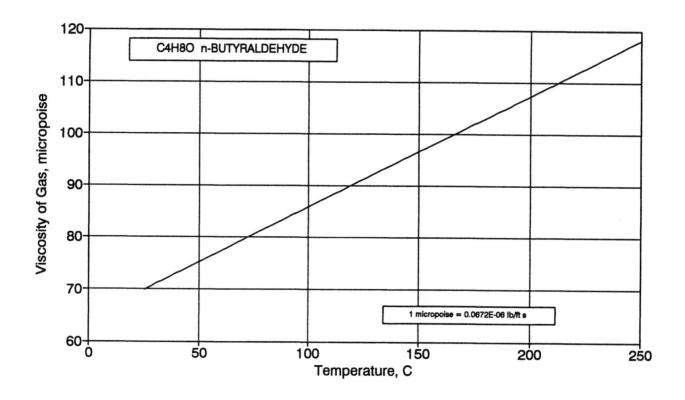

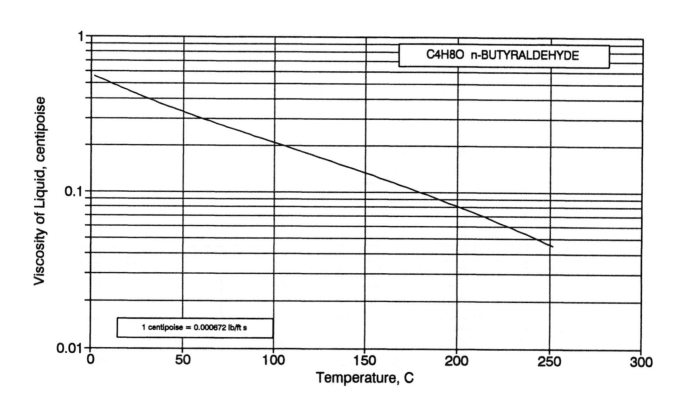

266

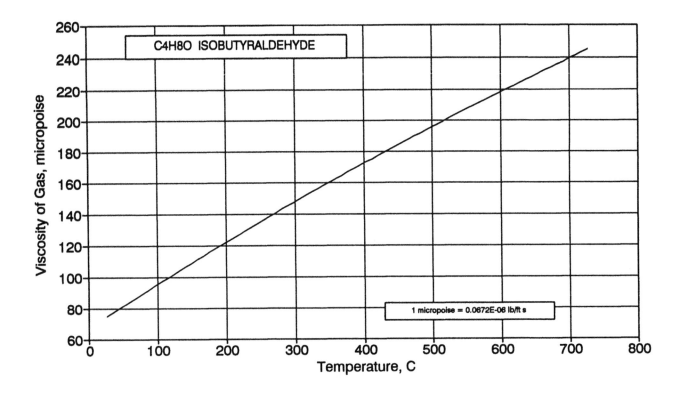

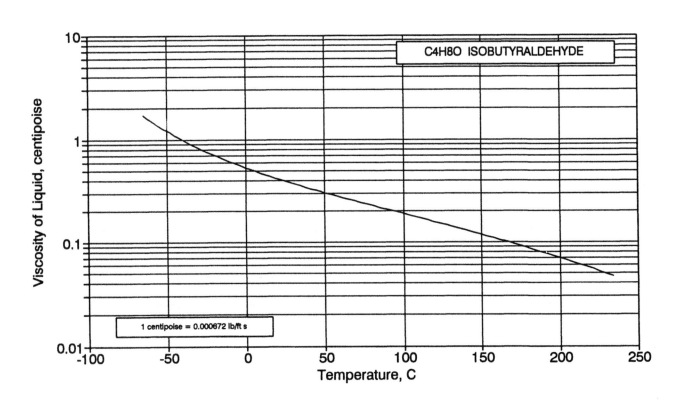

267

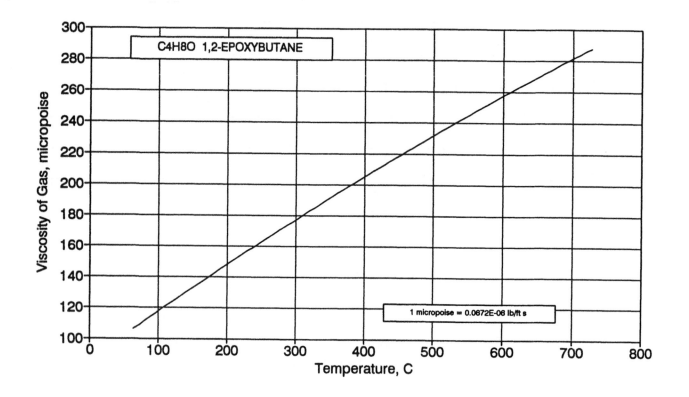

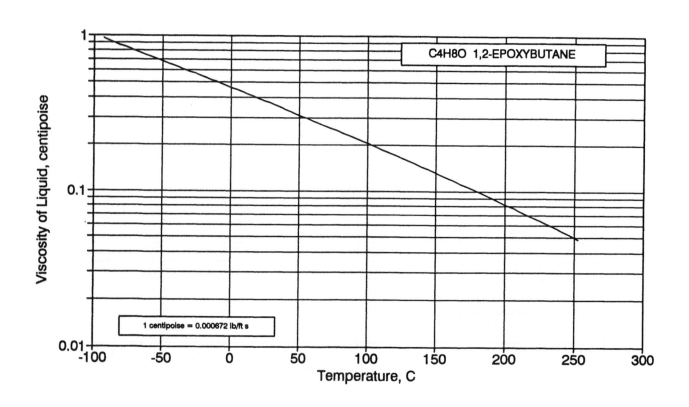

268

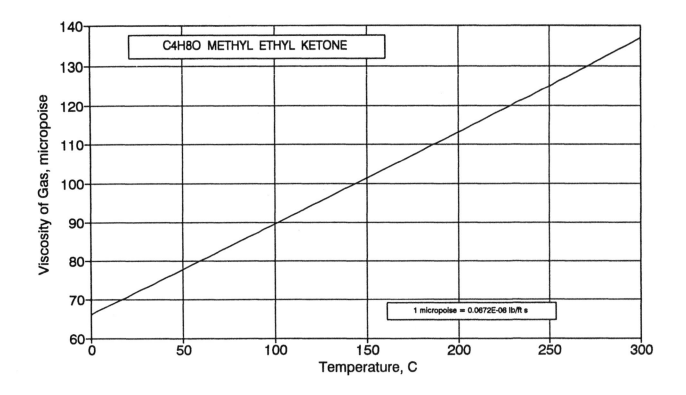

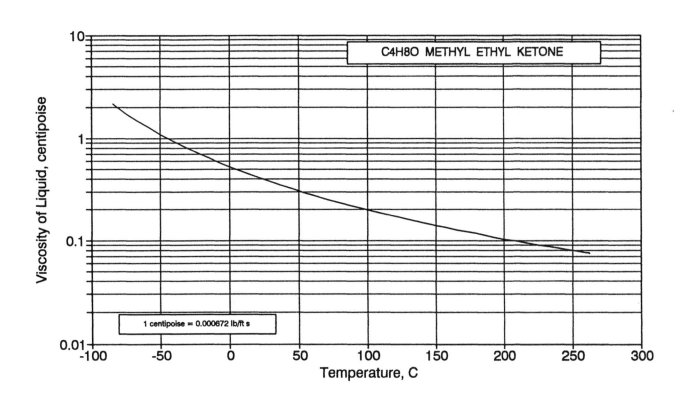

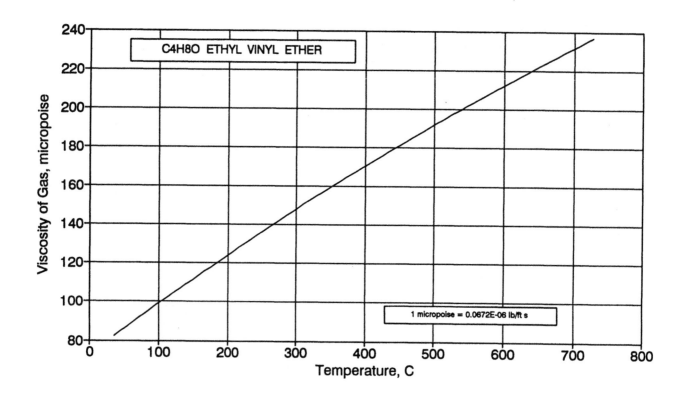

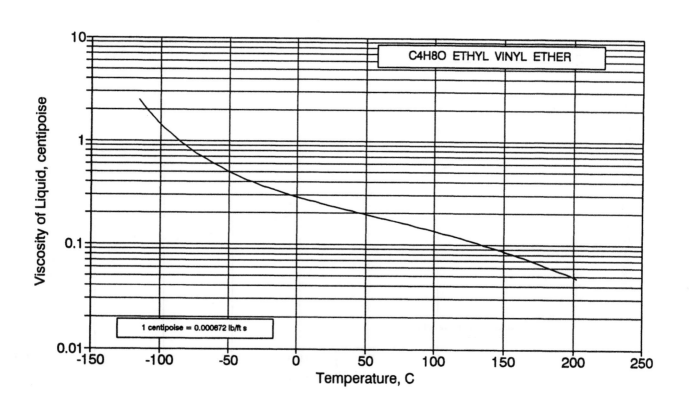

270

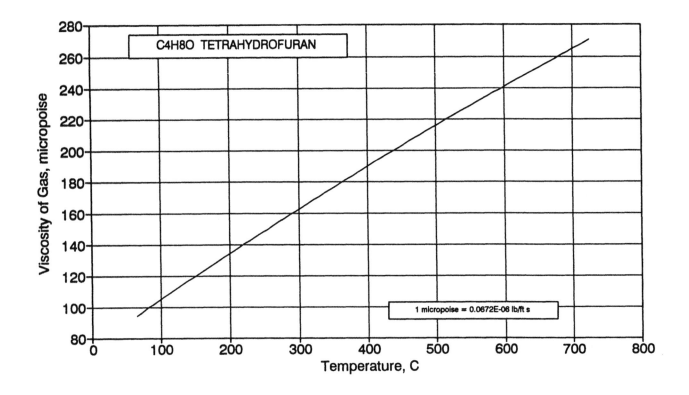

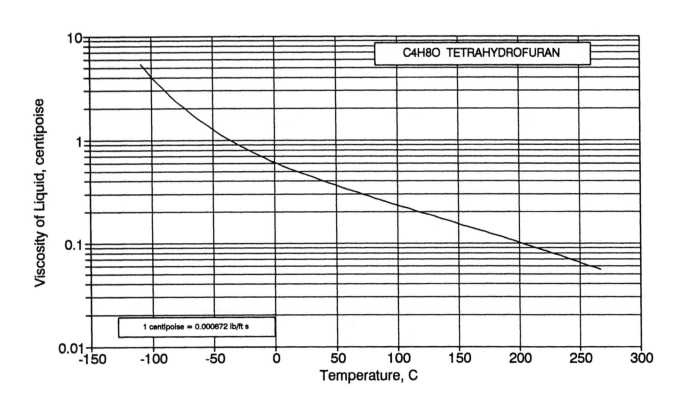

271

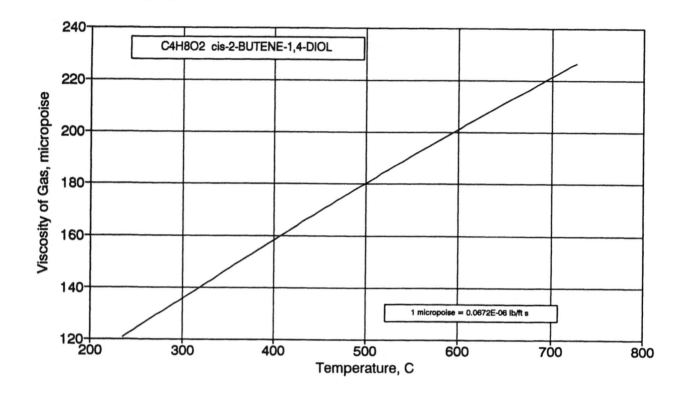

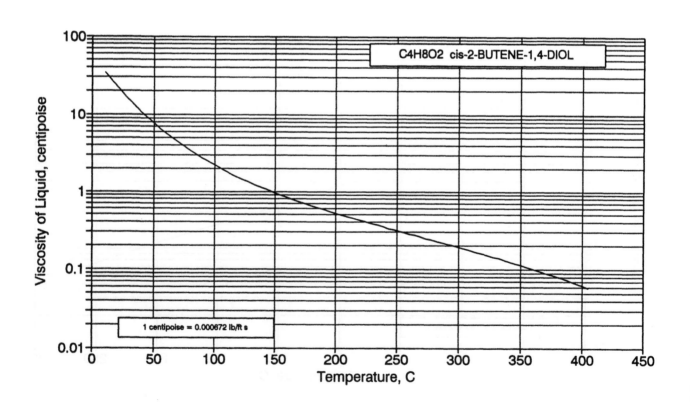

272

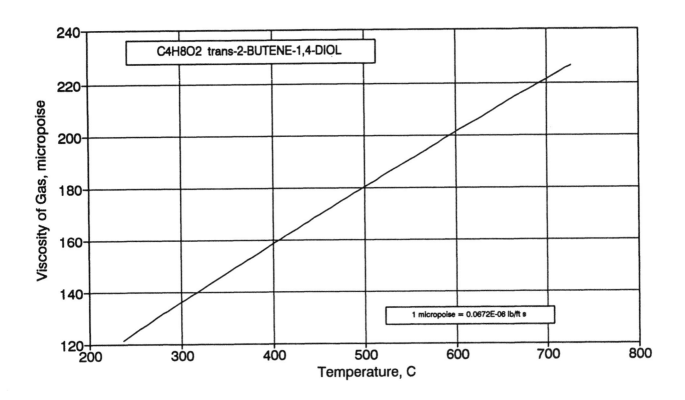

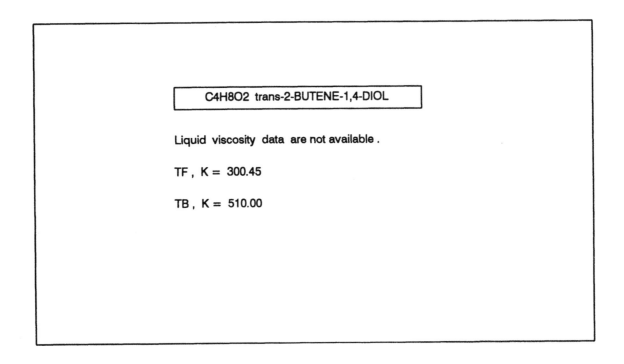

C4H8O2 trans-2-BUTENE-1,4-DIOL

Liquid viscosity data are not available .

TF , K = 300.45

TB , K = 510.00

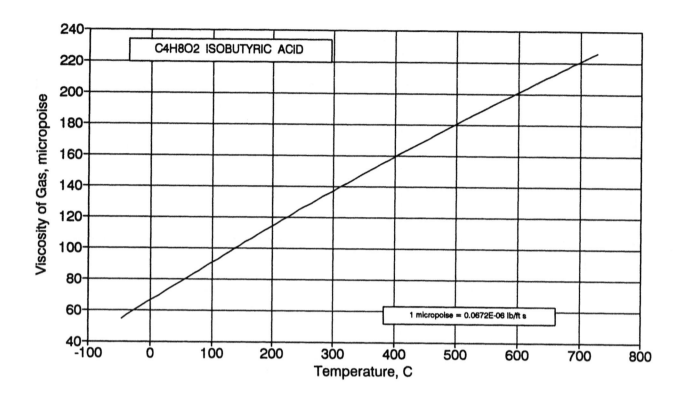

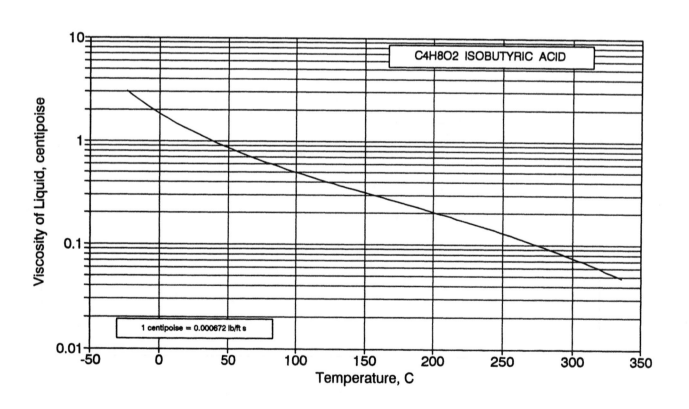

274

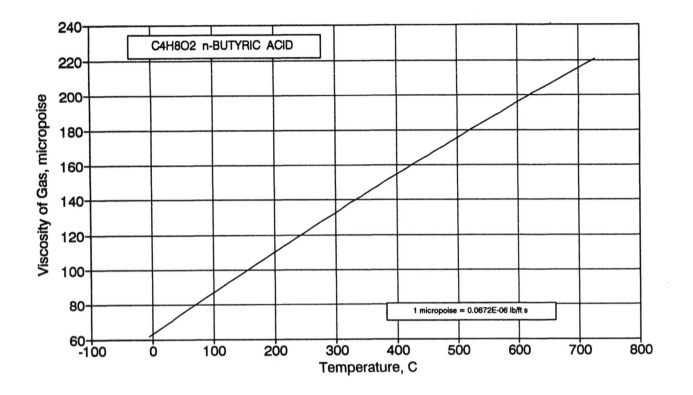

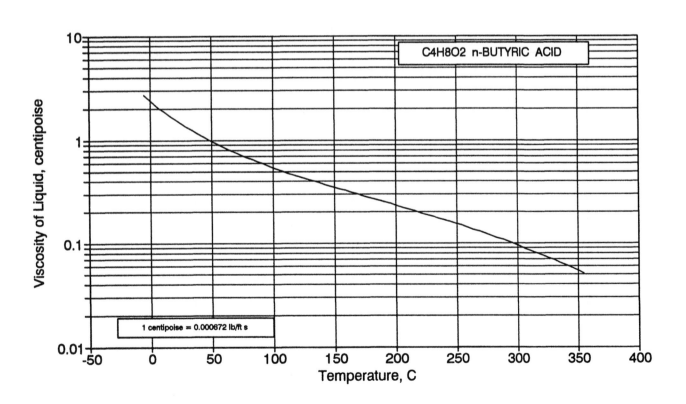

275

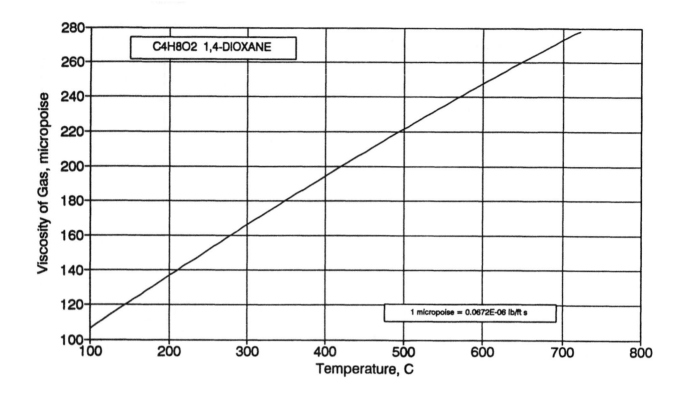

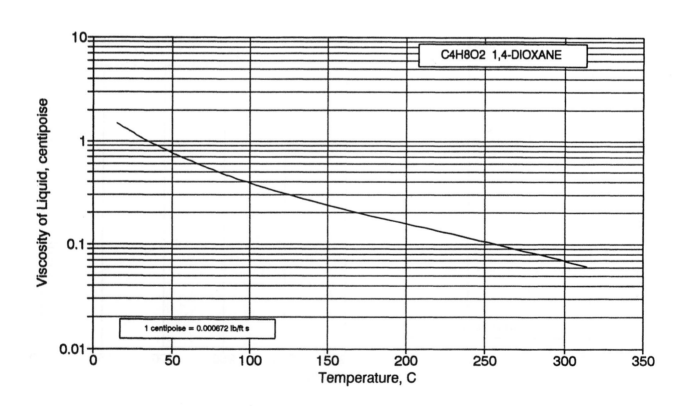

276

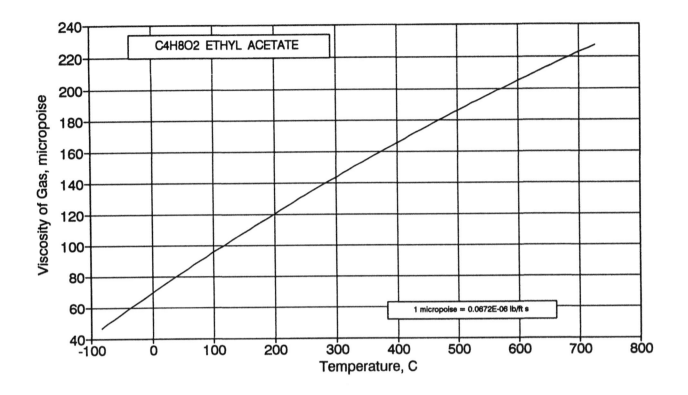

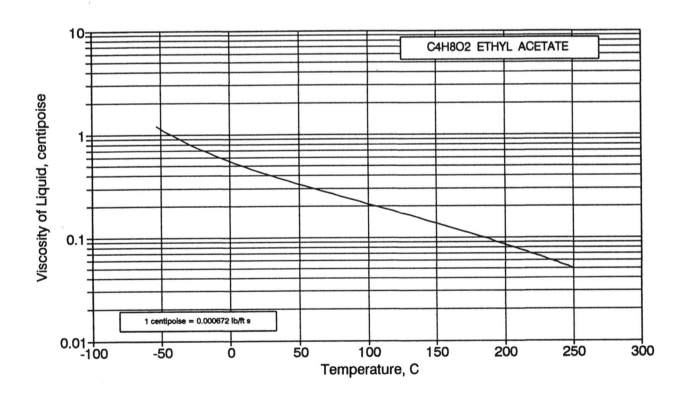

277

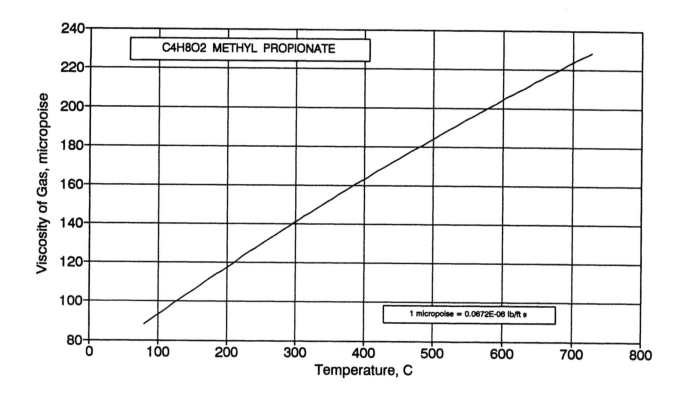

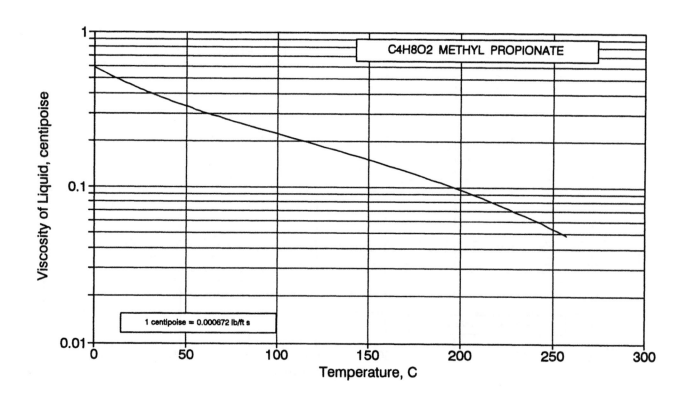

278

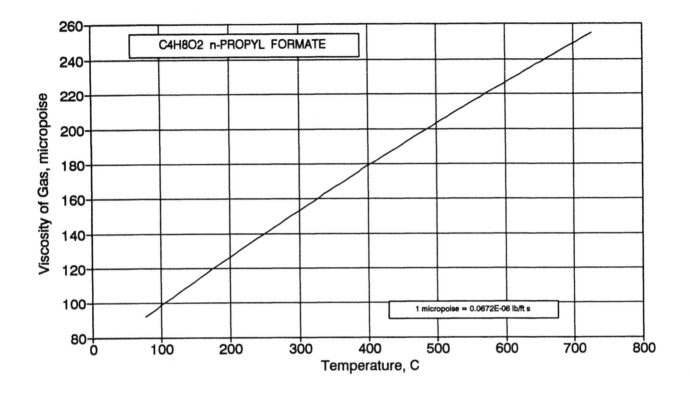

1 micropoise = 0.0672E-06 lb/ft s

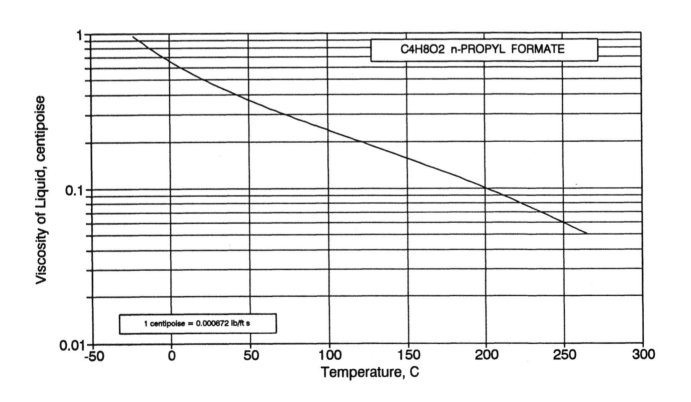

1 centipoise = 0.000672 lb/ft s

279

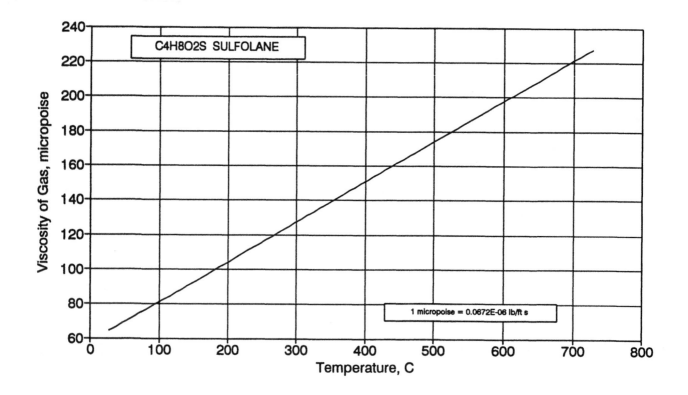

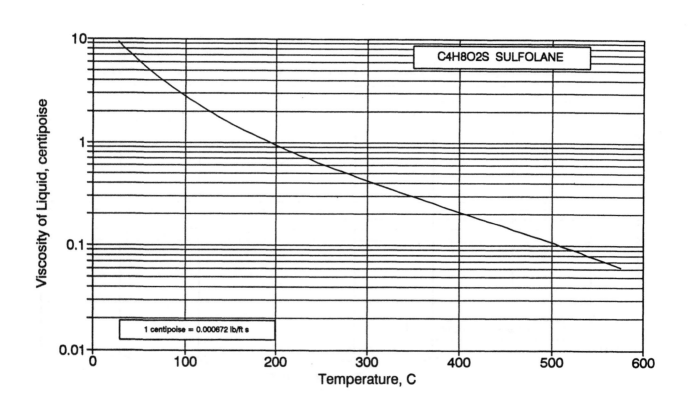

280

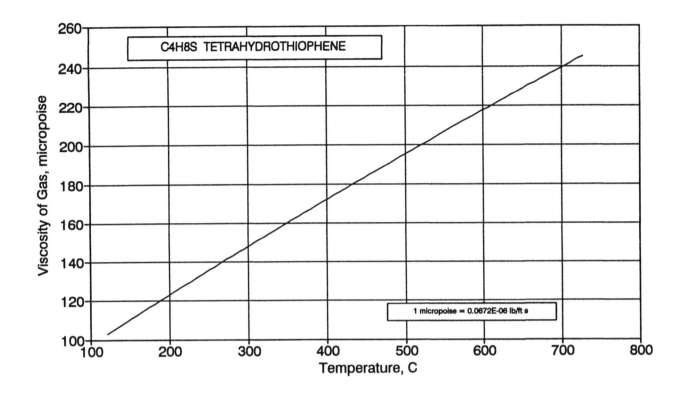

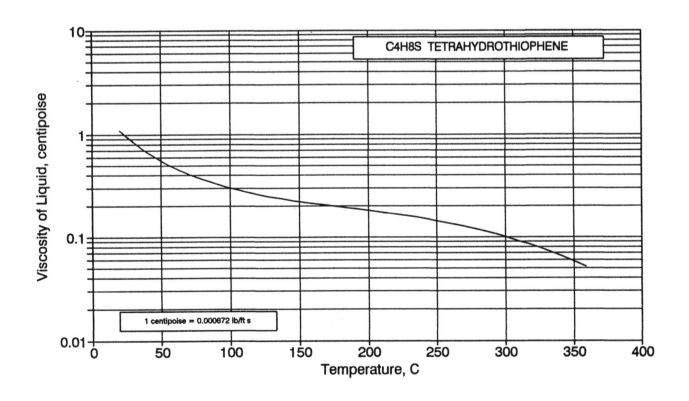

281

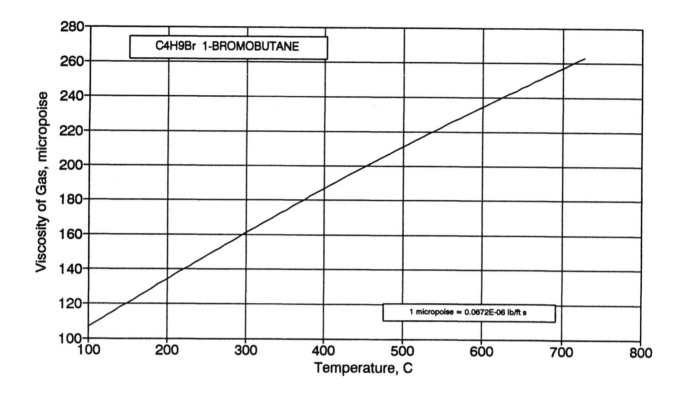

C4H9Br 1-BROMOBUTANE

1 micropoise = 0.0672E-06 lb/ft s

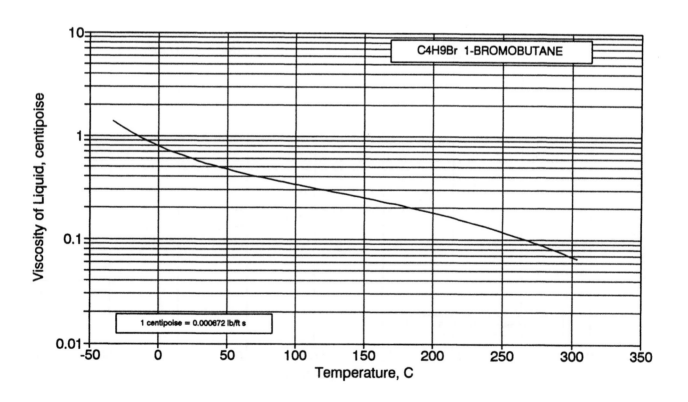

C4H9Br 1-BROMOBUTANE

1 centipoise = 0.000672 lb/ft s

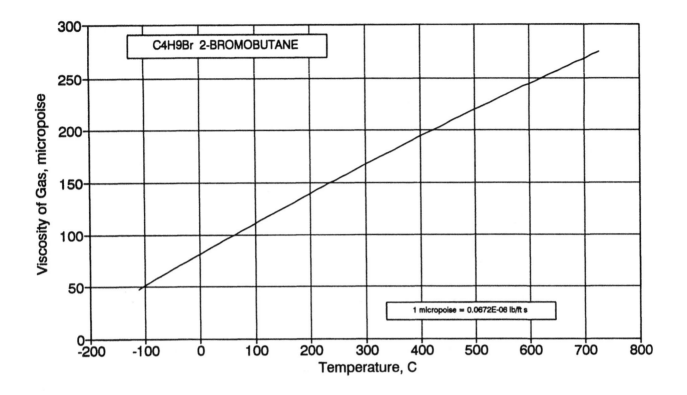

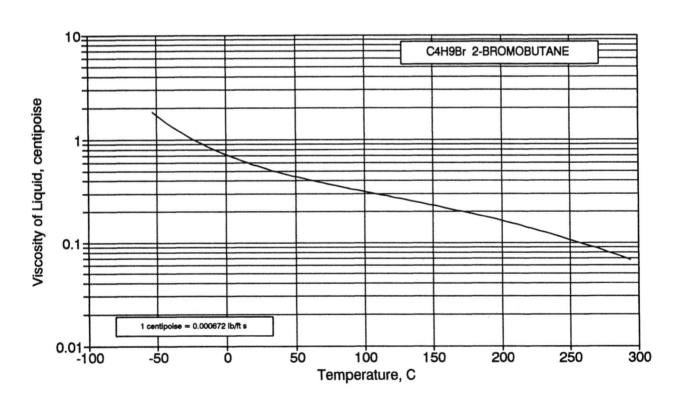

283

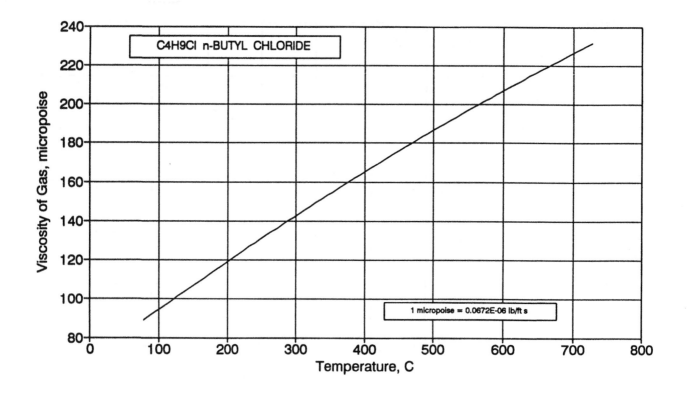

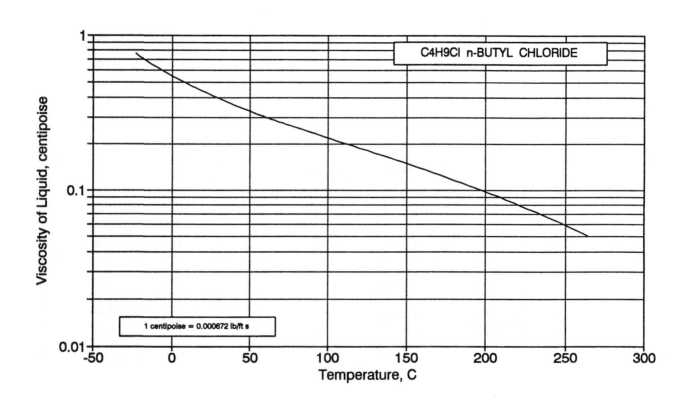

284

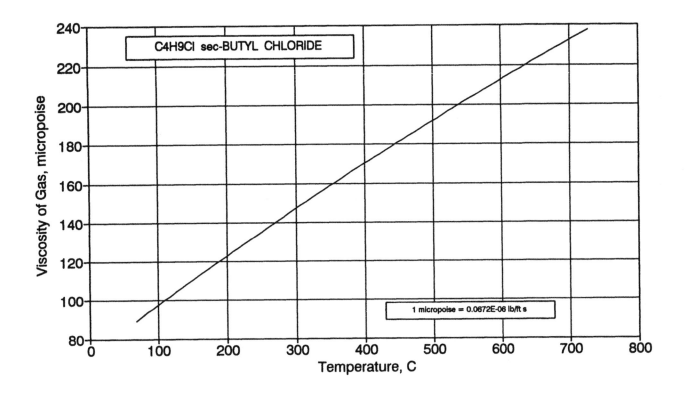

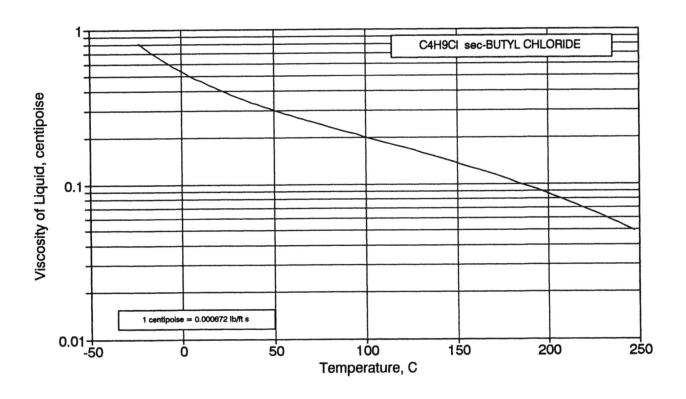

285

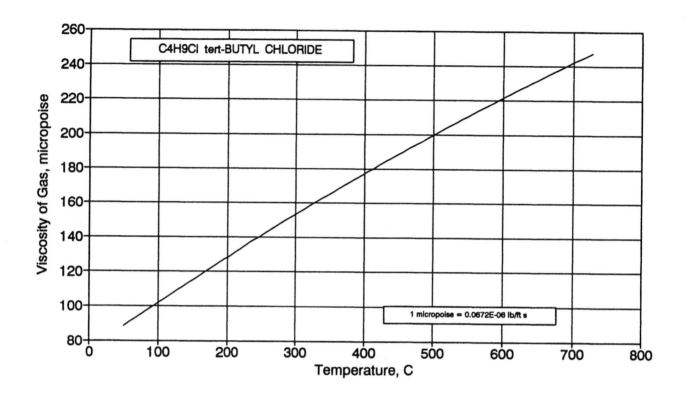

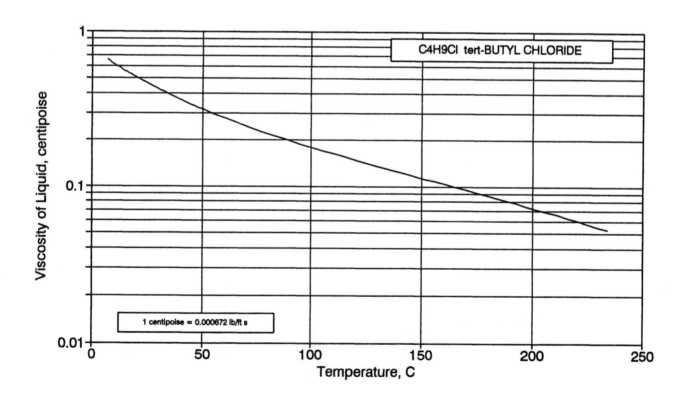

286

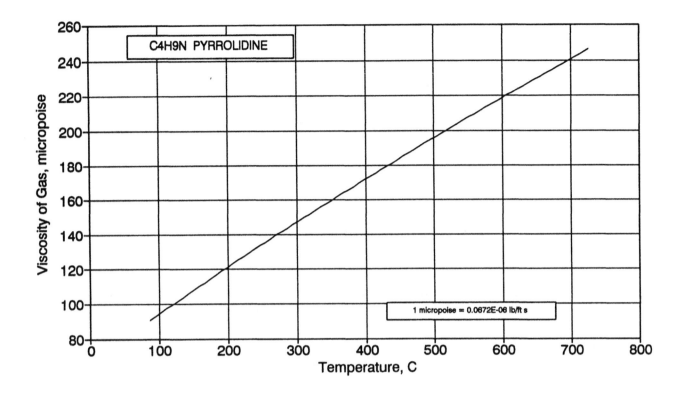

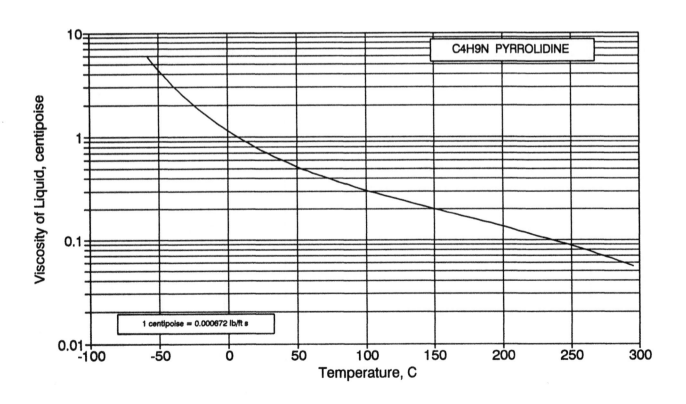

287

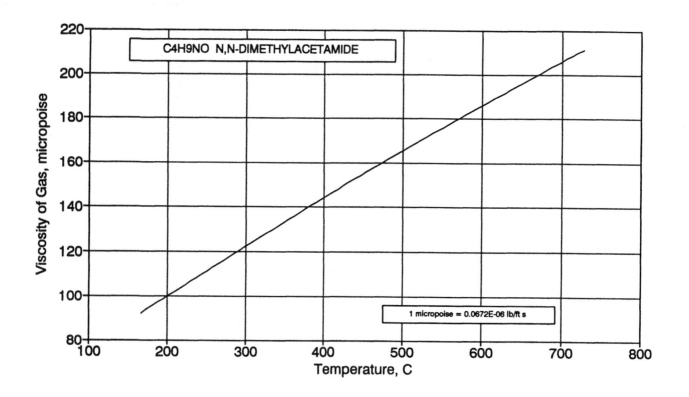

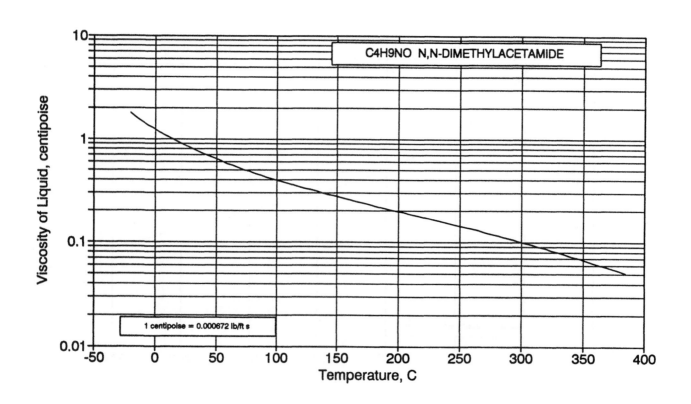

288

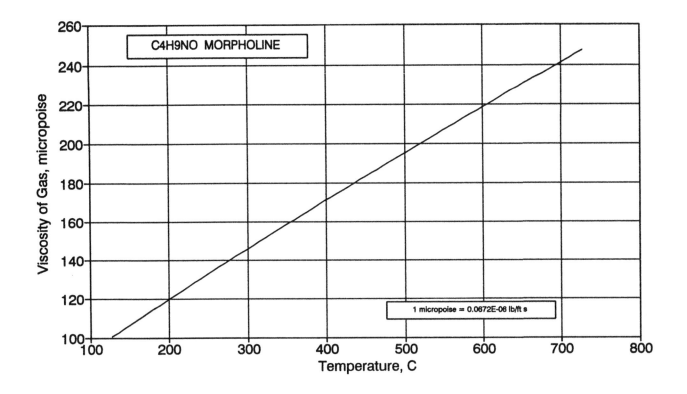

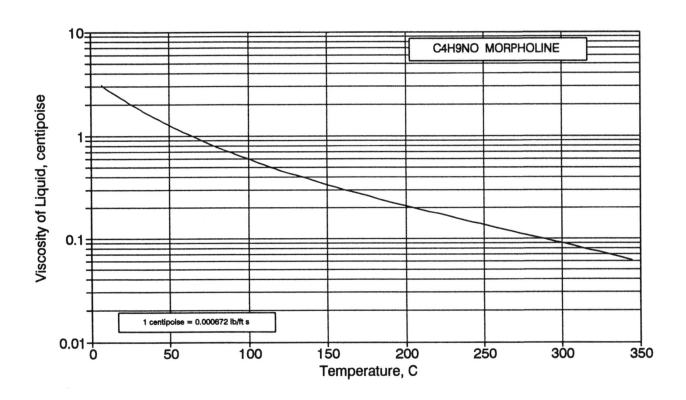

289

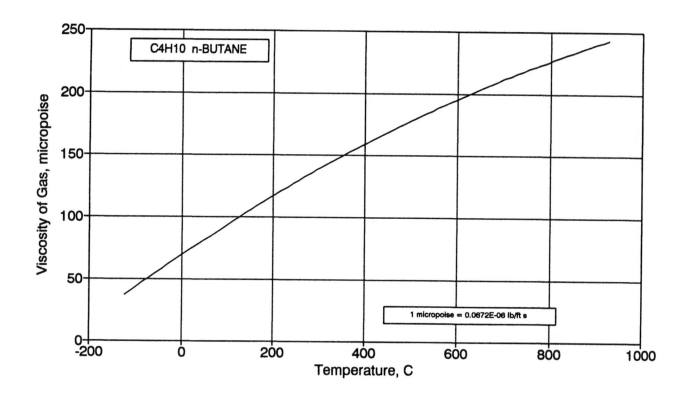

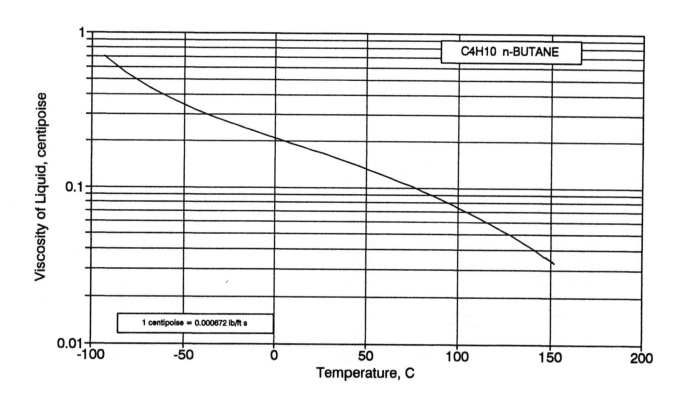

290

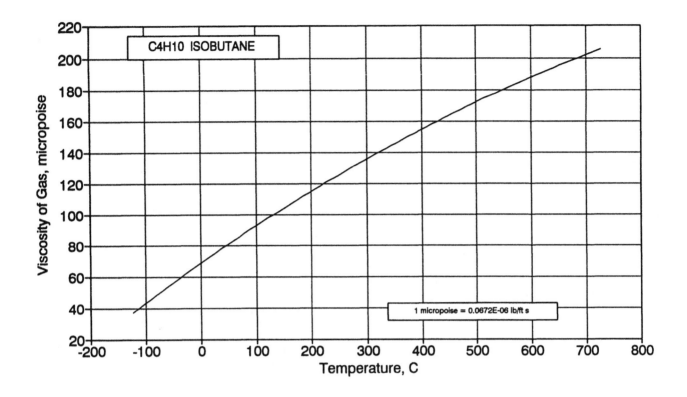

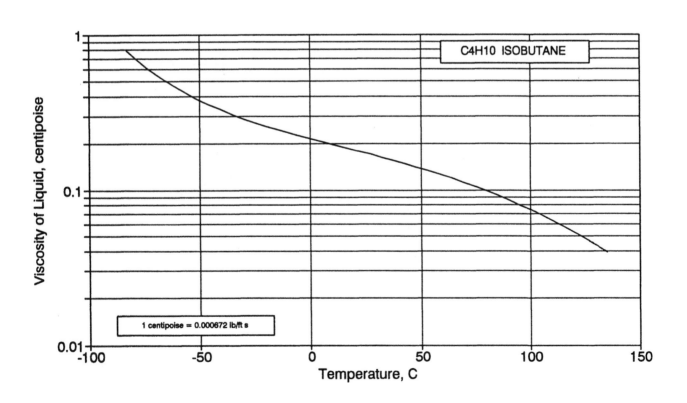

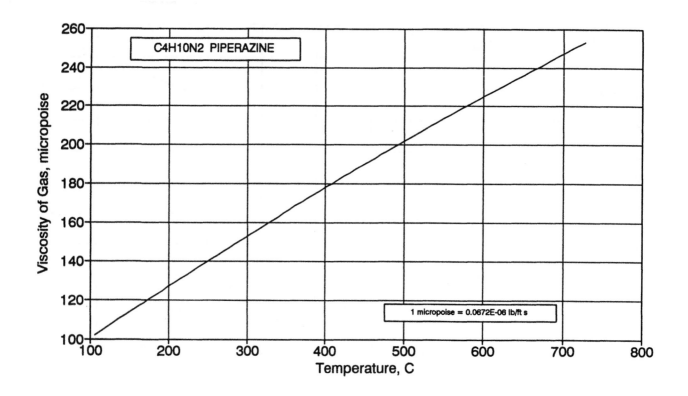

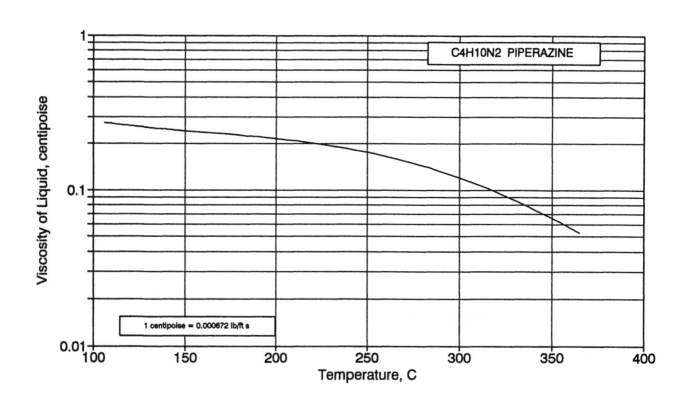

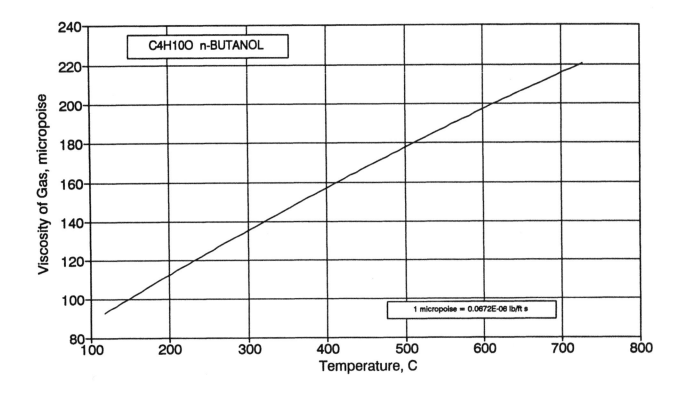

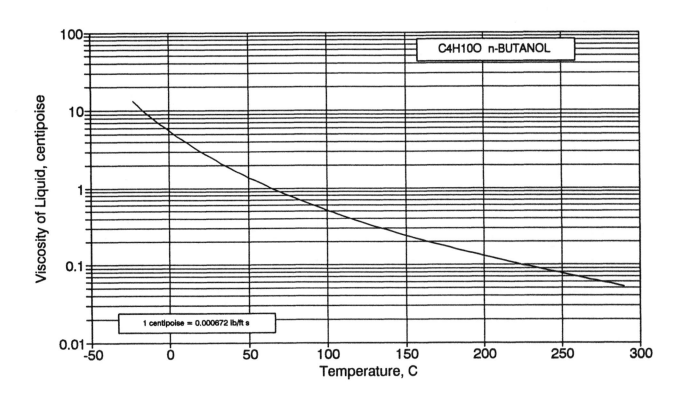

293

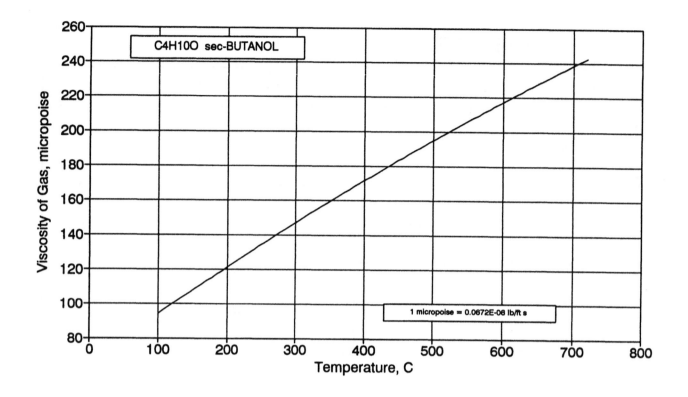

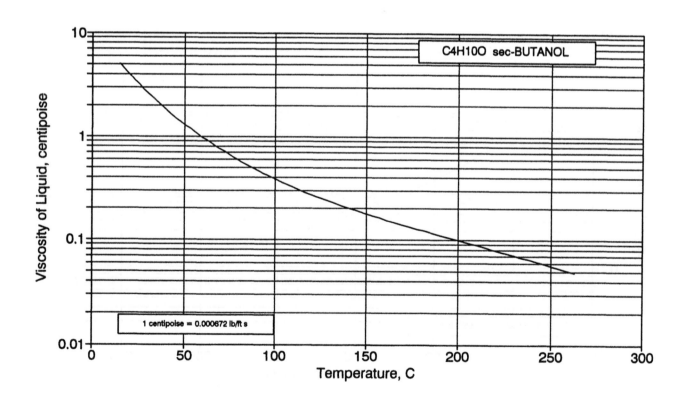

294

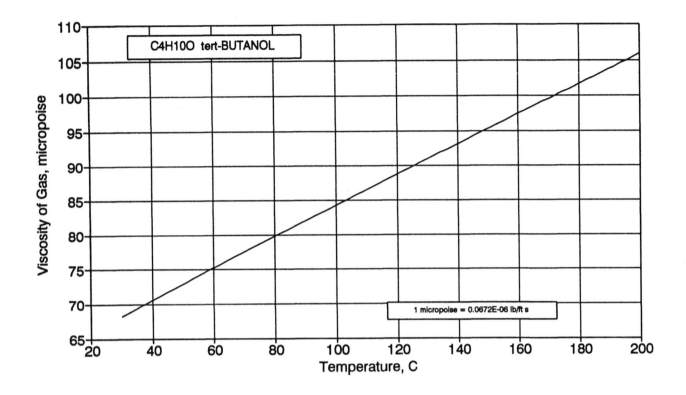

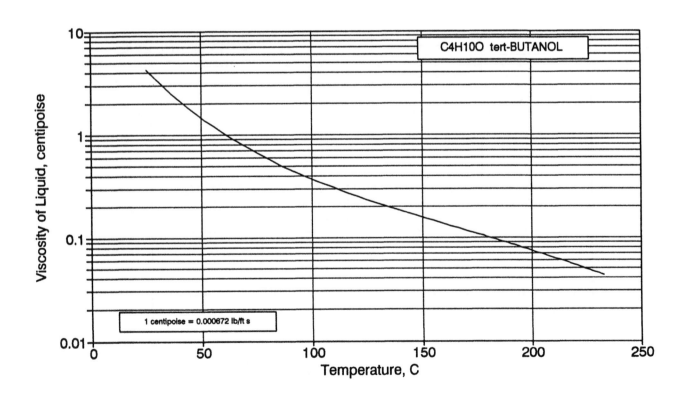

295

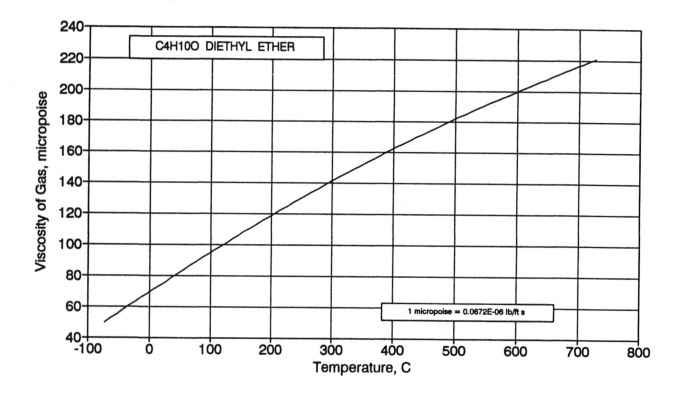

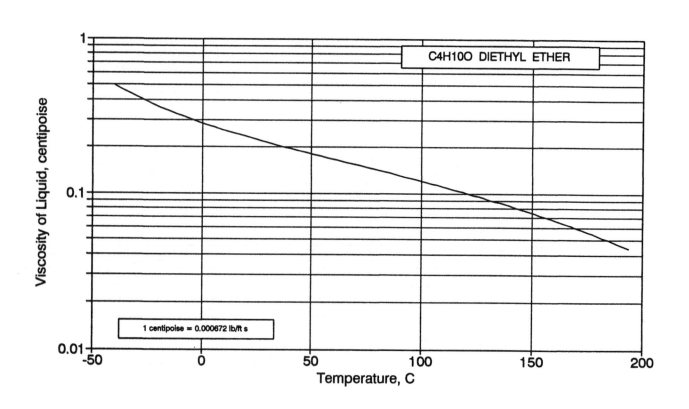

296

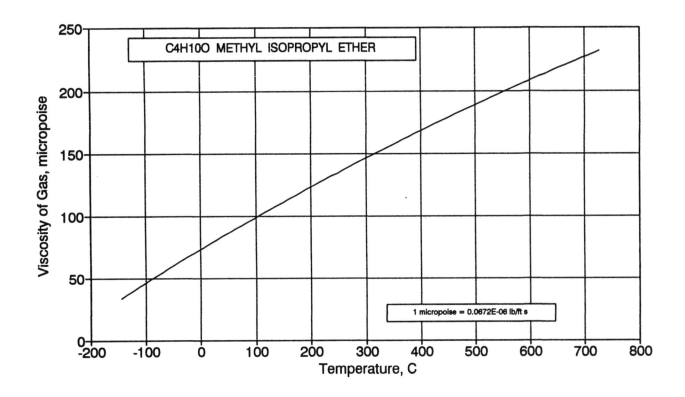

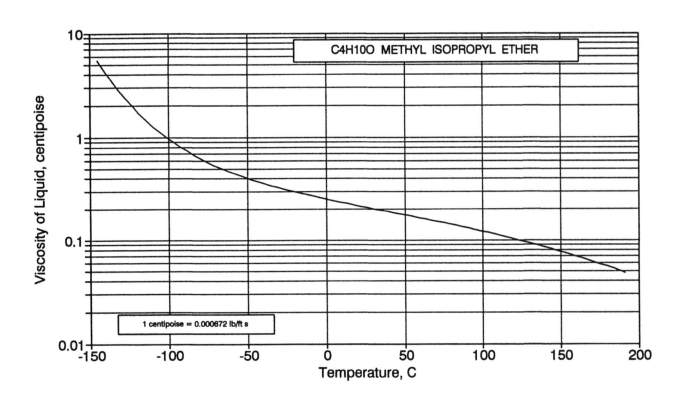

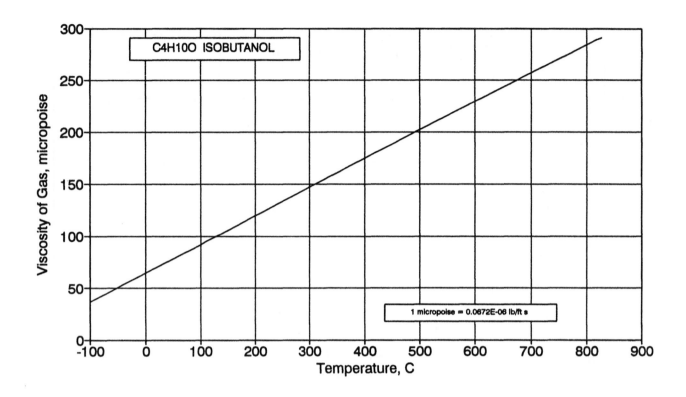

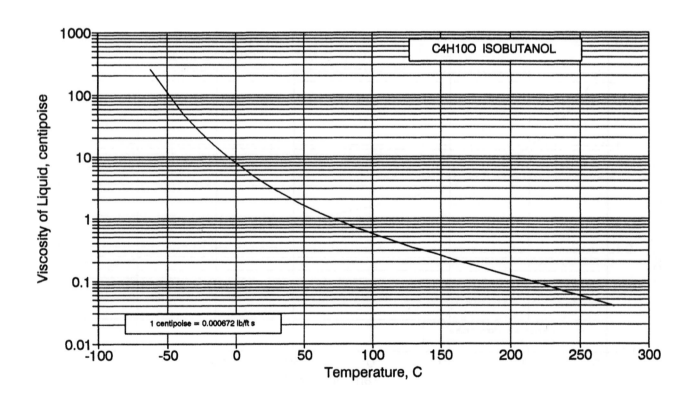

298

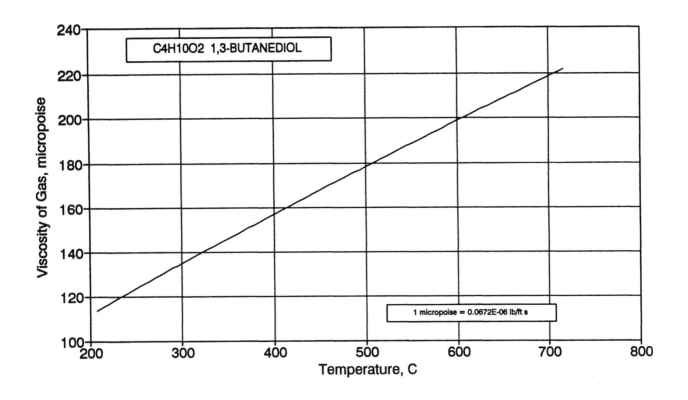

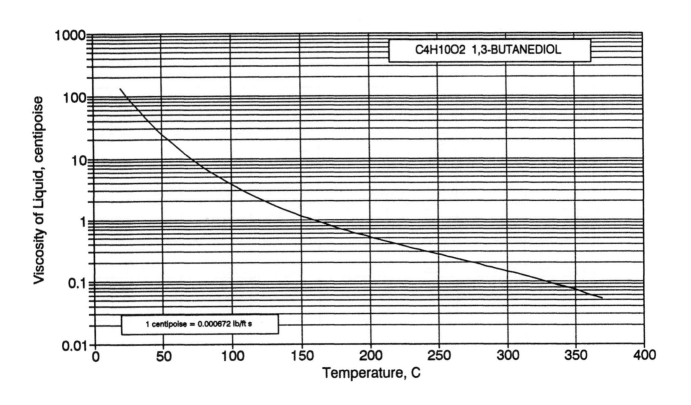

299

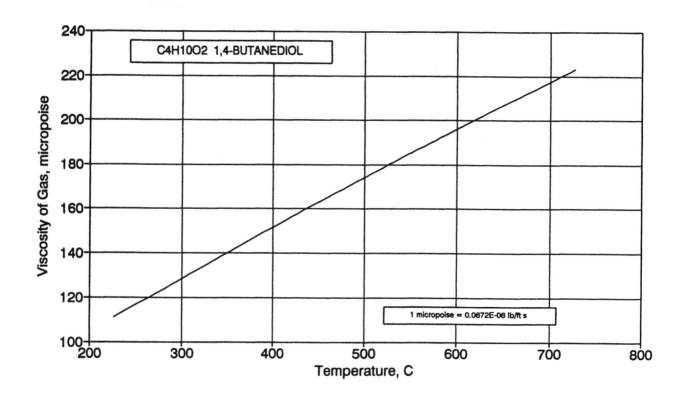

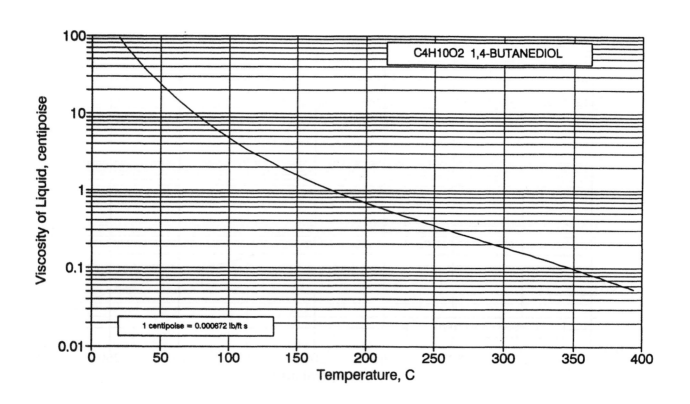

300

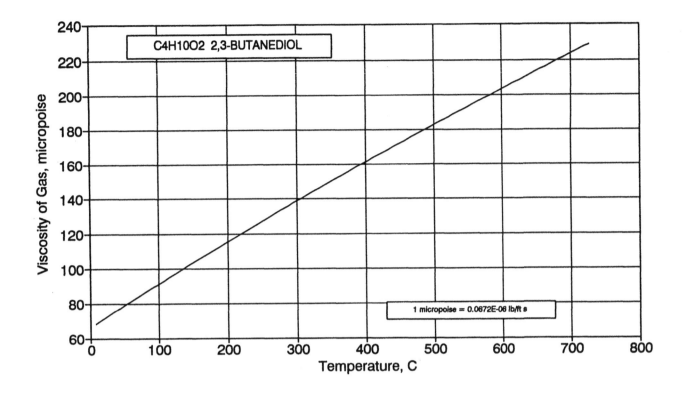

Viscosity of Gas, micropoise

C4H10O2 2,3-BUTANEDIOL

1 micropoise = 0.0672E-06 lb/ft s

Temperature, C

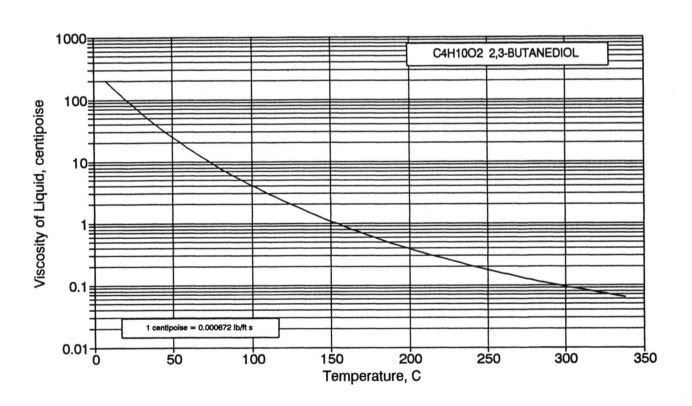

Viscosity of Liquid, centipoise

C4H10O2 2,3-BUTANEDIOL

1 centipoise = 0.000672 lb/ft s

Temperature, C

301

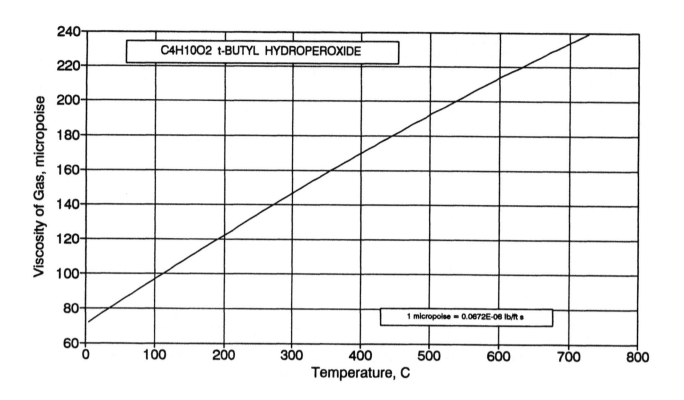

C4H10O2 t-BUTYL HYDROPEROXIDE

1 micropoise = 0.0672E-06 lb/ft s

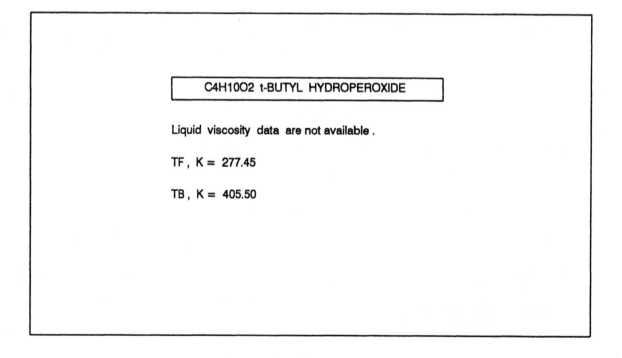

C4H10O2 t-BUTYL HYDROPEROXIDE

Liquid viscosity data are not available.

TF, K = 277.45

TB, K = 405.50

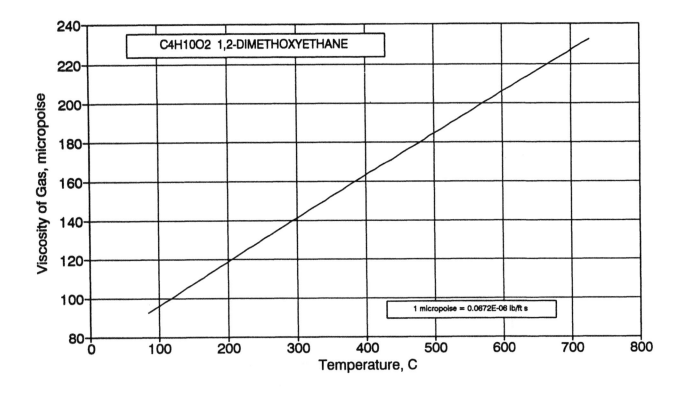

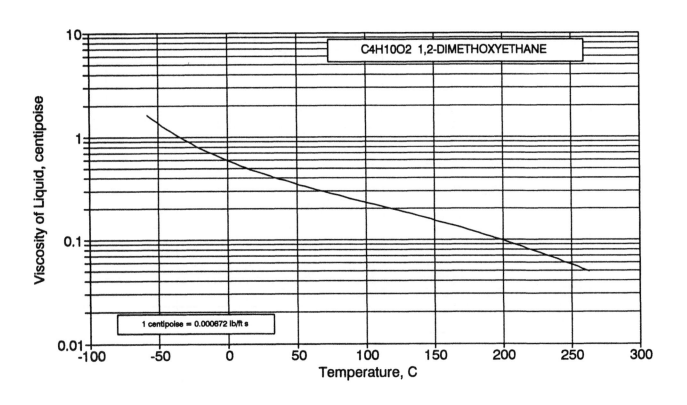

303

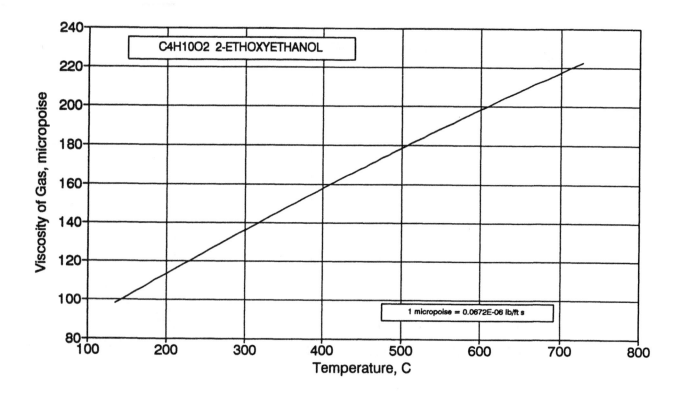

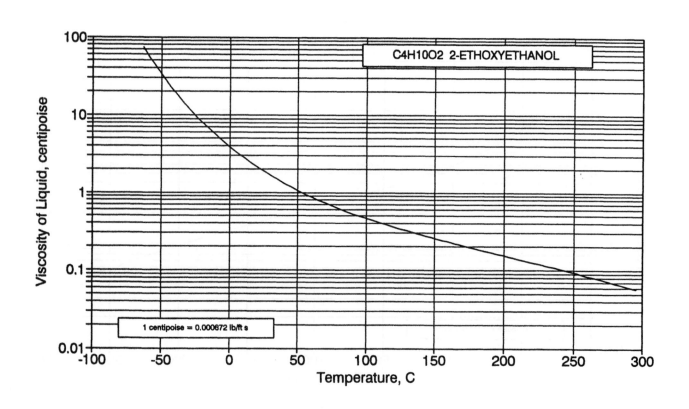

304

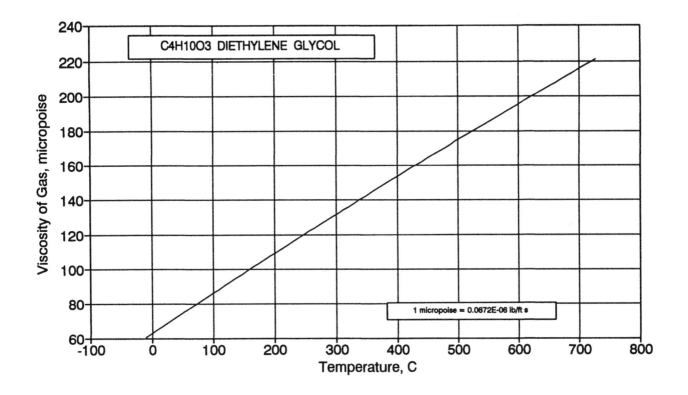

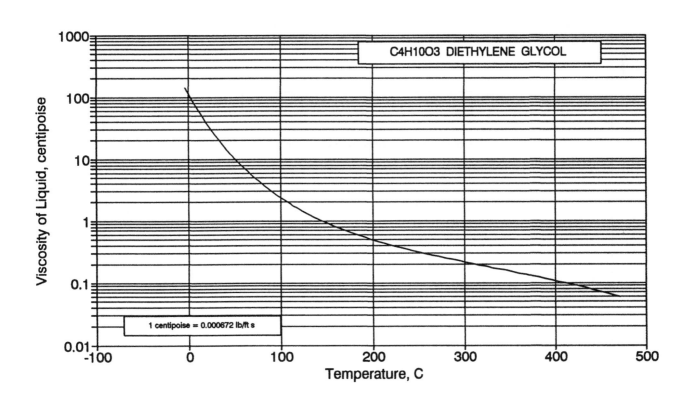

305

C4H10O4S DIETHYL SULFATE

Gas viscosity data are not available.

TF , K = 248.00

TB , K = 483.00

C4H10O4S DIETHYL SULFATE

Liquid viscosity data are not available.

TF , K = 248.00

TB , K = 483.00

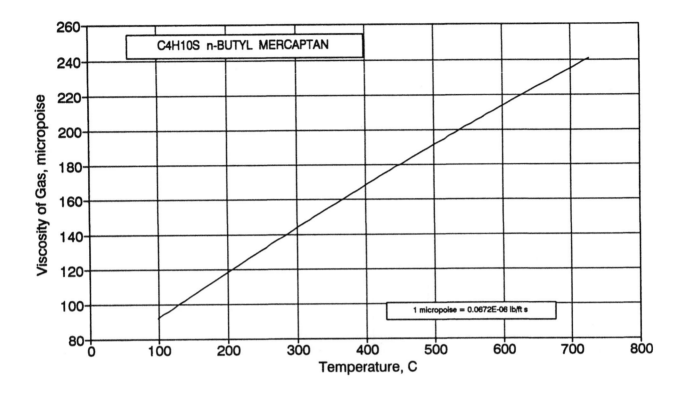

1 micropoise = 0.0672E-06 lb/ft s

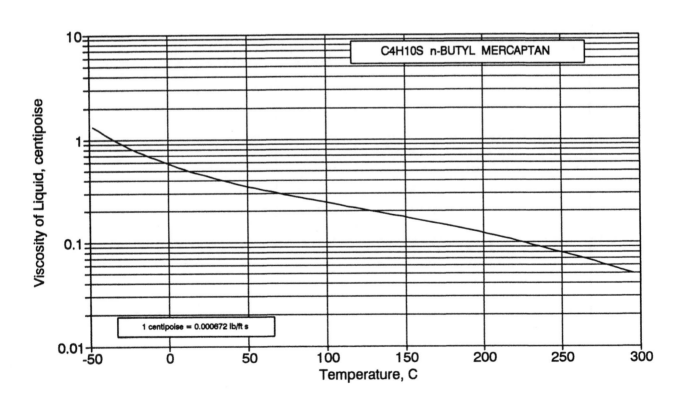

1 centipoise = 0.000672 lb/ft s

307

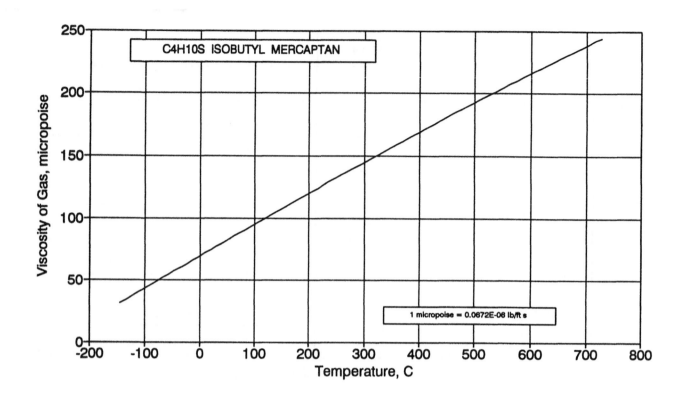

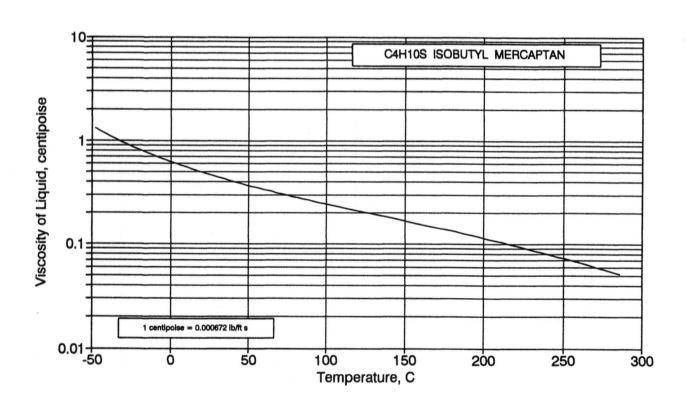

308

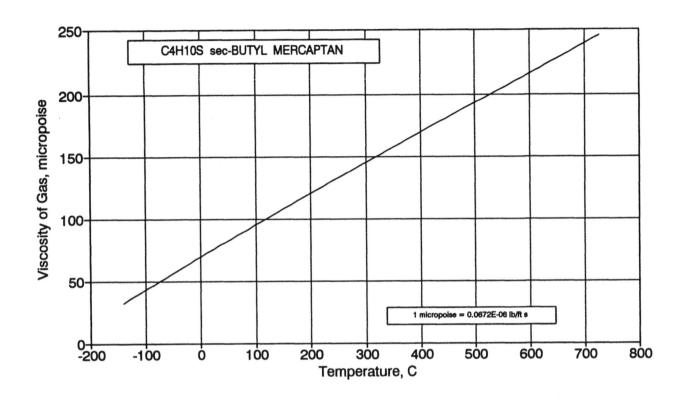

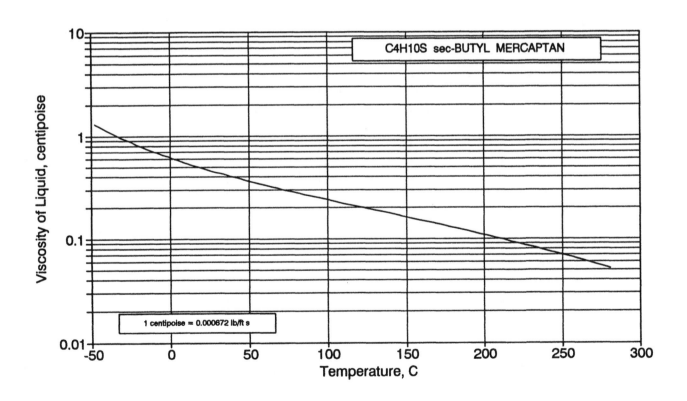

309

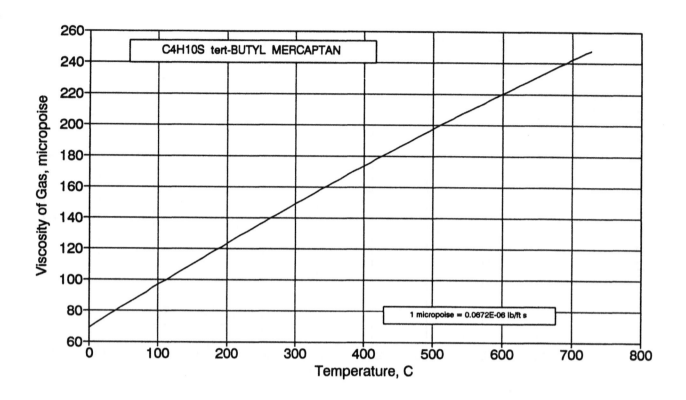

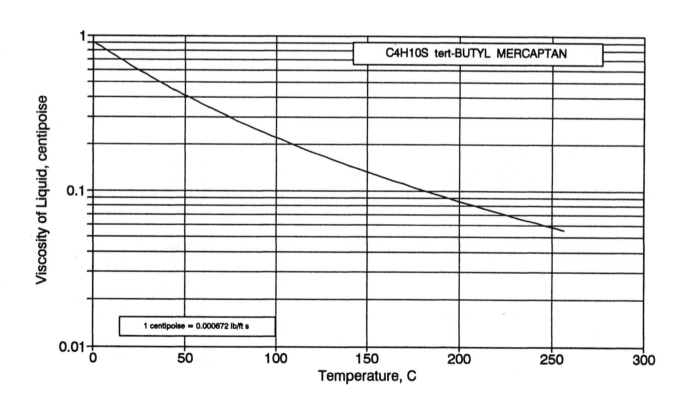

310

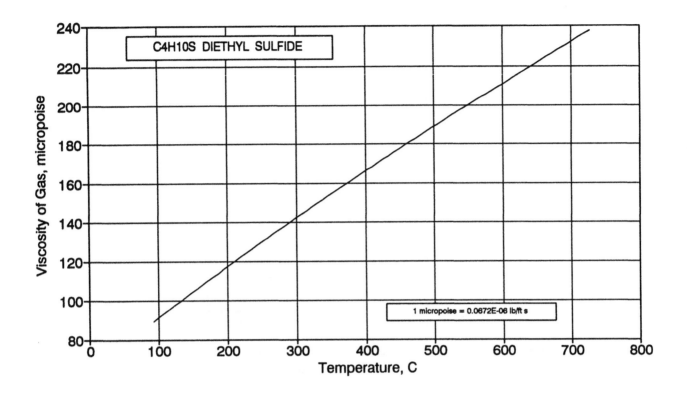

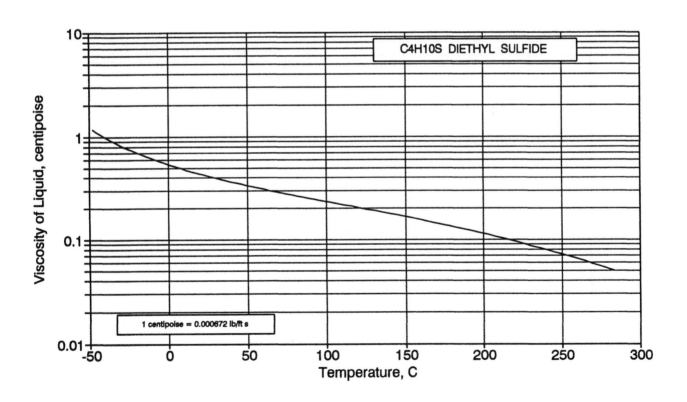

311

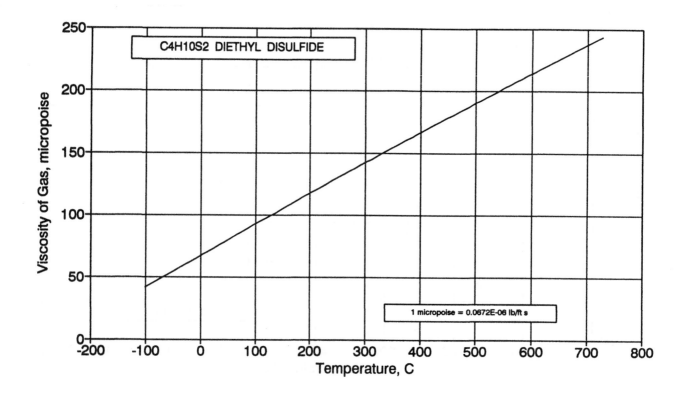

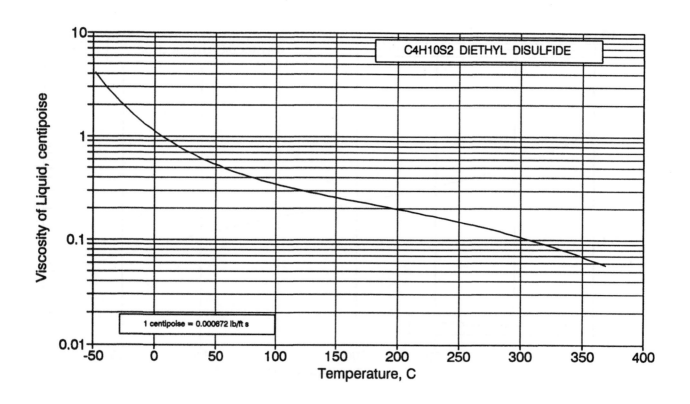

312

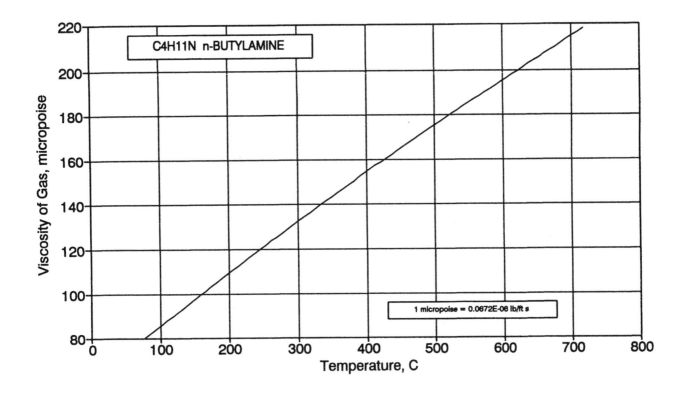

C4H11N n-BUTYLAMINE

1 micropoise = 0.0672E-06 lb/ft s

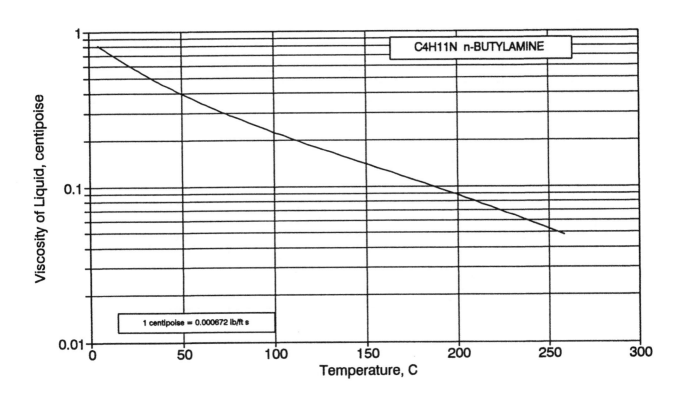

C4H11N n-BUTYLAMINE

1 centipoise = 0.000672 lb/ft s

313

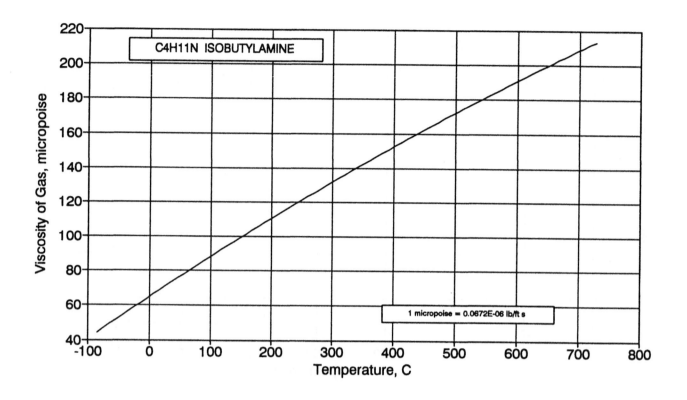

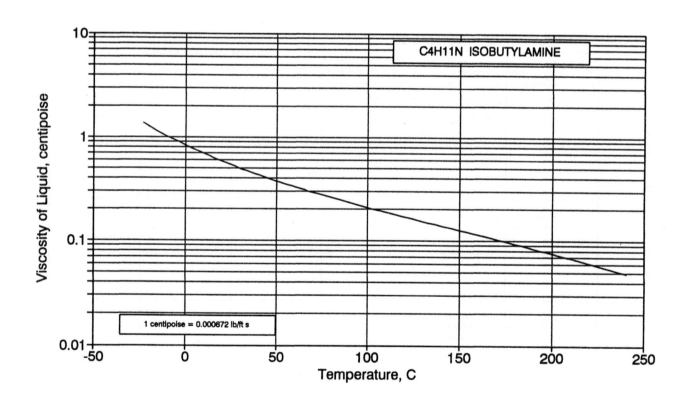

314

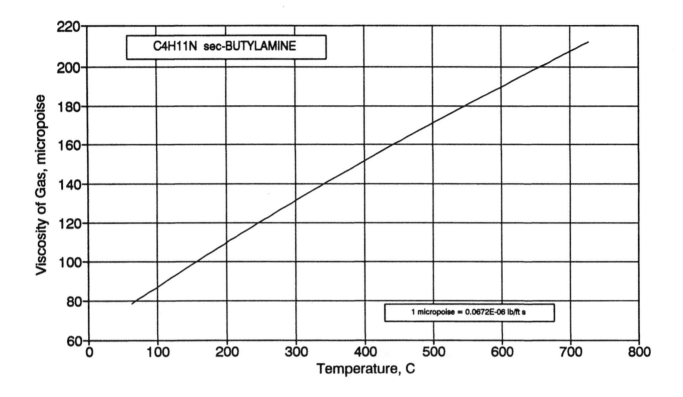

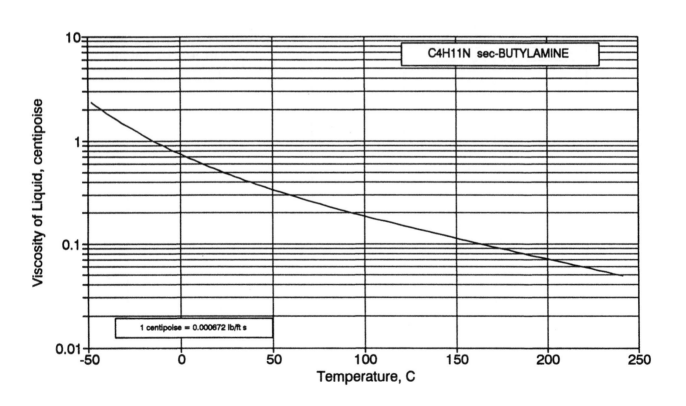

315

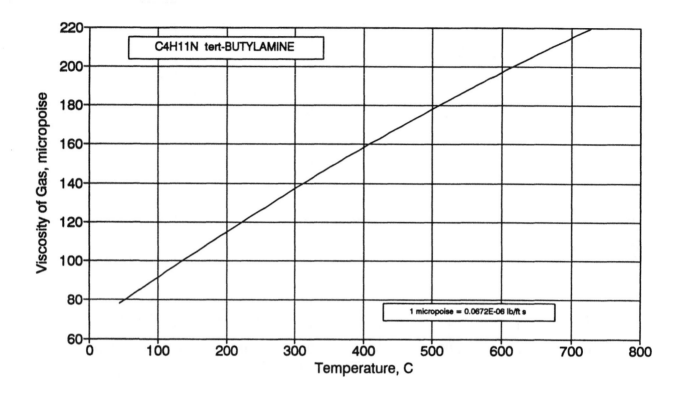

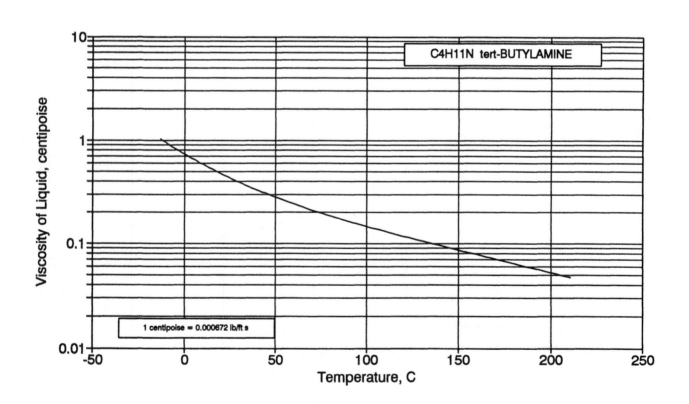

316

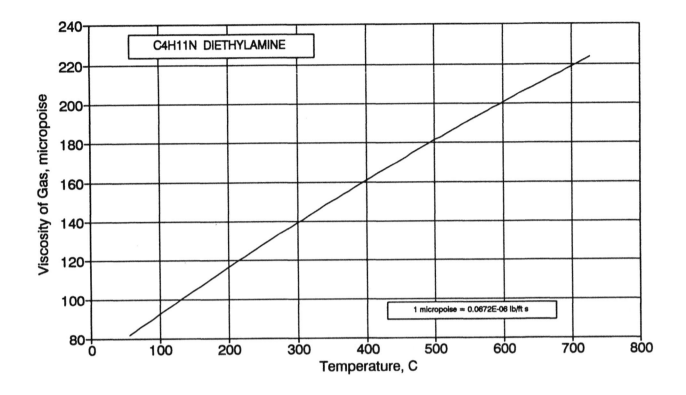

C4H11N DIETHYLAMINE

Viscosity of Gas, micropoise

1 micropoise = 0.0672E-06 lb/ft s

Temperature, C

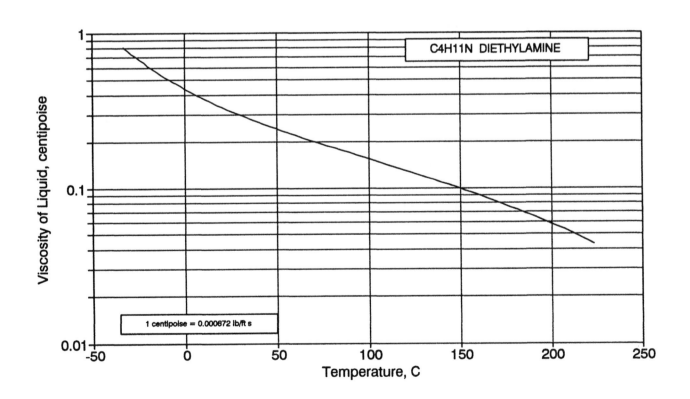

C4H11N DIETHYLAMINE

Viscosity of Liquid, centipoise

1 centipoise = 0.000672 lb/ft s

Temperature, C

317

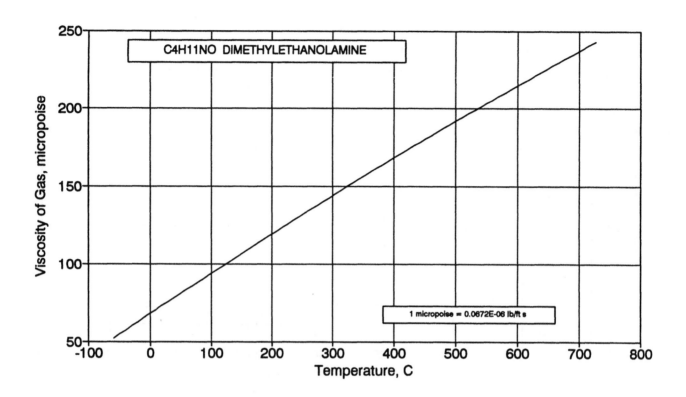

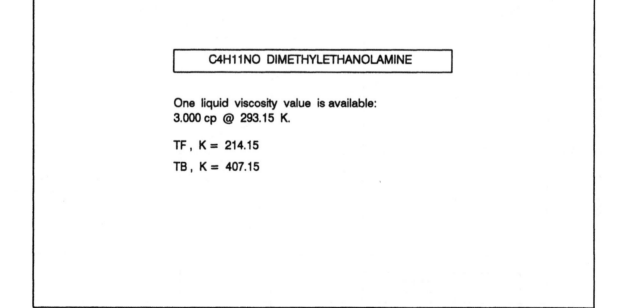

C4H11NO DIMETHYLETHANOLAMINE

One liquid viscosity value is available:
3.000 cp @ 293.15 K.

TF , K = 214.15

TB , K = 407.15

318

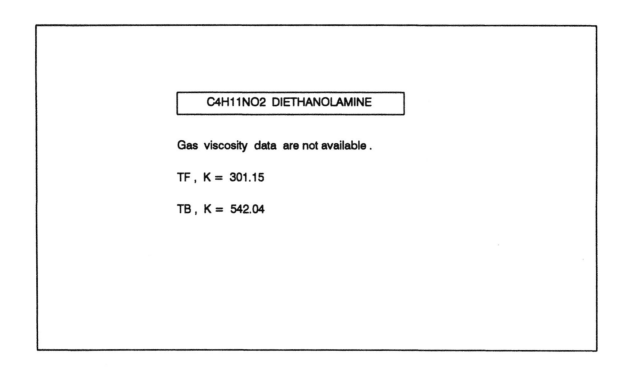

C4H11NO2 DIETHANOLAMINE

Gas viscosity data are not available .

TF , K = 301.15

TB , K = 542.04

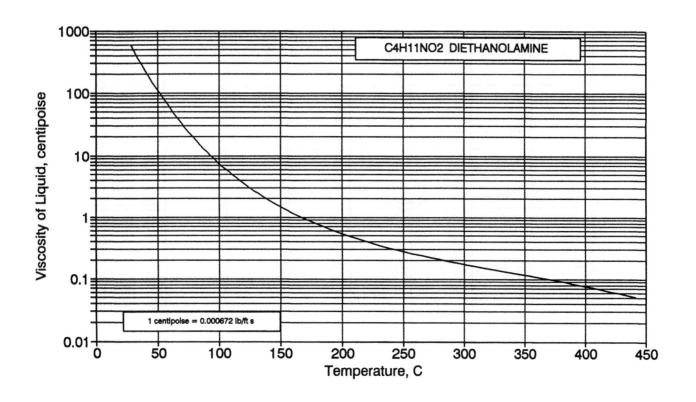

1 centipoise = 0.000672 lb/ft s

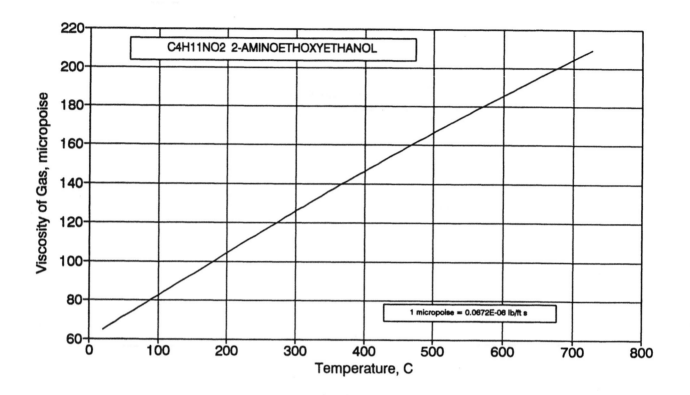

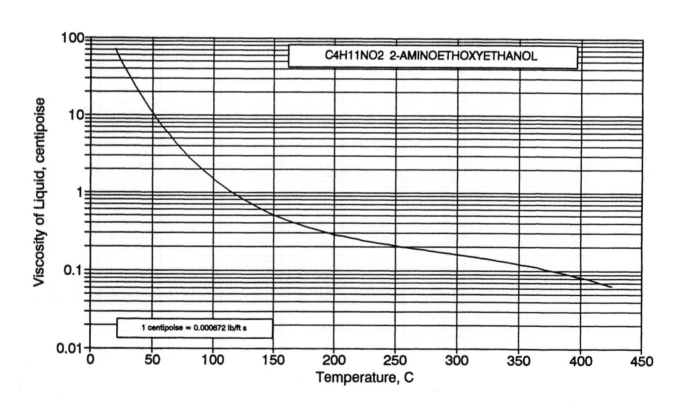

320

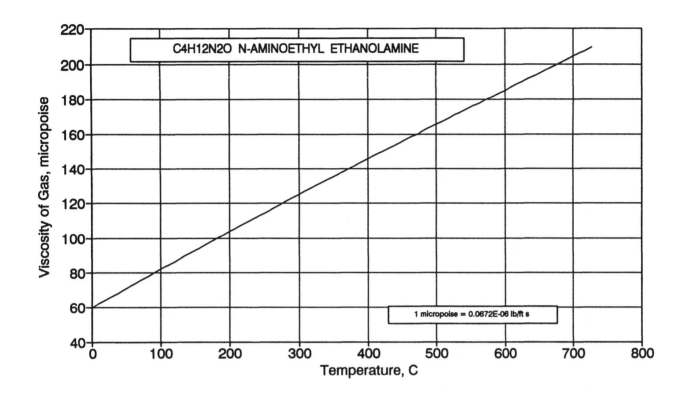

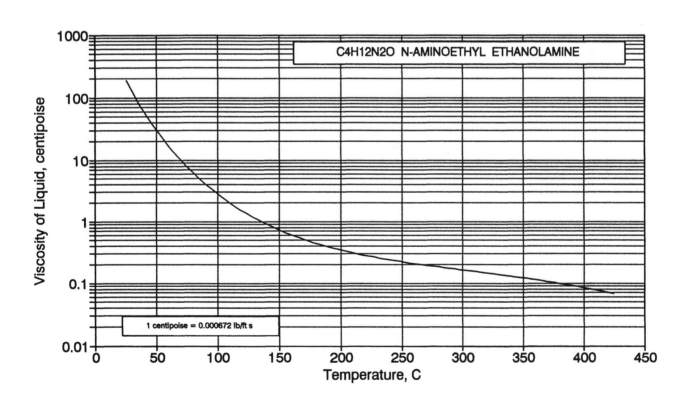

321

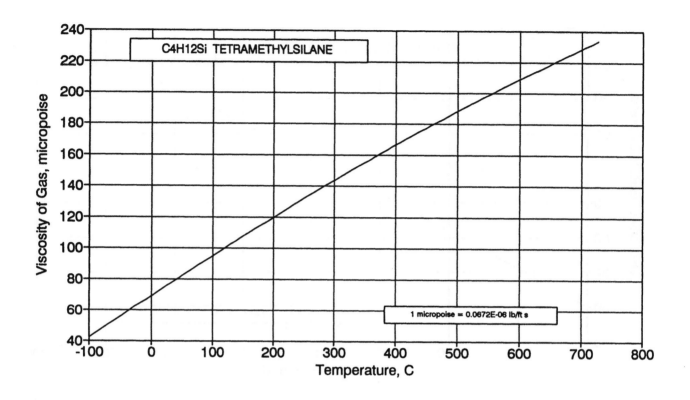

C4H12Si TETRAMETHYLSILANE

1 micropoise = 0.0672E-06 lb/ft s

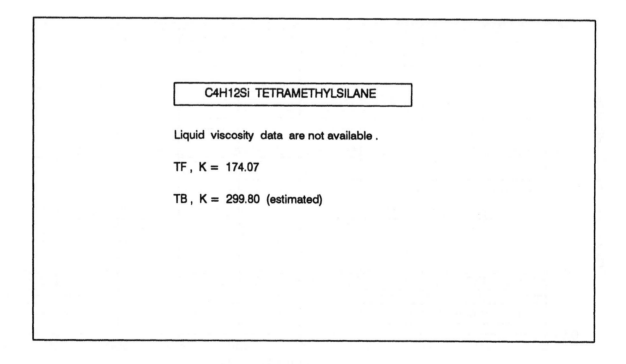

C4H12Si TETRAMETHYLSILANE

Liquid viscosity data are not available .

TF , K = 174.07

TB , K = 299.80 (estimated)

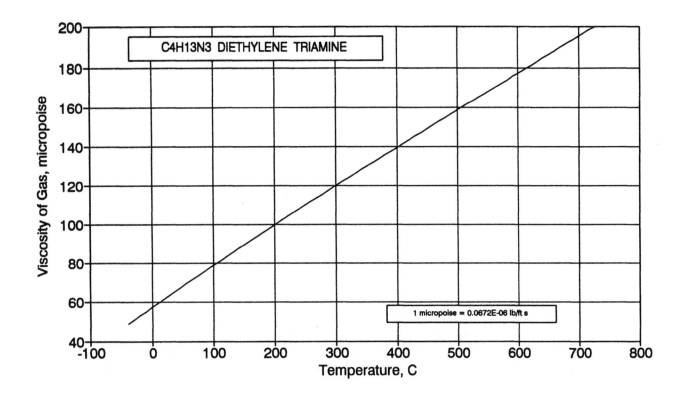

C4H13N3 DIETHYLENE TRIAMINE

1 micropoise = 0.0672E-06 lb/ft s

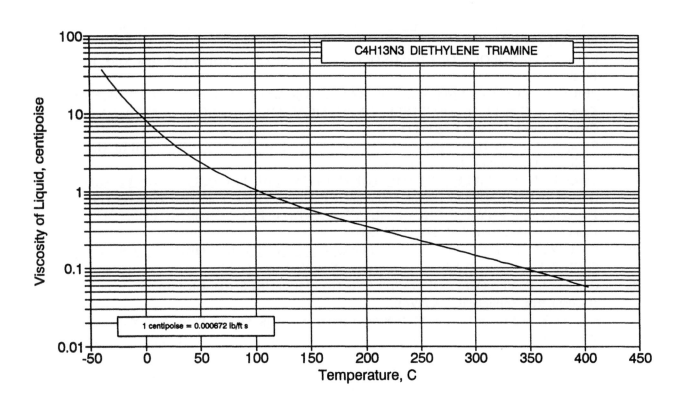

C4H13N3 DIETHYLENE TRIAMINE

1 centipoise = 0.000672 lb/ft s

REFERENCES

1. API Research Project No. 44, <u>SELECTED VALUES OF PHYSICAL AND THERMODYNAMIC PROPERTIES OF HYDROCARBONS AND RELATED COMPOUNDS</u>, Carnegie Press, Carnegie Institute of Technology, Pittsburgh, PA (1953).

2. <u>SELECTED VALUES OF PROPERTIES OF HYDROCARBONS AND RELATED COMPOUNDS</u>, Thermodynamics Research Center, TAMU, College Station, TX (1977, 1984).

3. <u>SELECTED VALUES OF PROPERTIES OF CHEMICAL COMPOUNDS</u>, Thermodynamics Research Center, TAMU, College Station, TX (1977, 1987).

4. <u>TECHNICAL DATA BOOK - PETROLEUM REFINING</u>, Vol. I and II, American Petroleum Institute, Washington, DC (1972, 1977, 1982).

5. Daubert, T. E. and R. P. Danner, <u>DATA COMPILATION OF PROPERTIES OF PURE COMPOUNDS</u>, Parts 1, 2, 3 and 4, Supplements 1 and 2, DIPPR Project, AIChE, New York, NY (1985-1992).

6. Ambrose, D., <u>VAPOUR-LIQUID CRITICAL PROPERTIES</u>, National Physical Laboratory, Teddington, England, NPL Report Chem 107 (Feb., 1980).

7. Simmrock, K. H., R. Janowsky and A. Ohnsorge, <u>CRITICAL DATA OF PURE SUBSTANCES</u>, Vol. II, Parts 1 and 2, Dechema Chemistry Data Series, 6000 Frankfurt/Main, Germany (1986).

8. <u>INTERNATIONAL CRITICAL TABLES</u>, McGraw-Hill, New York, NY (1926).

9. Maxwell, J. B., <u>DATA BOOK ON HYDROCARBONS</u>, D. Van Nostrand, Princeton. NJ (1958).

10. Egloff, G., <u>PHYSICAL CONSTANTS OF HYDROCARBONS</u>, Vols. 1-6, Reinhold Publishing Corp., New York, NY (1939-1947).

11. Braker, W. and A. L. Mossman, <u>MATHESON GAS DATA BOOK</u>, 6th ed., Matheson Gas Products, Secaucaus, NJ (1980).

12. <u>CRC HANDBOOK OF CHEMISTRY AND PHYSICS</u>, 66th - 74th eds., CRC Press, Inc., Boca Raton, FL (1985-1993).

13. <u>LANGE'S HANDBOOK OF CHEMISTRY</u>, 13th and 14th eds., McGraw-Hill, New York, NY (1985, 1992).

14. <u>PERRY'S CHEMICAL ENGINEERING HANDBOOK</u>, 6th ed., McGraw-Hill, New York, NY (1984).

15. Kaye, G. W. C. and T. H. Laby, <u>TABLES OF PHYSICAL AND CHEMICAL CONSTANTS</u>, Longman Group Limited, London, England (1973).

16. Raznjevic, Kuzman, <u>HANDBOOK OF THERMODYNAMIC TABLES AND CHARTS</u>, Hemisphere Publishing Corporation, New York, NY (1976).

17. Vargaftik, N. B., <u>TABLES ON THE THERMOPHYSICAL PROPERTIES OF LIQUIDS AND GASES</u>, 2nd ed., English translation, Hemisphere Publishing Corporation, New York, NY (1975, 1983).

18. Timmermans, J., <u>PHYSICO-CHEMICAL CONSTANTS OF PURE ORGANIC COMPOUNDS</u>, Vol. 1 and 2, Elsevier, New York, NY (1950,1965).

19. Ho, C. Y., P. E. Liley, T. Makita and Y. Tanaka, <u>PROPERTIES OF INORGANIC AND ORGANIC FLUIDS</u>, Hemisphere Publishing Corporation, New York, NY (1988).

20. Golubev, I. F., <u>VISCOSITY OF GASES AND GAS MIXTURES</u>, translated from Russian, US Dept. of Commerce, Springfield, VA (1970).

21. Stephan, K. and Lucas K., <u>VISCOSITY OF DENSE FLUIDS</u>, Plenum Press, New York, NY (1979).

22. Viswanath, D. S. and G. Natarajan, <u>DATA BOOK ON THE VISCOSITY OF LIQUIDS</u>, Hemisphere Publishing Corporation, New York, NY (1989).

23. Landolt-Bornstein, <u>ZAHLENWERTE UND FUNKIONEN ANS PHYSIK, CHEMEI, ASTRONOMIE UND TECHNIK</u>, 6th ed., Springer-Verlag, Berlin, Germany (1972).

24. Touloukian, Y. S., P. E. Liley and S. C. Saxena, <u>THERMOPHYSICAL PROPERTIES OF MATTER</u>,

Vol. 11 - Viscosity, IFI/Plenum Press, New York, NY (1974).

25. Lyman, W. J., W. F. Reehl and D. H. Rosenblatt, HANDBOOK OF CHEMICAL PROPERTY ESTIMATION METHODS, McGraw-Hill, New York, NY (1982).

26. Reid, R. C., J. M. Prausnitz and B. E. Poling, THE PROPERTIES OF GASES AND LIQUIDS, 3rd ed. (R. C. Reid and T. K. Sherwood), 4th ed., McGraw-Hill, New York, NY (1977, 1987).

27. Kirk, R. E. and D. F. Othmer, editors, ENCYCLOPEDIA OF CHEMICAL TECHNOLOGY, 3rd ed., Vols. 1-24, John Wiley and Sons, Inc., New York, NY (1978-1984).

28. Beaton, C. F. and G. F. Hewitt, PHYSICAL PROPERTY DATA FOR THE DESIGN ENGINEER, Hemisphere Publishing Corporation, New York, NY (1989).

29. Yaws, C. L., PHYSICAL PROPERTIES, McGraw-Hill, New York, NY (1977).

30. Yaws, C. L., THERMODYNAMIC AND PHYSICAL PROPERTY DATA, Gulf Publishing Co., Houston, TX (1992).

31. Yaws, C. L. and R. W. Gallant, PHYSICAL PROPERTIES OF HYDROCARBONS, Vols. 1 (2nd ed.), 2 (3rd ed.) and 3 (1st ed.), Gulf Publishing Co., Houston, TX (1992, 1993, 1993).

Appendix A

COEFFICIENTS FOR LIQUID VISCOSITY EQUATION[*]

Carl L. Yaws, Xiaoyan Lin and Li Bu
Lamar University, Beaumont, Texas

$$\log_{10} \eta_{liq} = A + B/T + C\ T + D\ T^2 \qquad (\eta_{liq} - \text{centipoise}, T - K)$$

NO	FORMULA	NAME	A	B	C	D	TMIN	TMAX	η_{liq} @ 25 C
1	CBrClF2	BROMOCHLORODIFLUOROMETHANE	-5.3203	5.2567E+02	1.8649E-02	-2.7670E-05	160	426	0.349
2	CBrCl3	BROMOTRICHLOROMETHANE	-2.4807	6.5384E+02	2.4106E-03	-2.9900E-06	252	606	1.463
3	CBrF3	BROMOTRIFLUOROMETHANE	-17.3630	1.4685E+03	6.7833E-02	-9.6658E-05	170	340	0.156
4	CBr2F2	DIBROMODIFLUOROMETHANE	-2.3825	3.4361E+02	6.1053E-03	-9.9243E-06	220	478	0.511
5	CClF3	CHLOROTRIFLUOROMETHANE	-15.2276	1.1396E+03	6.6727E-02	-1.0913E-04	170	302	0.061
6	CClN	CYANOGEN CHLORIDE	-34.0692	4.3759E+03	8.8017E-02	-8.1600E-05	267	449	0.395
7	CCl2F2	DICHLORODIFLUOROMETHANE	-14.1271	1.2812E+03	5.1192E-02	-6.8214E-05	170	385	0.234
8	CCl2O	PHOSGENE	-5.9900	8.9328E+02	1.2942E-02	-1.4515E-05	253	455	0.375
9	CCl3F	TRICHLOROFLUOROMETHANE	-8.7050	9.7314E+02	2.6505E-02	-3.1615E-05	170	471	0.448
10	CCl4	CARBON TETRACHLORIDE	-6.4564	1.0379E+03	1.4021E-02	-1.4107E-05	265	556	0.894
11	CF2O	CARBONYL FLUORIDE	15.3424	-1.0000E+03	-6.0892E-02	5.3802E-05	162	297	--
12	CF4	CARBON TETRAFLUORIDE	-8.1062	4.7871E+02	5.0987E-02	-1.3379E-04	90	228	--
13	CHBr3	TRIBROMOMETHANE	-3.3401	7.2801E+02	5.4337E-03	-5.0412E-06	281	696	1.878
14	CHClF2	CHLORODIFLUOROMETHANE	-10.8934	9.7972E+02	3.8730E-02	-5.2025E-05	170	369	0.207
15	CHCl2F	DICHLOROFLUOROMETHANE	-9.3552	9.7434E+02	2.9611E-02	-3.6067E-05	170	452	0.343
16	CHCl3	CHLOROFORM	-4.7831	6.9902E+02	1.0929E-02	-1.2244E-05	210	536	0.539
17	CHF3	TRIFLUOROMETHANE	-21.3082	1.5503E+03	9.5945E-02	-1.5489E-04	170	299	0.054
18	CHN	HYDROGEN CYANIDE	-12.0812	1.3183E+03	3.5234E-02	-4.0185E-05	260	457	0.188
19	CH2BrCl	BROMOCHLOROMETHANE	-5.2060	7.7028E+02	1.1905E-02	-1.2455E-05	185	557	0.660
20	CH2Br2	DIBROMOMETHANE	-5.4865	8.6380E+02	1.2128E-02	-1.1614E-05	230	611	0.987
21	CH2Cl2	DICHLOROMETHANE	-5.1043	6.8653E+02	1.2459E-02	-1.4540E-05	208	510	0.417
22	CH2F2	DIFLUOROMETHANE	-0.9739	1.3685E+02	2.7097E-03	-1.3376E-05	200	352	0.127
23	CH2I2	DIIODOMETHANE	-1.4610	5.0743E+02	1.1256E-03	-1.8470E-06	279	747	2.584
24	CH2O	FORMALDEHYDE	-6.3673	6.5848E+02	1.9414E-02	-2.7279E-05	193	408	0.160
25	CH2O2	FORMIC ACID	-4.2125	9.7953E+02	5.5520E-03	-5.7723E-06	281	580	1.641
26	CH3Br	METHYL BROMIDE	-9.5533	1.0306E+03	2.8322E-02	-3.1920E-05	193	467	0.324
27	CH3Cl	METHYL CHLORIDE	-7.3473	8.5395E+02	1.9485E-02	-2.3484E-05	249	416	0.173
28	CH3Cl3Si	METHYL TRICHLOROSILANE	-5.0787	7.1071E+02	1.2214E-02	-1.4351E-05	225	517	0.469
29	CH3F	METHYL FLUORIDE	-7.1229	5.0487E+02	3.0365E-02	-5.4345E-05	131	318	0.062
30	CH3I	METHYL IODIDE	-9.3737	1.1262E+03	2.5513E-02	-2.6102E-05	250	528	0.490
31	CH3NO	FORMAMIDE	-10.3646	1.9650E+03	1.8169E-02	-1.2609E-05	276	771	3.329
32	CH3NO2	NITROMETHANE	-7.1521	1.0567E+03	1.5983E-02	-1.5345E-05	245	588	0.621
33	CH4	METHANE	-7.3801	3.1925E+02	4.7934E-02	-1.4120E-04	91	191	--
34	CH4Cl2Si	METHYL DICHLOROSILANE	-7.2040	8.6650E+02	1.9542E-02	-2.2621E-05	275	483	0.330
35	CH4O	METHANOL	-9.0562	1.2542E+03	2.2383E-02	-2.3538E-05	230	513	0.539
36	CH4O3S	METHANESULFONIC ACID	---	---	---	---	--	--	--
37	CH4S	METHYL MERCAPTAN	-3.8298	4.4874E+02	9.7914E-03	-1.3437E-05	150	470	0.251
38	CH5ClSi	METHYL CHLOROSILANE	-12.7087	1.3825E+03	3.7365E-02	-4.2205E-05	220	442	0.207
39	CH5N	METHYLAMINE	-9.4670	9.8286E+02	2.8918E-02	-3.5672E-05	180	430	0.191
40	CH6Si	METHYL SILANE	-2.3551	1.5725E+02	8.7028E-03	-2.0236E-05	180	353	0.093
41	CN4O8	TETRANITROMETHANE	1.5505	4.2561E+02	-1.2276E-02	1.0222E-05	287	540	1.685
42	CO	CARBON MONOXIDE	-1.1224	5.7858E+01	-4.9174E-03	8.2233E-06	69	133	--
43	COS	CARBONYL SULFIDE	-2.7939	3.0912E+02	7.0546E-03	-1.3233E-05	134	379	0.148
44	CO2	CARBON DIOXIDE	-19.4921	1.5948E+03	7.9274E-02	-1.2025E-04	219	304	0.064
45	CS2	CARBON DISULFIDE	-9.1108	1.1216E+03	2.3216E-02	-2.2648E-05	235	552	0.363
46	C2BrF3	BROMOTRIFLUOROETHYLENE	12.0582	-5.7499E+02	-5.2673E-02	5.9107E-05	220	432	0.478
47	C2Br2F4	1,2-DIBROMOTETRAFLUOROETHANE	1.4494	1.4794E+01	-5.6542E-03	6.9522E-07	227	488	0.750
48	C2ClF3	CHLOROTRIFLUOROETHYLENE	-14.1754	2.2800E+03	2.2548E-02	-1.0524E-05	220	379	0.181
49	C2ClF5	CHLOROPENTAFLUOROETHANE	-21.2668	1.8746E+03	8.1455E-02	-1.1282E-04	190	353	0.190
50	C2Cl2F4	1,2-DICHLOROTETRAFLUOROETHANE	-13.2921	1.3844E+03	4.4079E-02	-5.5267E-05	179	419	0.381
51	C2Cl3F3	1,1,2-TRICHLOROTRIFLUOROETHANE	-1.8516	4.1245E+02	3.2446E-03	-7.4593E-06	237	487	0.686
52	C2Cl4	TETRACHLOROETHYLENE	-7.4654	1.1063E+03	1.6888E-02	-1.5458E-05	251	620	0.805
53	C2Cl4F2	1,1,2,2-TETRACHLORODIFLUOROETHANE	-4.4583	1.0258E+03	5.4537E-03	-5.1203E-06	299	551	--
54	C2Cl4O	TRICHLOROACETYL CHLORIDE	-3.2061	7.3624E+02	4.6638E-03	-5.5161E-06	273	590	1.457
55	C2Cl6	HEXACHLOROETHANE	-5.0439	8.4518E+02	9.8272E-03	-8.6017E-06	460	698	--
56	C2F4	TETRAFLUOROETHYLENE	18.2028	-1.3315E+03	-6.6186E-02	5.3425E-05	174	306	0.057
57	C2F6	HEXAFLUOROETHANE	-14.9996	1.0011E+03	7.0996E-02	-1.2231E-04	173	293	--

[*] A computer program, containing coefficients for all compounds, is available for a nominal fee. The computer program is in ASCII which can be accessed by other software.

NO	FORMULA	NAME	A	B	C	D	TMIN	TMAX	η_{liq} @ 25 C
58	C2HBrClF3	HALOTHANE	2.8293	5.5081E+01	-1.4157E-02	1.2453E-05	223	521	0.795
59	C2HClF2	2-CHLORO-1,1-DIFLUOROETHYLENE	13.0230	-7.5837E+02	-5.6827E-02	6.5532E-05	200	401	0.230
60	C2HCl3	TRICHLOROETHYLENE	-5.5389	7.8313E+02	1.2849E-02	-1.3292E-05	250	571	0.546
61	C2HCl3O	DICHLOROACETYL CHLORIDE	-18.9816	2.5504E+03	4.5842E-02	-3.9281E-05	298	579	0.560
62	C2HCl3O	TRICHLOROACETALDEHYDE	-3.9777	7.8167E+02	6.8624E-03	-7.6177E-06	225	565	1.030
63	C2HCl5	PENTACHLOROETHANE	-2.2339	6.8941E+02	1.5280E-03	-2.1471E-06	244	665	2.204
64	C2HF3O2	TRIFLUOROACETIC ACID	7.9900	-5.8864E+02	-2.6509E-02	2.0573E-05	258	491	0.872
65	C2HF5	PENTAFLUOROETHANE	5.4452	-1.8170E+02	-3.0751E-02	3.7720E-05	170	342	0.105
66	C2H2	ACETYLENE	-0.0709	2.8381E+01	-4.6617E-03	3.1151E-06	193	308	0.082
67	C2H2Br4	1,1,2,2-TETRABROMOETHANE	-12.1956	2.4476E+03	2.0163E-02	-1.2385E-05	300	824	--
68	C2H2Cl2	1,1-DICHLOROETHYLENE	-2.8187	4.7865E+02	5.0534E-03	-7.6546E-06	151	482	0.410
69	C2H2Cl2	cis-1,2-DICHLOROETHYLENE	-5.4151	7.2994E+02	1.3225E-02	-1.4921E-05	208	527	0.446
70	C2H2Cl2	trans-1,2-DICHLOROETHYLENE	-7.5792	9.4638E+02	1.9835E-02	-2.1586E-05	223	508	0.389
71	C2H2Cl2O	CHLOROACETYL CHLORIDE	-6.4016	1.1523E+03	1.2139E-02	-1.1268E-05	251	581	1.204
72	C2H2Cl2O	DICHLOROACETALDEHYDE	-10.5217	1.8824E+03	1.9229E-02	-1.5230E-05	223	555	1.483
73	C2H2Cl2O2	DICHLOROACETIC ACID	-22.2651	3.8370E+03	4.3150E-02	-3.0014E-05	287	686	6.330
74	C2H2Cl3F	1,1,1-TRICHLOROFLUOROETHANE	-1.1714	4.5197E+02	-9.0334E-04	-8.6437E-07	220	565	0.996
75	C2H2Cl4	1,1,1,2-TETRACHLOROETHANE	-2.5068	6.5745E+02	2.4956E-03	-3.2075E-06	203	624	1.436
76	C2H2Cl4	1,1,2,2-TETRACHLOROETHANE	-3.5146	7.7224E+02	5.5571E-03	-5.8524E-06	273	645	1.629
77	C2H2F2	1,1-DIFLUOROETHYLENE	12.1301	-5.3605E+02	-7.1633E-02	1.1068E-04	129	303	0.065
78	C2H2F4	1,1,1,2-TETRAFLUOROETHANE	-14.4406	1.2563E+03	5.2393E-02	-6.9771E-05	172	380	0.156
79	C2H2O	KETENE	4.3585	-1.4009E+02	-2.5364E-02	3.0329E-05	130	370	0.105
80	C2H2O4	OXALIC ACID	---	---	---	---	--	--	--
81	C2H3Br	VINYL BROMIDE	-7.3663	8.1930E+02	2.2026E-02	-2.6270E-05	180	473	0.411
82	C2H3Cl	VINYL CHLORIDE	-1.1063	2.1454E+02	-8.5045E-04	-1.3519E-06	130	432	0.174
83	C2H3ClF2	1-CHLORO-1,1-DIFLUOROETHANE	-16.8159	1.5115E+03	6.1839E-02	-8.0549E-05	200	410	0.339
84	C2H3ClO	ACETYL CHLORIDE	-10.9887	1.3155E+03	2.9872E-02	-3.1074E-05	275	508	0.369
85	C2H3ClO	CHLOROACETALDEHYDE	-35.7780	5.0921E+03	7.9981E-02	-6.1723E-05	293	555	0.458
86	C2H3ClO2	CHLOROACETIC ACID	-8.5505	1.8169E+03	1.3987E-02	-1.0355E-05	333	686	--
87	C2H3ClO2	METHYL CHLOROFORMATE	-8.3513	1.1333E+03	2.0644E-02	-2.1205E-05	192	525	0.525
88	C2H3Cl3	1,1,1-TRICHLOROETHANE	-3.9096	7.0709E+02	7.5847E-03	-9.1662E-06	243	545	0.810
89	C2H3Cl3	1,1,2-TRICHLOROETHANE	-3.2716	6.8810E+02	4.8932E-03	-5.4671E-06	237	602	1.021
90	C2H3F	VINYL FLUORIDE	-1.1547	1.6886E+02	3.4703E-04	-7.2507E-06	113	328	0.074
91	C2H3F3	1,1,1-TRIFLUOROETHANE	-2.2123	2.4044E+02	7.7406E-04	1.6026E-07	162	346	0.069
92	C2H3N	ACETONITRILE	-2.9528	4.1475E+02	6.4299E-03	-9.1660E-06	288	546	0.347
93	C2H3NO	METHYL ISOCYANATE	-2.3960	3.2790E+02	6.4147E-03	-1.0316E-05	273	505	0.250
94	C2H4	ETHYLENE	-4.5611	3.0811E+02	1.8030E-02	-3.8145E-05	105	282	--
95	C2H4Br2	1,1-DIBROMOETHANE	-4.1302	7.5049E+02	7.8279E-03	-7.5923E-06	210	628	1.112
96	C2H4Br2	1,2-DIBROMOETHANE	-5.4223	1.0377E+03	9.6953E-03	-8.3417E-06	283	650	1.612
97	C2H4Cl2	1,1-DICHLOROETHANE	-3.8388	5.9046E+02	8.0953E-03	-9.9210E-06	176	523	0.471
98	C2H4Cl2	1,2-DICHLOROETHANE	-0.1656	2.7576E+02	-3.3493E-03	1.4093E-06	245	561	0.769
99	C2H4Cl2O	BIS(CHLOROMETHYL)ETHER	-2.4635	5.7215E+02	3.2060E-03	-4.6903E-06	232	579	0.987
100	C2H4F2	1,1-DIFLUOROETHANE	-37.3585	3.4509E+03	1.3343E-01	-1.6427E-04	243	387	0.248
101	C2H4F2	1,2-DIFLUOROETHANE	-10.3352	1.1272E+03	2.9491E-02	-3.2763E-05	215	476	0.212
102	C2H4O	ACETALDEHYDE	-6.6171	6.8123E+02	1.9979E-02	-2.5563E-05	260	461	0.225
103	C2H4O	ETHYLENE OXIDE	-5.7794	6.7020E+02	1.5686E-02	-1.9462E-05	190	469	0.260
104	C2H4O2	ACETIC ACID	-3.8937	7.8482E+02	6.6650E-03	-7.5606E-06	290	593	1.132
105	C2H4O2	METHYL FORMATE	-8.0637	1.0137E+03	2.0884E-02	-2.2997E-05	250	487	0.330
106	C2H5Br	BROMOETHANE	-5.3844	6.7418E+02	1.4140E-02	-1.6501E-05	155	504	0.422
107	C2H5Cl	ETHYL CHLORIDE	-4.4279	5.1891E+02	1.2035E-02	-1.6620E-05	150	460	0.265
108	C2H5ClO	2-CHLOROETHANOL	-10.3253	1.8994E+03	1.9820E-02	-1.6723E-05	250	585	2.939
109	C2H5F	ETHYL FLUORIDE	-5.4713	4.8451E+02	1.8733E-02	-2.9742E-05	130	375	0.124
110	C2H5I	ETHYL IODIDE	-10.4954	1.3679E+03	2.6346E-02	-2.4827E-05	273	561	0.551
111	C2H5N	ETHYLENEIMINE	-4.7646	7.3673E+02	9.5205E-03	-1.0431E-05	250	537	0.415
112	C2H5NO	ACETAMIDE	-15.0576	3.0478E+03	2.4646E-02	-1.5506E-05	354	761	--
113	C2H5NO	N-METHYLFORMAMIDE	-6.9983	1.2046E+03	1.2046E-02	-9.2805E-06	269	721	1.651
114	C2H5NO2	NITROETHANE	-3.7814	6.3484E+02	7.5441E-03	-8.5933E-06	200	593	0.681
115	C2H6	ETHANE	-4.2694	2.8954E+02	1.7111E-02	-3.6092E-05	98	305	0.039
116	C2H6AlCl	DIMETHYLALUMINUM CHLORIDE	---	---	---	---	(@ 30 C)		11.000
117	C2H6O	DIMETHYL ETHER	-7.4844	5.8392E+02	2.7815E-02	-4.0433E-05	132	400	0.149
118	C2H6O	ETHANOL	-6.4406	1.1176E+03	1.3721E-02	-1.5465E-05	240	516	1.057
119	C2H6OS	DIMETHYL SULFOXIDE	-3.6341	8.5487E+02	4.8721E-03	-4.4070E-06	292	726	1.968
120	C2H6O2	ETHYLENE GLYCOL	-16.9728	3.1886E+03	3.2537E-02	-2.4480E-05	261	645	17.647
121	C2H6O4S	DIMETHYL SULFATE	---	---	---	---	--	--	--
122	C2H6S	DIMETHYL SULFIDE	-7.8503	9.4330E+02	2.0507E-02	-2.2249E-05	273	503	0.282
123	C2H6S	ETHYL MERCAPTAN	-3.0781	4.2896E+02	6.4239E-03	-9.0060E-06	125	499	0.299
124	C2H6S2	DIMETHYL DISULFIDE	-6.8447	1.0217E+03	1.4599E-02	-1.3298E-05	188	606	0.566
125	C2H7N	DIMETHYLAMINE	-11.5558	1.2126E+03	3.4999E-02	-4.1253E-05	240	438	0.190

* A computer program, containing coefficients for all compounds, is available for a nominal fee. The computer program is in ASCII which can be accessed by other software.

$$\log_{10} \eta_{liq} = A + B/T + C\,T + D\,T^2 \qquad (\eta_{liq} - \text{centipoise}, \; T - K)$$

NO	FORMULA	NAME	A	B	C	D	TMIN	TMAX	η_{liq} @ 25 C
126	C2H7N	ETHYLAMINE	-7.0668	9.0544E+02	1.7675E-02	-2.0701E-05	192	456	0.251
127	C2H7NO	MONOETHANOLAMINE	-13.1818	2.8596E+03	2.0826E-02	-1.4230E-05	288	638	22.572
128	C2H8N2	ETHYLENEDIAMINE	-18.3052	2.9617E+03	3.7865E-02	-2.9650E-05	303	593	--
129	C2H8Si	DIMETHYL SILANE	---	---	---	---	--	--	--
130	C2N2	CYANOGEN	-79.0878	1.0824E+04	1.8829E-01	-1.5317E-04	269	400	0.550
131	C3F6	HEXAFLUOROPROPYLENE	17.0240	-9.7487E+02	-7.8331E-02	9.8244E-05	160	368	0.136
132	C3F6O	HEXAFLUOROACETONE	7.0539	-2.5115E+02	-3.6300E-02	4.2474E-05	148	357	0.146
133	C3F8	OCTAFLUOROPROPANE	4.8406	-2.8807E+01	-2.9852E-02	3.6536E-05	125	345	0.123
134	C3H2N2	MALONONITRILE	-15.5463	2.8953E+03	2.7321E-02	-1.8425E-05	305	715	--
135	C3H3Cl	PROPARGYL CHLORIDE	-2.0693	3.0412E+02	4.3357E-03	-7.1639E-06	293	541	0.404
136	C3H3N	ACRYLONITRILE	-6.3470	8.1502E+02	1.5664E-02	-1.7275E-05	240	535	0.332
137	C3H3NO	OXAZOLE	---	---	---	---	--	--	--
138	C3H4	METHYLACETYLENE	-8.4493	7.7571E+02	2.8408E-02	-3.8708E-05	170	402	0.152
139	C3H4	PROPADIENE	-4.0226	3.5646E+02	1.3512E-02	-2.3072E-05	173	393	0.141
140	C3H4Cl2	2,3-DICHLOROPROPENE	-1.9545	5.4794E+02	7.6738E-04	-1.9049E-06	192	577	0.876
141	C3H4O	ACROLEIN	-5.5517	7.0871E+02	1.4056E-02	-1.6788E-05	223	506	0.334
142	C3H4O	PROPARGYL ALCOHOL	-12.5948	2.0520E+03	2.5944E-02	-2.1497E-05	221	580	1.294
143	C3H4O2	ACRYLIC ACID	-15.9215	2.4408E+03	3.4383E-02	-2.7677E-05	293	615	1.137
144	C3H4O2	beta-PROPIOLACTONE	-6.6127	1.0109E+03	1.2263E-02	-9.3813E-06	240	686	0.398
145	C3H4O2	VINYL FORMATE	-4.4042	5.9523E+02	1.0606E-02	-1.3477E-05	200	498	0.360
146	C3H4O3	ETHYLENE CARBONATE	-8.5203	1.3364E+03	1.7273E-02	-1.2911E-05	311	790	--
147	C3H4O3	PYRUVIC ACID	-17.9536	2.9060E+03	3.7391E-02	-2.8775E-05	287	635	2.418
148	C3H5Cl	2-CHLOROPROPENE	-2.1425	3.7179E+02	2.8199E-03	-5.4004E-06	136	478	0.292
149	C3H5Cl	3-CHLOROPROPENE	-6.0433	8.0040E+02	1.4186E-02	-1.5455E-05	250	514	0.314
150	C3H5ClO	alpha-EPICHLOROHYDRIN	-2.3159	5.6442E+02	2.7982E-03	-4.1693E-06	223	610	1.100
151	C3H5ClO2	METHYL CHLOROACETATE	-6.3947	1.0712E+03	1.3156E-02	-1.2533E-05	241	600	1.015
152	C3H5ClO2	ETHYL CHLOROFORMATE	-12.4292	1.5619E+03	3.3551E-02	-3.4501E-05	192	508	0.557
153	C3H5Cl3	1,2,3-TRICHLOROPROPANE	-1.7913	6.4440E+02	3.8924E-04	-1.4969E-06	258	652	2.254
154	C3H5N	PROPIONITRILE	-5.6142	8.0233E+02	1.2446E-02	-1.3286E-05	250	564	0.404
155	C3H5NO	ACRYLAMIDE	2.7157	2.5375E+02	-1.1286E-02	7.4172E-06	358	710	--
156	C3H5NO	HYDRACRYLONITRILE	-13.0827	1.9718E+03	2.6820E-02	-2.0038E-05	227	690	0.557
157	C3H5NO	LACTONITRILE	-13.0054	2.6255E+03	2.1406E-02	-1.4674E-05	233	643	7.554
158	C3H5N3O9	NITROGLYCERINE	-30.0495	5.6062E+03	5.2649E-02	-3.2609E-05	286	680	35.683
159	C3H6	CYCLOPROPANE	-3.2541	3.2192E+02	9.9766E-03	-1.8191E-05	146	398	0.152
160	C3H6	PROPYLENE	-5.1758	4.2982E+02	1.8611E-02	-3.1662E-05	90	365	0.100
161	C3H6Cl2	1,1-DICHLOROPROPANE	-3.5820	6.1920E+02	6.6598E-03	-7.8836E-06	200	560	0.602
162	C3H6Cl2	1,2-DICHLOROPROPANE	-2.8218	5.9927E+02	3.7360E-03	-4.8006E-06	173	572	0.751
163	C3H6Cl2	1,3-DICHLOROPROPANE	-2.8361	6.0541E+02	4.0515E-03	-5.0439E-06	174	603	0.900
164	C3H6O	ACETONE	-7.2126	9.0305E+02	1.8385E-02	-2.0353E-05	223	508	0.308
165	C3H6O	ALLYL ALCOHOL	-11.8248	1.9173E+03	2.5034E-02	-2.2322E-05	281	545	1.217
166	C3H6O	METHYL VINYL ETHER	-5.8282	5.4577E+02	1.8540E-02	-2.5458E-05	151	437	0.185
167	C3H6O	n-PROPIONALDEHYDE	-9.8172	1.2714E+03	2.4587E-02	-2.5572E-05	280	496	0.319
168	C3H6O	1,2-PROPYLENE OXIDE	-7.2842	9.7539E+02	1.7425E-02	-1.9160E-05	200	482	0.302
169	C3H6O	1,3-PROPYLENE OXIDE	-10.8675	1.2901E+03	2.7838E-02	-2.7265E-05	255	520	0.217
170	C3H6O2	ETHYL FORMATE	-6.3477	8.5383E+02	1.5404E-02	-1.7222E-05	245	508	0.378
171	C3H6O2	METHYL ACETATE	-7.0933	9.3074E+02	1.7481E-02	-1.9038E-05	250	507	0.353
172	C3H6O2	PROPIONIC ACID	-5.0177	8.7365E+02	1.0302E-02	-1.0883E-05	252	604	1.039
173	C3H6O2S	3-MERCAPTOPROPIONIC ACID	-18.6383	3.2116E+03	3.5095E-02	-2.3670E-05	291	729	3.112
174	C3H6O3	LACTIC ACID	---	---	---	---	--	--	--
175	C3H6O3	METHOXYACETIC ACID	-16.1627	2.5857E+03	3.2785E-02	-2.4094E-05	281	691	1.389
176	C3H6O3	TRIOXANE	---	---	---	---	(@ 70 C)		8.120
177	C3H7Br	1-BROMOPROPANE	-6.3524	8.3429E+02	1.5963E-02	-1.6934E-05	200	544	0.501
178	C3H7Br	2-BROMOPROPANE	-11.4833	1.4831E+03	2.9203E-02	-2.8268E-05	250	532	0.484
179	C3H7Cl	ISOPROPYL CHLORIDE	-8.3740	1.0583E+03	2.1296E-02	-2.2963E-05	250	489	0.305
180	C3H7Cl	n-PROPYL CHLORIDE	-6.4801	8.5514E+02	1.5738E-02	-1.7463E-05	250	503	0.337
181	C3H7I	ISOPROPYL IODIDE	-6.6205	9.4055E+02	1.5487E-02	-1.5127E-05	225	578	0.641
182	C3H7I	n-PROPYL IODIDE	-7.8304	1.1203E+03	1.8135E-02	-1.6845E-05	240	593	0.687
183	C3H7N	ALLYLAMINE	-6.2650	9.0217E+02	1.4036E-02	-1.5335E-05	185	505	0.382
184	C3H7N	PROPYLENEIMINE	-12.0978	1.5277E+03	2.7981E-02	-2.4453E-05	229	529	0.157
185	C3H7NO	N,N-DIMETHYLFORMAMIDE	-5.3292	8.9547E+02	1.0559E-02	-1.0088E-05	240	647	0.842
186	C3H7NO	N-METHYLACETAMIDE	-12.0806	2.3614E+03	2.0484E-02	-1.3984E-05	301	718	--
187	C3H7NO2	1-NITROPROPANE	-6.5870	1.0437E+03	1.4009E-02	-1.3397E-05	240	605	0.794
188	C3H7NO2	2-NITROPROPANE	-3.5405	6.1915E+02	6.7366E-03	-7.8310E-06	182	594	0.706
189	C3H8	PROPANE	-3.1759	2.9712E+02	9.5453E-03	-1.8781E-05	85	370	0.099
190	C3H8O	ISOPROPANOL	-0.7009	8.4150E+02	-8.6068E-03	8.2964E-06	187	508	1.963
191	C3H8O	METHYL ETHYL ETHER	-29.8385	3.0253E+03	9.3761E-02	-1.0176E-04	233	438	0.165
192	C3H8O	n-PROPANOL	-3.7702	9.9151E+02	4.0836E-03	-5.4586E-06	220	537	1.939
193	C3H8O2	2-METHOXYETHANOL	-15.1858	2.4914E+03	3.0625E-02	-2.4291E-05	250	564	1.386

* A computer program, containing coefficients for all compounds, is available for a nominal fee. The computer program is in ASCII which can be accessed by other software.

$$\log_{10} \eta_{liq} = A + B/T + C\ T + D\ T^2 \quad (\eta_{liq} - centipoise,\ T - K)$$

NO	FORMULA	NAME	A	B	C	D	TMIN	TMAX	η_{liq} @ 25 C
194	C3H8O2	METHYLAL	-3.9255	5.1395E+02	1.0108E-02	-1.4478E-05	168	481	0.335
195	C3H8O2	1,2-PROPYLENE GLYCOL	-29.4920	5.2456E+03	5.8169E-02	-4.2343E-05	233	626	47.965
196	C3H8O2	1,3-PROPYLENE GLYCOL	-7.9787	1.9800E+03	1.1850E-02	-9.3205E-06	246	658	23.266
197	C3H8O3	GLYCEROL	-18.2152	4.2305E+03	2.8705E-02	-1.8648E-05	293	723	749.242
198	C3H8S	n-PROPYLMERCAPTAN	-4.1495	5.8084E+02	9.1407E-03	-1.0824E-05	160	536	0.365
199	C3H8S	ISOPROPYL MERCAPTAN	-2.7739	4.3005E+02	5.2653E-03	-7.6377E-06	143	517	0.363
200	C3H9N	n-PROPYLAMINE	-4.9620	7.9761E+02	9.5072E-03	-1.0844E-05	190	497	0.384
201	C3H9N	ISOPROPYLAMINE	-3.6769	6.8343E+02	3.4740E-03	-1.5653E-06	273	472	0.325
202	C3H9N	TRIMETHYLAMINE	-3.9726	4.4221E+02	1.0657E-02	-1.6134E-05	200	433	0.179
203	C3H9NO	1-AMINO-2-PROPANOL	-31.3874	5.5522E+03	5.9514E-02	-4.0875E-05	275	614	22.147
204	C3H9NO	3-AMINO-1-PROPANOL	-26.7477	4.7518E+03	5.0500E-02	-3.4638E-05	284	649	14.701
205	C3H9NO	METHYLETHANOLAMINE	-10.5527	2.3490E+03	1.5490E-02	-1.0492E-05	269	630	10.267
206	C3H9O4P	TRIMETHYL PHOSPHATE	-6.0196	1.2096E+03	9.3929E-03	-6.3749E-06	240	764	1.867
207	C3H10N2	1,2-PROPANEDIAMINE	-13.3632	2.2430E+03	2.6624E-02	-2.1292E-05	237	587	1.604
208	C3H10Si	TRIMETHYL SILANE	---	---	---	---	--	--	--
209	C4Cl4S	TETRACHLOROTHIOPHENE	---	---	---	---	(@ 30 C)		3.318
210	C4Cl6	HEXACHLORO-1,3-BUTADIENE	-0.1976	5.7506E+02	-3.8913E-03	2.1134E-06	252	741	5.739
211	C4F8	OCTAFLUORO-2-BUTENE	-1.9732	4.6632E+02	3.5001E-05	-2.1423E-08	174	392	0.398
212	C4F8	OCTAFLUOROCYCLOBUTANE	-2.0637	4.9325E+02	3.8652E-05	-2.0508E-08	245	388	0.398
213	C4F10	DECAFLUOROBUTANE	-2.0446	6.0604E+02	-2.3464E-05	1.3468E-08	210	386	0.960
214	C4H2O3	MALEIC ANHYDRIDE	-1.0811	5.5616E+02	-1.2536E-03	4.1553E-07	326	721	--
215	C4H4	VINYLACETYLENE	---	---	---	---	--	--	--
216	C4H4N2	SUCCINONITRILE	-1.1564	5.9038E+02	-6.9203E-04	2.1818E-07	331	770	--
217	C4H4O	FURAN	-3.6715	5.5958E+02	7.4026E-03	-9.7493E-06	220	490	0.351
218	C4H4O2	DIKETENE	---	---	---	---	(@ 25 C)		7.200
219	C4H4O3	SUCCINIC ANHYDRIDE	-1.7466	5.0881E+02	-9.6474E-05	2.7683E-08	393	811	--
220	C4H4O4	FUMARIC ACID	-3.0627	1.1080E+03	6.7047E-04	-1.6982E-07	560	771	--
221	C4H4O4	MALEIC ACID	-52.9808	9.9837E+03	9.2110E-02	-5.4257E-05	403	773	--
222	C4H4S	THIOPHENE	-6.5928	1.0206E+03	1.3826E-02	-1.3089E-05	273	579	0.616
223	C4H5Cl	CHLOROPRENE	-1.8266	4.2271E+02	2.5963E-05	-1.2880E-08	174	525	0.396
224	C4H5N	trans-CROTONITRILE	-5.6497	8.2828E+02	1.2430E-02	-1.2891E-05	222	586	0.488
225	C4H5N	cis-CROTONITRILE	-4.1893	6.5278E+02	8.7208E-03	-1.0178E-05	201	568	0.496
226	C4H5N	METHACRYLONITRILE	-9.9472	1.3376E+03	2.3416E-02	-2.2227E-05	237	554	0.351
227	C4H5N	PYRROLE	-7.8869	1.3872E+03	1.4837E-02	-1.2263E-05	250	640	1.257
228	C4H5N	VINYLACETONITRILE	-4.3272	6.6740E+02	8.9861E-03	-1.0067E-05	186	584	0.496
229	C4H5NO2	METHYL CYANOACETATE	-8.0928	1.5829E+03	1.4063E-02	-1.0951E-05	260	687	2.727
230	C4H6	1,2-BUTADIENE	-3.5214	4.0958E+02	7.8148E-03	-1.0143E-05	181	444	0.191
231	C4H6	1,3-BUTADIENE	0.3772	7.9658E+01	-5.8889E-03	2.9221E-06	250	425	0.141
232	C4H6	DIMETHYLACETYLENE	---	---	---	---	--	--	--
233	C4H6	ETHYLACETYLENE	-6.9326	7.0136E+02	2.1415E-02	-2.7943E-05	193	443	0.209
234	C4H6Cl2	1,3-DICHLORO-trans-2-BUTENE	-6.0849	1.0588E+03	1.1364E-02	-1.0201E-05	276	618	0.886
235	C4H6Cl2	1,4-DICHLORO-cis-2-BUTENE	-6.4731	1.0734E+03	1.2744E-02	-1.1246E-05	225	640	0.845
236	C4H6Cl2	1,4-DICHLORO-trans-2-BUTENE	-9.9634	1.5750E+03	2.0497E-02	-1.6731E-05	274	646	0.877
237	C4H6Cl2	3,4-DICHLORO-1-BUTENE	-3.9073	7.5340E+02	6.5569E-03	-7.0911E-06	212	589	0.879
238	C4H6O	trans-CROTONALDEHYDE	-5.2285	7.3316E+02	1.1862E-02	-1.2777E-05	197	571	0.428
239	C4H6O	2,5-DIHYDROFURAN	-3.0034	4.5961E+02	2.8049E-03	-1.9102E-06	273	542	0.160
240	C4H6O	DIVINYL ETHER	-7.2676	7.4786E+02	2.1778E-02	-2.6886E-05	172	463	0.221
241	C4H6O	METHACROLEIN	-2.0243	4.1152E+02	2.2500E-03	-4.4668E-06	192	530	0.426
242	C4H6O2	2-BUTYNE-1,4-DIOL	---	---	---	---	--	--	--
243	C4H6O2	gamma-BUTYROLACTONE	---	---	---	---	(@ 25 C)		1.717
244	C4H6O2	cis-CROTONIC ACID	-13.7877	2.2066E+03	2.8706E-02	-2.2671E-05	289	647	1.434
245	C4H6O2	trans-CROTONIC ACID	-23.6229	3.8043E+03	4.8126E-02	-3.4867E-05	345	666	--
246	C4H6O2	METHACRYLIC ACID	-4.2607	8.4654E+02	7.6050E-03	-7.7782E-06	295	643	1.428
247	C4H6O2	METHYL ACRYLATE	-12.1755	1.6859E+03	2.8551E-02	-2.6324E-05	273	536	0.448
248	C4H6O2	VINYL ACETATE	-9.0671	1.1863E+03	2.2663E-02	-2.3208E-05	250	524	0.403
249	C4H6O3	ACETIC ANHYDRIDE	-17.3580	2.3611E+03	4.2734E-02	-3.8202E-05	265	569	0.806
250	C4H6O4	SUCCINIC ACID	-46.7860	9.6164E+03	7.4588E-02	-4.0794E-05	461	806	--
251	C4H6O5	DIGLYCOLIC ACID	-20.3046	4.5696E+03	2.8803E-02	-1.4915E-05	421	820	--
252	C4H6O5	MALIC ACID	-3.4665	1.2435E+03	-1.4020E-03	4.0266E-07	403	781	--
253	C4H6O6	TARTARIC ACID	---	---	---	---	--	--	--
254	C4H7N	n-BUTYRONITRILE	-3.4336	5.8884E+02	6.3219E-03	-7.7312E-06	161	582	0.548
255	C4H7N	ISOBUTYRONITRILE	-4.9492	7.8421E+02	9.9363E-03	-1.0712E-05	202	565	0.491
256	C4H7NO	ACETONE CYANOHYDRIN	---	---	---	---	--	--	--
257	C4H7NO	2-METHACRYLAMIDE	-2.3289	1.0046E+03	2.8308E-04	-8.1490E-07	384	741	--
258	C4H7NO	3-METHOXYPROPIONITRILE	-7.4121	1.1519E+03	1.5220E-02	-1.3361E-05	210	638	0.633
259	C4H7NO	2-PYRROLIDONE	-14.3974	2.9106E+03	2.3755E-02	-1.4858E-05	298	792	13.386
260	C4H8	1-BUTENE	-4.9218	4.9503E+02	1.4390E-02	-2.0853E-05	160	420	0.150
261	C4H8	cis-2-BUTENE	-6.3837	6.8524E+02	1.8356E-02	-2.3949E-05	200	439	0.181

* A computer program, containing coefficients for all compounds, is available for a nominal fee. The computer program is in ASCII which can be accessed by other software.

$$\log_{10} \eta_{liq} = A + B/T + C\,T + D\,T^2 \qquad (\eta_{liq} - \text{centipoise}, \; T - K)$$

NO	FORMULA	NAME	A	B	C	D	TMIN	TMAX	η_{liq} @ 25 C
262	C4H8	trans-2-BUTENE	-7.9461	8.1334E+02	2.4583E-02	-3.2036E-05	200	429	0.183
263	C4H8	CYCLOBUTANE	-3.3295	3.8110E+02	9.4773E-03	-1.5146E-05	182	460	0.268
264	C4H8	ISOBUTENE	-5.1190	3.4126E+02	1.9893E-02	-3.1057E-05	240	418	0.157
265	C4H8Cl2	1,4-DICHLOROBUTANE	-2.2175	5.7867E+02	2.4157E-03	-3.6109E-06	236	641	1.326
266	C4H8O	n-BUTYRALDEHYDE	-4.6882	6.8181E+02	1.0648E-02	-1.2871E-05	275	525	0.426
267	C4H8O	ISOBUTYRALDEHYDE	-4.9534	7.1084E+02	1.1385E-02	-1.3812E-05	208	507	0.396
268	C4H8O	1,2-EPOXYBUTANE	0.1130	3.6921E+01	-1.3654E-03	-2.8143E-06	180	526	0.380
269	C4H8O	METHYL ETHYL KETONE	-0.8761	2.9257E+02	-1.9833E-03	9.4354E-07	189	536	0.396
270	C4H8O	ETHYL VINYL ETHER	-5.7800	6.6206E+02	1.5513E-02	-1.9072E-05	157	475	0.235
271	C4H8O	TETRAHYDROFURAN	-2.7860	4.7681E+02	4.9173E-03	-6.8815E-06	165	540	0.465
272	C4H8O2	cis-2-BUTENE-1,4-DIOL	-13.7194	2.7695E+03	2.4403E-02	-1.7709E-05	284	678	18.676
273	C4H8O2	trans-2-BUTENE-1,4-DIOL	---	---	---	---	--	--	--
274	C4H8O2	ISOBUTYRIC ACID	-6.5789	1.1041E+03	1.3998E-02	-1.3690E-05	250	609	1.205
275	C4H8O2	n-BUTYRIC ACID	-7.9846	1.3636E+03	1.6315E-02	-1.4511E-05	268	628	1.456
276	C4H8O2	1,4-DIOXANE	-7.5724	1.3813E+03	1.3556E-02	-1.1464E-05	288	587	1.210
277	C4H8O2	ETHYL ACETATE	-3.6861	5.5228E+02	8.0018E-03	-1.0439E-05	220	523	0.421
278	C4H8O2	METHYL PROPIONATE	-9.4254	1.2618E+03	2.3072E-02	-2.3116E-05	274	531	0.428
279	C4H8O2	n-PROPYL FORMATE	-5.9614	8.7189E+02	1.3431E-02	-1.4452E-05	250	538	0.482
280	C4H8O2S	SULFOLANE	-2.6246	9.0825E+02	2.7366E-03	-2.7441E-06	301	849	--
281	C4H8S	TETRAHYDROTHIOPHENE	-20.0809	3.0608E+03	4.2486E-02	-3.2296E-05	293	632	0.958
282	C4H9Br	1-BROMOBUTANE	-8.1638	1.1342E+03	1.9376E-02	-1.8526E-05	240	577	0.590
283	C4H9Br	2-BROMOBUTANE	-6.8783	9.5838E+02	1.6202E-02	-1.6077E-05	220	567	0.546
284	C4H9Cl	n-BUTYL CHLORIDE	-5.5792	7.8795E+02	1.2714E-02	-1.3909E-05	250	537	0.415
285	C4H9Cl	sec-BUTYL CHLORIDE	-9.9788	1.3237E+03	2.4470E-02	-2.4390E-05	250	521	0.388
286	C4H9Cl	tert-BUTYL CHLORIDE	-10.2240	1.5480E+03	2.1671E-02	-1.9828E-05	280	507	0.464
287	C4H9N	PYRROLIDINE	-7.7495	1.2386E+03	1.6051E-02	-1.4879E-05	215	569	0.737
288	C4H9NO	N,N-DIMETHYLACETAMIDE	-4.6530	8.3643E+02	8.3151E-03	-7.8272E-06	253	658	0.863
289	C4H9NO	MORPHOLINE	-5.0308	1.1150E+03	7.3378E-03	-6.5955E-06	280	618	2.044
290	C4H10	n-BUTANE	-6.8590	6.7393E+02	2.1973E-02	-3.0686E-05	180	425	0.168
291	C4H10	ISOBUTANE	-13.4207	1.3131E+03	4.4329E-02	-5.5793E-05	190	408	0.174
292	C4H10N2	PIPERAZINE	-19.4532	2.7745E+03	4.3501E-02	-3.4220E-05	379	638	--
293	C4H10O	n-BUTANOL	-5.3970	1.3256E+03	6.2223E-03	-5.5062E-06	250	563	2.600
294	C4H10O	sec-BUTANOL	-20.6736	3.5493E+03	4.0352E-02	-3.0937E-05	288	536	3.249
295	C4H10O	tert-BUTANOL	-35.2655	5.4737E+03	7.7742E-02	-6.3499E-05	298	506	4.243
296	C4H10O	DIETHYL ETHER	-8.5060	1.0020E+03	2.2753E-02	-2.5780E-05	233	467	0.222
297	C4H10O	METHYL ISOPROPYL ETHER	-5.1341	5.5878E+02	1.4090E-02	-1.8212E-05	128	465	0.210
298	C4H10O	ISOBUTANOL	-11.9687	2.1770E+03	2.3767E-02	-2.1427E-05	211	548	3.268
299	C4H10O2	1,3-BUTANEDIOL	-24.0994	4.6841E+03	4.4009E-02	-3.0862E-05	293	643	97.473
300	C4H10O2	1,4-BUTANEDIOL	-14.3559	3.1423E+03	2.4253E-02	-1.7544E-05	293	667	71.594
301	C4H10O2	2,3-BUTANEDIOL	-6.0949	2.1560E+03	2.8597E-03	-1.0083E-06	281	611	79.334
302	C4H10O2	t-BUTYL HYDROPEROXIDE	---	---	---	---	--	--	--
303	C4H10O2	1,2-DIMETHOXYETHANE	-5.7926	8.1316E+02	1.3579E-02	-1.5006E-05	215	536	0.446
304	C4H10O2	2-ETHOXYETHANOL	-10.1080	1.8207E+03	1.9143E-02	-1.6143E-05	210	569	1.868
305	C4H10O3	DIETHYLENE GLYCOL	-14.7942	3.1502E+03	2.3543E-02	-1.4786E-05	270	745	29.973
306	C4H10O4S	DIETHYL SULFATE	---	---	---	---	--	--	--
307	C4H10S	n-BUTYL MERCAPTAN	-6.3227	8.9344E+02	1.4252E-02	-1.4396E-05	225	569	0.440
308	C4H10S	ISOBUTYL MERCAPTAN	-4.3170	6.6031E+02	8.9698E-03	-1.0136E-05	225	559	0.469
309	C4H10S	sec-BUTYL MERCAPTAN	-4.1544	6.4242E+02	8.5141E-03	-9.8036E-06	225	554	0.465
310	C4H10S	tert-BUTYL MERCAPTAN	-1.7534	5.2688E+02	-6.2337E-04	-5.9282E-07	274	530	0.596
311	C4H10S	DIETHYL SULFIDE	-6.1635	8.4801E+02	1.4306E-02	-1.4925E-05	225	557	0.416
312	C4H10S2	DIETHYL DISULFIDE	-8.6196	1.3497E+03	1.7682E-02	-1.4750E-05	225	642	0.738
313	C4H11N	n-BUTYLAMINE	-7.6575	1.2106E+03	1.5854E-02	-1.5448E-05	278	532	0.571
314	C4H11N	ISOBUTYLAMINE	-6.3000	1.0270E+03	1.2630E-02	-1.3256E-05	250	514	0.539
315	C4H11N	sec-BUTYLAMINE	-4.7287	8.5139E+02	7.7286E-03	-8.3765E-06	225	514	0.486
316	C4H11N	tert-BUTYLAMINE	-8.3011	1.3490E+03	1.5924E-02	-1.5016E-05	260	484	0.433
317	C4H11N	DIETHYLAMINE	-9.2189	1.2133E+03	2.2638E-02	-2.3633E-05	240	497	0.316
318	C4H11NO	DIMETHYLETHANOLAMINE	---	---	---	---	(@ 20 C)		3.000
319	C4H11NO2	DIETHANOLAMINE	-27.9385	5.9547E+03	4.4120E-02	-2.5871E-05	301	715	--
320	C4H11NO2	2-AMINOETHOXYETHANOL	-30.0369	5.6570E+03	5.2553E-02	-3.2716E-05	293	699	49.793
321	C4H12N2O	N-AMINOETHYL ETHANOLAMINE	-32.0709	6.2908E+03	5.4172E-02	-3.2693E-05	298	698	187.811
322	C4H12Si	TETRAMETHYLSILANE	---	---	---	---	--	--	--
323	C4H13N3	DIETHYLENE TRIAMINE	-6.9597	1.4821E+03	1.1547E-02	-9.3746E-06	234	676	4.177

* A computer program, containing coefficients for all compounds, is available for a nominal fee. The computer program is in ASCII which can be accessed by other software.

330

Appendix B

COEFFICIENTS FOR GAS VISCOSITY EQUATION[*]

Carl L. Yaws, Xiaoyan Lin and Li Bu
Lamar University, Beaumont, Texas

$$\eta_{gas} = A + B\,T + C\,T^2 \qquad (\eta_{gas} - \text{micropoise}, \; T - K)$$

NO	FORMULA	NAME	A	B	C	TMIN	TMAX	η_{gas} @ 25 C
1	CBrClF2	BROMOCHLORODIFLUOROMETHANE	-4.3690	4.8220E-01	-1.0333E-04	233	1000	130.2
2	CBrCl3	BROMOTRICHLOROMETHANE	-10.2744	3.9477E-01	-6.3126E-05	252	1000	101.8
3	CBrF3	BROMOTRIFLUOROMETHANE	-4.9136	6.0829E-01	-2.2428E-04	230	500	156.5
4	CBr2F2	DIBROMODIFLUOROMETHANE	-9.4009	5.0711E-01	-1.0180E-04	296	1000	132.7
5	CClF3	CHLOROTRIFLUOROMETHANE	21.2289	3.9296E-01	6.5918E-05	230	500	144.2
6	CClN	CYANOGEN CHLORIDE	-5.2355	3.1918E-01	-6.6139E-05	286	1000	84.0
7	CCl2F2	DICHLORODIFLUOROMETHANE	13.9862	3.7629E-01	-1.3267E-05	250	575	125.0
8	CCl2O	PHOSGENE	-37.2351	5.2032E-01	-1.9868E-05	200	900	116.1
9	CCl3F	TRICHLOROFLUOROMETHANE	-3.3913	4.0268E-01	-2.9161E-05	248	997	114.1
10	CCl4	CARBON TETRACHLORIDE	-7.7453	3.9481E-01	-1.1150E-04	280	800	100.1
11	CF2O	CARBONYL FLUORIDE	0.0033	4.2190E-01	-1.1407E-04	189	1000	115.7
12	CF4	CARBON TETRAFLUORIDE	10.8962	5.9490E-01	-1.7559E-04	145	900	172.7
13	CHBr3	TRIBROMOMETHANE	-17.0263	3.4203E-01	-3.9741E-05	273	573	81.4
14	CHClF2	CHLORODIFLUOROMETHANE	-9.5169	5.0379E-01	-1.3232E-04	232	500	128.9
15	CHCl2F	DICHLOROFLUOROMETHANE	-3.0160	4.2016E-01	-7.7903E-05	280	500	115.3
16	CHCl3	CHLOROFORM	-4.3915	3.7309E-01	-5.1696E-05	250	700	102.2
17	CHF3	TRIFLUOROMETHANE	-10.4947	5.6948E-01	-1.7587E-04	191	700	143.7
18	CHN	HYDROGEN CYANIDE	-8.4856	9.0368E-02	7.9146E-05	300	425	--
19	CH2BrCl	BROMOCHLOROMETHANE	-10.7977	4.5078E-01	-7.8817E-05	341	1000	--
20	CH2Br2	DIBROMOMETHANE	-9.5466	4.6829E-01	-6.9599E-05	370	1000	--
21	CH2Cl2	DICHLOROMETHANE	-20.3721	4.3745E-01	-7.7549E-05	273	993	103.2
22	CH2F2	DIFLUOROMETHANE	-3.8067	5.2411E-01	-1.2877E-04	222	1000	141.0
23	CH2I2	DIIODOMETHANE	-28.1747	5.0959E-01	-6.8132E-05	288	573	117.7
24	CH2O	FORMALDEHYDE	-6.4389	4.4802E-01	-1.0130E-04	254	1000	118.1
25	CH2O2	FORMIC ACID	-13.1393	2.7486E-01	1.9189E-04	363	386	--
26	CH3Br	METHYL BROMIDE	-27.7403	5.5901E-01	-7.0942E-05	260	440	132.6
27	CH3Cl	METHYL CHLORIDE	-1.3735	3.8627E-01	-4.8650E-05	230	700	109.5
28	CH3Cl3Si	METHYL TRICHLOROSILANE	-9.3434	3.6088E-01	-5.1146E-05	195	1000	93.7
29	CH3F	METHYL FLUORIDE	-0.9005	4.7265E-01	-1.2197E-04	195	1000	129.2
30	CH3I	METHYL IODIDE	-21.6979	5.4959E-01	-9.1954E-05	300	1000	--
31	CH3NO	FORMAMIDE	---	---	---	--	--	--
32	CH3NO2	NITROMETHANE	-20.2060	3.2616E-01	-4.3498E-05	374	1000	--
33	CH4	METHANE	3.8435	4.0112E-01	-1.4303E-04	91	850	110.7
34	CH4Cl2Si	METHYL DICHLOROSILANE	-8.8851	3.6879E-01	-5.9741E-05	183	1000	95.8
35	CH4O	METHANOL	-14.2360	3.8935E-01	-6.2762E-05	240	1000	96.3
36	CH4O3S	METHANESULFONIC ACID	---	---	---	--	--	--
37	CH4S	METHYL MERCAPTAN	-39.3801	4.6695E-01	-6.2465E-05	273	473	94.3
38	CH5ClSi	METHYL CHLOROSILANE	-6.6913	3.5509E-01	-6.0616E-05	139	1000	93.8
39	CH5N	METHYLAMINE	-5.3340	3.4181E-01	-7.4297E-05	267	1000	90.0
40	CH6Si	METHYL SILANE	-5.4289	3.4287E-01	-7.7044E-05	116	1000	89.9
41	CN4O8	TETRANITROMETHANE	-15.7515	3.4934E-01	-5.9980E-05	287	1000	83.1
42	CO	CARBON MONOXIDE	23.8114	5.3944E-01	-1.5411E-04	68	1250	170.9
43	COS	CARBONYL SULFIDE	-11.4897	4.9840E-01	-9.9991E-05	134	1000	128.2
44	CO2	CARBON DIOXIDE	11.8109	4.9838E-01	-1.0851E-04	195	1500	150.8
45	CS2	CARBON DISULFIDE	-7.7001	3.6594E-01	-2.5416E-05	273	583	99.1
46	C2BrF3	BROMOTRIFLUOROETHYLENE	-7.5408	4.8895E-01	-1.0519E-04	271	1000	128.9
47	C2Br2F4	1,2-DIBROMOTETRAFLUOROETHANE	-7.1322	4.5248E-01	-8.5233E-05	320	1000	--
48	C2ClF3	CHLOROTRIFLUOROETHYLENE	-3.9477	4.5063E-01	-1.0575E-04	245	1000	121.0
49	C2ClF5	CHLOROPENTAFLUOROETHANE	3.7363	4.4326E-01	-1.0051E-04	250	500	127.0
50	C2Cl2F4	1,2-DICHLOROTETRAFLUOROETHANE	26.6416	2.7226E-01	8.7050E-05	230	500	115.6
51	C2Cl3F3	1,1,2-TRICHLOROTRIFLUOROETHANE	32.9953	2.5082E-01	-5.4159E-05	237	1000	103.0
52	C2Cl4	TETRACHLOROETHYLENE	-10.8414	3.1582E-01	2.2833E-05	251	570	85.4
53	C2Cl4F2	1,1,2,2-TETRACHLORODIFLUOROETHANE	-8.8607	3.6067E-01	-6.3826E-05	366	1000	--
54	C2Cl4O	TRICHLOROACETYL CHLORIDE	-19.5535	4.0312E-01	-6.2634E-05	391	991	--
55	C2Cl6	HEXACHLOROETHANE	-19.9606	3.6258E-01	-4.5006E-05	460	1000	--
56	C2F4	TETRAFLUOROETHYLENE	-0.6618	5.1057E-01	-1.3502E-04	198	1000	139.6
57	C2F6	HEXAFLUOROETHANE	0.9720	5.3473E-01	-1.4324E-04	195	1000	147.7

[*] A computer program, containing coefficients for all compounds, is available for a nominal fee. The computer program is in ASCII which can be accessed by other software.

$$\eta_{gas} = A + B\,T + C\,T^2 \qquad (\eta_{gas} - \text{micropoise}, \ T - K)$$

NO	FORMULA	NAME	A	B	C	TMIN	TMAX	η_{gas} @ 25 C
58	C2HBrClF3	HALOTHANE	-7.3577	4.2906E-01	-7.5755E-05	223	1000	113.8
59	C2HClF2	2-CHLORO-1,1-DIFLUOROETHYLENE	-4.7752	4.2835E-01	-9.6753E-05	255	1000	114.3
60	C2HCl3	TRICHLOROETHYLENE	-23.6213	4.2164E-01	-6.8948E-05	333	990	--
61	C2HCl3O	DICHLOROACETYL CHLORIDE	-19.4991	4.0357E-01	-6.4804E-05	382	992	--
62	C2HCl3O	TRICHLOROACETALDEHYDE	-9.1630	3.7547E-01	-4.3356E-05	216	1000	98.9
63	C2HCl5	PENTACHLOROETHANE	-4.8758	3.0910E-01	-3.8437E-05	244	1000	83.9
64	C2HF3O2	TRIFLUOROACETIC ACID	-6.8229	4.0112E-01	-7.5762E-05	258	1000	106.0
65	C2HF5	PENTAFLUOROETHANE	-4.5043	5.2704E-01	-1.3533E-04	170	1000	140.6
66	C2H2	ACETYLENE	-11.5572	4.2363E-01	-1.4174E-04	193	500	102.1
67	C2H2Br4	1,1,2,2-TETRABROMOETHANE	-8.3089	3.6748E-01	-3.5912E-05	517	1000	--
68	C2H2Cl2	1,1-DICHLOROETHYLENE	-8.0777	3.8859E-01	-7.9107E-05	305	1000	--
69	C2H2Cl2	cis-1,2-DICHLOROETHYLENE	-7.5739	3.6605E-01	-6.7144E-05	334	1000	--
70	C2H2Cl2	trans-1,2-DICHLOROETHYLENE	-7.1249	3.6505E-01	-6.8692E-05	321	1000	--
71	C2H2Cl2O	CHLOROACETYL CHLORIDE	-18.2047	3.8542E-01	-5.7860E-05	350	1000	
72	C2H2Cl2O	DICHLOROACETALDEHYDE	-5.3627	3.2751E-01	-5.2553E-05	223	1000	87.6
73	C2H2Cl2O2	DICHLOROACETIC ACID	-5.5075	3.0753E-01	-3.7358E-05	287	1000	82.9
74	C2H2Cl3F	1,1,1-TRICHLOROFLUOROETHANE	-4.8526	3.4986E-01	-5.4139E-05	173	1000	94.6
75	C2H2Cl4	1,1,1,2-TETRACHLOROETHANE	-9.9634	3.3385E-01	-5.3254E-05	404	1000	--
76	C2H2Cl4	1,1,2,2-TETRACHLOROETHANE	-20.6051	3.7770E-01	-5.4670E-05	418	998	--
77	C2H2F2	1,1-DIFLUOROETHYLENE	2.5368	4.6454E-01	-1.1805E-04	188	1000	130.5
78	C2H2F4	1,1,1,2-TETRAFLUOROETHANE	-5.0829	5.0162E-01	-1.1800E-04	172	1000	134.0
79	C2H2O	KETENE	-10.9239	4.1236E-01	-9.3098E-05	200	1000	103.7
80	C2H2O4	OXALIC ACID	-6.8421	2.8812E-01	-3.1538E-05	463	1000	--
81	C2H3Br	VINYL BROMIDE	-9.3086	4.6189E-01	-9.4500E-05	289	1000	120.0
82	C2H3Cl	VINYL CHLORIDE	-6.0669	3.9013E-01	-8.3970E-05	260	1000	102.8
83	C2H3ClF2	1-CHLORO-1,1-DIFLUOROETHANE	-6.4501	4.3456E-01	-9.8180E-05	263	1000	114.4
84	C2H3ClO	ACETYL CHLORIDE	1.0116	3.1562E-01	-6.2753E-06	277	761	94.6
85	C2H3ClO	CHLOROACETALDEHYDE	-6.3829	3.3296E-01	-5.6014E-05	293	1000	87.9
86	C2H3ClO2	CHLOROACETIC ACID	-6.4890	3.1233E-01	-3.8880E-05	333	1000	--
87	C2H3ClO2	METHYL CHLOROFORMATE	-4.7415	3.3821E-01	-5.7423E-05	192	1000	91.0
88	C2H3Cl3	1,1,1-TRICHLOROETHANE	-19.2155	4.0308E-01	-6.8618E-05	347	997	--
89	C2H3Cl3	1,1,2-TRICHLOROETHANE	-8.2930	3.3989E-01	-5.3678E-05	387	987	--
90	C2H3F	VINYL FLUORIDE	-1.9542	4.2566E-01	-1.0977E-04	201	1000	115.2
91	C2H3F3	1,1,1-TRIFLUOROETHANE	-3.5390	4.9228E-01	-1.2372E-04	226	1000	132.2
92	C2H3N	ACETONITRILE	-1.3840	2.5204E-01	-2.4595E-05	308	750	--
93	C2H3NO	METHYL ISOCYANATE	-15.4337	3.2530E-01	-5.4292E-05	312	1000	--
94	C2H4	ETHYLENE	-3.9851	3.8726E-01	-1.1227E-04	150	1000	101.5
95	C2H4Br2	1,1-DIBROMOETHANE	-6.1365	4.1000E-01	-5.4252E-05	210	1000	111.3
96	C2H4Br2	1,2-DIBROMOETHANE	-21.7955	4.5011E-01	-5.9439E-05	405	995	--
97	C2H4Cl2	1,1-DICHLOROETHANE	-12.9909	4.0085E-01	-1.1779E-04	320	472	--
98	C2H4Cl2	1,2-DICHLOROETHANE	1.0252	3.1792E-01	-4.1853E-05	322	561	--
99	C2H4Cl2O	BIS(CHLOROMETHYL)ETHER	-5.2068	3.3051E-01	-5.0373E-05	232	1000	88.9
100	C2H4F2	1,1-DIFLUOROETHANE	-2.9486	4.3374E-01	-9.2979E-05	247	1000	118.1
101	C2H4F2	1,2-DIFLUOROETHANE	-5.9508	4.2281E-01	-8.0663E-05	215	1000	112.9
102	C2H4O	ACETALDEHYDE	0.0689	3.0246E-01	-4.2372E-05	294	1000	86.5
103	C2H4O	ETHYLENE OXIDE	-12.1799	3.7672E-01	-7.7599E-05	250	1000	93.2
104	C2H4O2	ACETIC ACID	-28.6600	2.3510E-01	2.2087E-04	366	523	--
105	C2H4O2	METHYL FORMATE	-16.0306	3.9071E-01	-7.1883E-05	300	1000	--
106	C2H5Br	BROMOETHANE	-10.1477	4.3036E-01	-8.1147E-05	312	1000	--
107	C2H5Cl	ETHYL CHLORIDE	0.4584	3.2827E-01	-1.2467E-05	213	523	97.2
108	C2H5ClO	2-CHLOROETHANOL	-2.5292	1.0612E-01	-1.7298E-05	402	1000	--
109	C2H5F	ETHYL FLUORIDE	-4.5694	3.9509E-01	-9.5177E-05	235	1000	104.8
110	C2H5I	ETHYL IODIDE	-17.0712	4.6084E-01	-6.7487E-05	300	1000	--
111	C2H5N	ETHYLENEIMINE	-12.2201	3.0761E-01	-4.7922E-05	329	1000	--
112	C2H5NO	ACETAMIDE	-17.0898	2.5005E-01	-2.3629E-05	494	994	--
113	C2H5NO	N-METHYLFORMAMIDE	-8.8530	3.2988E-01	-4.1861E-05	473	983	--
114	C2H5NO2	NITROETHANE	-16.7014	2.9762E-01	-3.7773E-05	387	1000	--
115	C2H6	ETHANE	0.5142	3.3449E-01	-7.1071E-05	150	1000	93.9
116	C2H6AlCl	DIMETHYLALUMINUM CHLORIDE	-7.2297	2.6575E-01	-2.7116E-05	252	1000	69.6
117	C2H6O	DIMETHYL ETHER	-4.2760	3.0262E-01	6.3528E-05	216	373	91.6
118	C2H6O	ETHANOL	1.4991	3.0741E-01	-4.4479E-05	200	1000	89.2
119	C2H6OS	DIMETHYL SULFOXIDE	-14.9024	2.7069E-01	-2.0800E-05	462	992	--
120	C2H6O2	ETHYLENE GLYCOL	-7.1778	3.1246E-01	-4.4028E-05	260	1000	82.1
121	C2H6O4S	DIMETHYL SULFATE	---	---	---	--	--	--
122	C2H6S	DIMETHYL SULFIDE	-14.0760	3.5306E-01	-6.2780E-05	310	990	--
123	C2H6S	ETHYL MERCAPTAN	-15.4317	3.5300E-01	-6.6454E-05	308	998	--
124	C2H6S2	DIMETHYL DISULFIDE	-15.4317	3.5300E-01	-6.6454E-05	308	998	--
125	C2H7N	DIMETHYLAMINE	-9.2752	2.8958E-01	-1.3875E-06	250	450	76.9

* A computer program, containing coefficients for all compounds, is available for a nominal fee. The computer program is in ASCII which can be accessed by other software.

$$\eta_{gas} = A + B\,T + C\,T^2 \quad (\eta_{gas} - \text{micropoise}, \; T - K)$$

NO	FORMULA	NAME	A	B	C	TMIN	TMAX	η_{gas} @ 25 C
126	C2H7N	ETHYLAMINE	-5.5381	3.0778E-01	-6.4363E-05	290	1000	80.5
127	C2H7NO	MONOETHANOLAMINE	-12.5920	2.8971E-01	-3.9470E-05	400	1000	--
128	C2H8N2	ETHYLENEDIAMINE	-4.8426	2.7125E-01	-4.0319E-05	284	1000	72.4
129	C2H8Si	DIMETHYL SILANE	-5.5550	3.1359E-01	-6.1179E-05	123	1000	82.5
130	C2N2	CYANOGEN	-0.2930	3.4461E-01	-1.1345E-05	252	600	101.4
131	C3F6	HEXAFLUOROPROPYLENE	-4.9707	4.8190E-01	-1.1694E-04	244	1000	128.3
132	C3F6O	HEXAFLUOROACETONE	-3.3193	4.0569E-01	-9.9426E-05	246	1000	108.8
133	C3F8	OCTAFLUOROPROPANE	-1.6563	4.8878E-01	-1.2050E-04	236	1000	133.4
134	C3H2N2	MALONONITRILE	-4.0961	2.1382E-01	-2.5089E-05	305	1000	--
135	C3H3Cl	PROPARGYL CHLORIDE	-7.0922	3.4137E-01	-5.9573E-05	293	1000	89.4
136	C3H3N	ACRYLONITRILE	-4.7825	2.4047E-01	-1.4526E-05	303	1000	--
137	C3H3NO	OXAZOLE	-6.5226	2.9548E-01	-3.4972E-05	189	1000	78.5
138	C3H4	METHYLACETYLENE	-9.6262	3.5605E-01	-1.1504E-04	173	573	86.3
139	C3H4	PROPADIENE	-12.4311	3.4616E-01	-8.6996E-05	175	600	83.0
140	C3H4Cl2	2,3-DICHLOROPROPENE	-4.5994	3.2783E-01	-4.8207E-05	192	1000	88.9
141	C3H4O	ACROLEIN	-16.9104	3.2167E-01	-5.2581E-05	326	996	--
142	C3H4O	PROPARGYL ALCOHOL	-9.3808	3.2585E-01	-5.6888E-05	387	1000	--
143	C3H4O2	ACRYLIC ACID	-6.5320	3.0600E-01	-4.6620E-05	287	1000	80.6
144	C3H4O2	beta-PROPIOLACTONE	-5.1650	3.4624E-01	-3.9574E-05	240	1000	94.5
145	C3H4O2	VINYL FORMATE	-9.5852	3.4817E-01	-5.4813E-05	200	1000	89.3
146	C3H4O3	ETHYLENE CARBONATE	-21.0193	2.8757E-01	-1.5204E-05	511	1000	--
147	C3H4O3	PYRUVIC ACID	-4.4961	2.6567E-01	-3.5453E-05	287	1000	71.6
148	C3H5Cl	2-CHLOROPROPENE	-7.5121	3.5502E-01	-7.1618E-05	296	1000	92.0
149	C3H5Cl	3-CHLOROPROPENE	-7.0348	3.3654E-01	-6.2098E-05	318	1000	--
150	C3H5ClO	alpha-EPICHLOROHYDRIN	-17.2208	3.5374E-01	-5.1027E-05	389	999	--
151	C3H5ClO2	METHYL CHLOROACETATE	-5.3181	3.0230E-01	-4.3113E-05	241	1000	81.0
152	C3H5ClO2	ETHYL CHLOROFORMATE	-4.1456	3.0221E-01	-5.3347E-05	192	1000	81.2
153	C3H5Cl3	1,2,3-TRICHLOROPROPANE	-5.6119	2.9957E-01	-3.9788E-05	430	1000	--
154	C3H5N	PROPIONITRILE	-15.4093	2.5623E-01	-3.0933E-05	371	991	--
155	C3H5NO	ACRYLAMIDE	-6.3594	2.7465E-01	-3.4477E-05	358	1000	--
156	C3H5NO	HYDRACRYLONITRILE	--	---	---	--	--	--
157	C3H5NO	LACTONITRILE	-3.6185	2.4863E-01	-3.1666E-05	233	1000	67.7
158	C3H5N3O9	NITROGLYCERINE	-7.5435	3.1769E-01	-2.0439E-05	286	1000	85.4
159	C3H6	CYCLOPROPANE	-9.5210	3.7037E-01	-1.3138E-04	240	446	89.2
160	C3H6	PROPYLENE	-7.2303	3.4180E-01	-9.4516E-05	193	1000	86.3
161	C3H6Cl2	1,1-DICHLOROPROPANE	-4.9657	3.1422E-01	-4.9988E-05	200	1000	84.3
162	C3H6Cl2	1,2-DICHLOROPROPANE	-16.8017	3.5263E-01	-5.6713E-05	370	1000	--
163	C3H6Cl2	1,3-DICHLOROPROPANE	-3.9470	3.0000E-01	-4.1205E-05	174	1000	81.8
164	C3H6O	ACETONE	-4.0554	2.6655E-01	-5.6936E-06	300	650	--
165	C3H6O	ALLYL ALCOHOL	-9.2708	3.2529E-01	-5.9894E-05	370	1000	--
166	C3H6O	METHYL VINYL ETHER	-6.6324	3.4394E-01	-7.6282E-05	279	1000	89.1
167	C3H6O	n-PROPIONALDEHYDE	-14.8853	3.2999E-01	-5.9490E-05	321	991	--
168	C3H6O	1,2-PROPYLENE OXIDE	-15.2855	3.6549E-01	-6.7583E-05	300	1000	--
169	C3H6O	1,3-PROPYLENE OXIDE	-16.3753	3.5336E-01	-6.1967E-05	321	991	--
170	C3H6O2	ETHYL FORMATE	-15.2946	3.5359E-01	-6.3604E-05	300	1000	--
171	C3H6O2	METHYL ACETATE	-14.7802	3.5569E-01	-6.4353E-05	200	1000	79.6
172	C3H6O2	PROPIONIC ACID	-9.4464	2.9112E-01	-4.2258E-05	252	1000	73.6
173	C3H6O2S	3-MERCAPTOPROPIONIC ACID	-5.2411	2.6958E-01	-1.2110E-05	291	1000	74.1
174	C3H6O3	LACTIC ACID	-5.5144	2.8171E-01	-4.1031E-05	291	1000	74.8
175	C3H6O3	METHOXYACETIC ACID	-5.1948	2.8270E-01	-3.4304E-05	281	1000	76.0
176	C3H6O3	TRIOXANE	-17.9878	3.7799E-01	-5.4524E-05	388	998	--
177	C3H7Br	1-BROMOPROPANE	-7.9577	3.6805E-01	-6.3823E-05	344	1000	--
178	C3H7Br	2-BROMOPROPANE	-11.7925	5.1644E-01	-9.3057E-05	333	1000	--
179	C3H7Cl	ISOPROPYL CHLORIDE	-5.8563	3.3081E-01	-6.3622E-05	309	1000	--
180	C3H7Cl	n-PROPYL CHLORIDE	-6.2676	3.1908E-01	-5.9816E-05	320	1000	--
181	C3H7I	ISOPROPYL IODIDE	-16.9143	4.1524E-01	-5.9631E-05	363	1000	--
182	C3H7I	n-PROPYL IODIDE	-18.3019	4.1521E-01	-5.7130E-05	376	1000	--
183	C3H7N	ALLYLAMINE	-5.4216	2.9789E-01	-5.5989E-05	326	1000	--
184	C3H7N	PROPYLENEIMINE	-6.6844	4.1385E-01	-7.1281E-05	229	1000	110.4
185	C3H7NO	N,N-DIMETHYLFORMAMIDE	-17.8283	2.7374E-01	-3.5679E-05	426	996	--
186	C3H7NO	N-METHYLACETAMIDE	-4.2015	2.7991E-01	-2.9560E-05	301	1000	--
187	C3H7NO2	1-NITROPROPANE	-13.3552	2.7955E-01	-3.4102E-05	404	1000	--
188	C3H7NO2	2-NITROPROPANE	-17.5225	2.9622E-01	-4.2088E-05	393	1000	--
189	C3H8	PROPANE	-5.4615	3.2722E-01	-1.0672E-04	193	750	82.6
190	C3H8O	ISOPROPANOL	-10.8592	3.0873E-01	-4.8098E-05	200	1000	76.9
191	C3H8O	METHYL ETHYL ETHER	-12.2183	3.5144E-01	-7.2869E-05	281	991	86.1
192	C3H8O	n-PROPANOL	-14.8939	3.2171E-01	-5.8021E-05	200	1000	75.9
193	C3H8O2	2-METHOXYETHANOL	-4.6203	2.9356E-01	-4.5367E-05	398	1000	--

* A computer program, containing coefficients for all compounds, is available for a nominal fee. The computer program is in ASCII which can be accessed by other software.

$$\eta_{gas} = A + B T + C T^2 \quad (\eta_{gas} - \text{micropoise}, T - K)$$

NO	FORMULA	NAME	A	B	C	TMIN	TMAX	η_{gas} @ 25 C
194	C3H8O2	METHYLAL	-6.3193	3.1872E-01	-6.3471E-05	315	975	--
195	C3H8O2	1,2-PROPYLENE GLYCOL	-15.6199	2.8898E-01	-3.2128E-05	461	991	--
196	C3H8O2	1,3-PROPYLENE GLYCOL	-18.0143	3.2093E-01	-4.1746E-05	450	1000	--
197	C3H8O3	GLYCEROL	-23.1186	2.8879E-01	-3.4277E-05	563	993	--
198	C3H8S	n-PROPYLMERCAPTAN	-5.7053	2.9923E-01	-3.5958E-05	160	1000	80.3
199	C3H8S	ISOPROPYL MERCAPTAN	-5.0449	3.0176E-01	-3.5890E-05	143	1000	81.7
200	C3H9N	n-PROPYLAMINE	-12.3390	2.9593E-01	-5.5438E-05	300	1000	--
201	C3H9N	ISOPROPYLAMINE	-5.4714	2.9085E-01	-5.8656E-05	306	1000	--
202	C3H9N	TRIMETHYLAMINE	-11.5453	3.2107E-01	-6.8081E-05	250	1000	78.1
203	C3H9NO	1-AMINO-2-PROPANOL	-9.3751	2.8556E-01	-4.6813E-05	275	1000	71.6
204	C3H9NO	3-AMINO-1-PROPANOL	-7.7696	2.7106E-01	-3.9489E-05	284	1000	69.5
205	C3H9NO	METHYLETHANOLAMINE	-5.2317	2.9002E-01	-4.0148E-05	269	1000	77.7
206	C3H9O4P	TRIMETHYL PHOSPHATE	---	---	---	--	--	--
207	C3H10N2	1,2-PROPANEDIAMINE	-4.5129	2.5972E-01	-3.9437E-05	237	1000	69.4
208	C3H10Si	TRIMETHYL SILANE	-5.6910	2.9660E-01	-5.3047E-05	137	1000	78.0
209	C4Cl4S	TETRACHLOROTHIOPHENE	-5.7734	2.9116E-01	-1.2711E-05	302	1000	--
210	C4Cl6	HEXACHLORO-1,3-BUTADIENE	-19.8648	3.3082E-01	-4.0579E-05	488	998	--
211	C4F8	OCTAFLUORO-2-BUTENE	-4.2686	4.4109E-01	-1.0018E-04	270	1000	118.3
212	C4F8	OCTAFLUOROCYCLOBUTANE	-1.7974	4.2747E-01	-9.2773E-05	267	1000	117.4
213	C4F10	DECAFLUOROBUTANE	-5.2095	4.5542E-01	-1.0643E-04	271	1000	121.1
214	C4H2O3	MALEIC ANHYDRIDE	-11.2193	2.9181E-01	-1.0579E-05	326	1000	--
215	C4H4	VINYLACETYLENE	---	---	---	--	--	--
216	C4H4N2	SUCCINONITRILE	-11.2392	2.2724E-01	-1.6566E-05	540	990	--
217	C4H4O	FURAN	-13.6959	3.5655E-01	-6.6378E-05	305	995	--
218	C4H4O2	DIKETENE	-10.6837	3.3267E-01	-5.5054E-05	267	1000	83.6
219	C4H8O3	SUCCINIC ANHYDRIDE	-10.1249	2.7767E-01	-7.1078E-06	393	1000	--
220	C4H4O4	FUMARIC ACID	-11.3326	2.6574E-01	-3.6422E-05	560	1000	--
221	C4H4O4	MALEIC ACID	-5.8080	2.5158E-01	-2.7666E-05	403	1000	--
222	C4H4S	THIOPHENE	-23.8148	3.6576E-01	-4.9330E-05	293	997	80.8
223	C4H5Cl	CHLOROPRENE	-6.1062	3.2925E-01	-5.7652E-05	333	1000	--
224	C4H5N	trans-CROTONITRILE	-4.4251	2.4447E-01	-3.8374E-05	222	1000	65.1
225	C4H5N	cis-CROTONITRILE	-3.9973	2.4281E-01	-3.8310E-05	201	1000	65.0
226	C4H5N	METHACRYLONITRILE	-14.8704	2.7470E-01	-3.9631E-05	237	1000	63.5
227	C4H5N	PYRROLE	-14.3527	2.9421E-01	-3.9026E-05	403	993	--
228	C4H5N	VINYLACETONITRILE	-3.4064	2.4325E-01	-3.5875E-05	186	1000	65.9
229	C4H5NO2	METHYL CYANOACETATE	-3.9515	2.3617E-01	-2.7783E-05	478	1000	--
230	C4H6	1,2-BUTADIENE	0.8785	2.7938E-01	-6.6959E-05	284	994	78.2
231	C4H6	1,3-BUTADIENE	10.2555	2.6833E-01	-4.1148E-05	250	650	86.6
232	C4H6	DIMETHYLACETYLENE	-6.2929	2.9596E-01	-7.0767E-05	300	990	--
233	C4H6	ETHYLACETYLENE	-11.0514	3.0621E-01	-6.9582E-05	213	573	74.1
234	C4H6Cl2	1,3-DICHLORO-trans-2-BUTENE	-5.2168	2.9435E-01	-4.1350E-05	276	1000	78.9
235	C4H6Cl2	1,4-DICHLORO-cis-2-BUTENE	-3.8749	2.8158E-01	-3.5458E-05	225	1000	76.9
236	C4H6Cl2	1,4-DICHLORO-trans-2-BUTENE	-9.4198	3.3823E-01	-5.0131E-05	429	1000	--
237	C4H6Cl2	3,4-DICHLORO-1-BUTENE	-4.4558	2.9664E-01	-4.4188E-05	212	1000	80.1
238	C4H6O	trans-CROTONALDEHYDE	-6.5632	2.8401E-01	-4.7535E-05	377	1000	--
239	C4H6O	2,5-DIHYDROFURAN	-5.3408	3.2501E-01	-5.3305E-05	273	1000	86.8
240	C4H6O	DIVINYL ETHER	-5.5847	3.2050E-01	-6.5196E-05	301	1000	--
241	C4H6O	METHACROLEIN	-14.1028	3.1016E-01	-5.0202E-05	341	991	--
242	C4H6O2	2-BUTYNE-1,4-DIOL	-9.9309	2.7828E-01	-4.0329E-05	511	1000	--
243	C4H6O2	gamma-BUTYROLACTONE	-15.6303	2.6618E-01	-2.4638E-05	477	1000	--
244	C4H6O2	cis-CROTONIC ACID	-4.2445	2.6784E-01	-3.4297E-05	289	1000	72.6
245	C4H6O2	trans-CROTONIC ACID	-5.4454	2.7036E-01	-3.4454E-05	345	1000	--
246	C4H6O2	METHACRYLIC ACID	-6.0790	2.8607E-01	-4.1281E-05	288	1000	75.5
247	C4H6O2	METHYL ACRYLATE	-11.1733	3.1448E-01	-6.8774E-05	200	1000	76.5
248	C4H6O2	VINYL ACETATE	-7.4617	3.0466E-01	-5.7544E-05	346	1000	--
249	C4H6O3	ACETIC ANHYDRIDE	-1.4853	2.8869E-01	-2.3391E-05	295	993	82.5
250	C4H6O4	SUCCINIC ACID	-8.4579	2.5493E-01	-3.7515E-05	461	1000	--
251	C4H6O5	DIGLYCOLIC ACID	-7.0363	2.4872E-01	-3.2693E-05	421	1000	--
252	C4H6O5	MALIC ACID	-3.7125	2.3442E-01	-2.2353E-05	403	1000	--
253	C4H6O6	TARTARIC ACID	-3.7273	2.2814E-01	-1.5516E-05	479	1000	--
254	C4H7N	n-BUTYRONITRILE	-2.7850	2.3017E-01	-3.3051E-05	161	1000	62.9
255	C4H7N	ISOBUTYRONITRILE	-5.8841	2.4813E-01	-4.2192E-05	377	977	--
256	C4H7NO	ACETONE CYANOHYDRIN	-4.4975	2.4466E-01	-3.2844E-05	252	1000	65.5
257	C4H7NO	2-METHACRYLAMIDE	-5.5780	2.3845E-01	-2.8600E-05	384	1000	--
258	C4H7NO	3-METHOXYPROPIONITRILE	-3.4066	2.3563E-01	-3.0376E-05	210	1000	64.1
259	C4H7NO	2-PYRROLIDONE	-11.9390	2.5385E-01	-1.8021E-05	518	998	--
260	C4H8	1-BUTENE	-9.1429	3.1562E-01	-8.4164E-05	175	800	77.5
261	C4H8	cis-2-BUTENE	-9.9231	3.2622E-01	-1.0258E-04	277	450	78.2

* A computer program, containing coefficients for all compounds, is available for a nominal fee. The computer program is in ASCII which can be accessed by other software.

334

$$\eta_{gas} = A + B\,T + C\,T^2 \quad (\eta_{gas} - \text{micropoise}, T - K)$$

NO	FORMULA	NAME	A	B	C	TMIN	TMAX	η_{gas} @ 25 C
262	C4H8	trans-2-BUTENE	-9.9231	3.2622E-01	-1.0258E-04	277	450	78.2
263	C4H8	CYCLOBUTANE	0.4011	3.0851E-01	-7.9599E-05	286	1000	85.3
264	C4H8	ISOBUTENE	-8.6298	3.2415E-01	-7.1963E-05	175	1000	81.6
265	C4H8Cl2	1,4-DICHLOROBUTANE	-7.5752	2.8464E-01	-4.3227E-05	427	1000	--
266	C4H8O	n-BUTYRALDEHYDE	7.6943	2.0543E-01	1.0683E-05	298	523	69.9
267	C4H8O	ISOBUTYRALDEHYDE	-14.2389	3.1427E-01	-5.5302E-05	300	1000	--
268	C4H8O	1,2-EPOXYBUTANE	-6.7099	3.5563E-01	-6.1245E-05	337	1000	--
269	C4H8O	METHYL ETHYL KETONE	3.0100	2.2899E-01	7.8953E-06	273	573	72.0
270	C4H8O	ETHYL VINYL ETHER	-3.3642	2.9504E-01	-5.5005E-05	309	1000	--
271	C4H8O	TETRAHYDROFURAN	-14.2222	3.3961E-01	-5.4608E-05	338	998	--
272	C4H8O2	cis-2-BUTENE-1,4-DIOL	-8.5121	2.7407E-01	-3.9067E-05	508	1000	--
273	C4H8O2	trans-2-BUTENE-1,4-DIOL	-7.7299	2.7278E-01	-3.8322E-05	510	1000	--
274	C4H8O2	ISOBUTYRIC ACID	-4.4380	2.6891E-01	-3.8739E-05	227	1000	72.3
275	C4H8O2	n-BUTYRIC ACID	-5.7809	2.6159E-01	-3.4903E-05	268	1000	69.1
276	C4H8O2	1,4-DIOXANE	-16.7012	3.4988E-01	-5.3736E-05	374	994	--
277	C4H8O2	ETHYL ACETATE	-9.2585	3.0725E-01	-7.1069E-05	190	1000	76.0
278	C4H8O2	METHYL PROPIONATE	-6.7395	2.8777E-01	-5.2551E-05	353	1000	--
279	C4H8O2	n-PROPYL FORMATE	-15.0043	3.2541E-01	-5.5609E-05	350	1000	--
280	C4H8O2S	SULFOLANE	-3.7493	2.2530E-01	6.1191E-06	301	1000	--
281	C4H8S	TETRAHYDROTHIOPHENE	-5.6120	2.9145E-01	-4.0805E-05	394	1000	--
282	C4H9Br	1-BROMOBUTANE	-6.9399	3.2469E-01	-5.4949E-05	375	1000	--
283	C4H9Br	2-BROMOBUTANE	-3.8557	3.2515E-01	-4.6707E-05	161	1000	88.9
284	C4H9Cl	n-BUTYL CHLORIDE	-5.8469	2.8831E-01	-5.0974E-05	352	1000	--
285	C4H9Cl	sec-BUTYL CHLORIDE	-7.4512	3.0318E-01	-5.8353E-05	341	1000	--
286	C4H9Cl	tert-BUTYL CHLORIDE	-5.2539	3.0857E-01	-5.5912E-05	324	1000	--
287	C4H9N	PYRROLIDINE	-14.4541	3.1056E-01	-4.9770E-05	360	1000	--
288	C4H9NO	N,N-DIMETHYLACETAMIDE	-14.9371	2.5775E-01	-3.1476E-05	439	1000	--
289	C4H9NO	MORPHOLINE	-15.4022	3.0711E-01	-4.4586E-05	401	1000	--
290	C4H10	n-BUTANE	-4.9462	2.9001E-01	-6.9665E-05	150	1200	75.3
291	C4H10	ISOBUTANE	-4.7305	2.9131E-01	-8.0995E-05	150	1000	74.9
292	C4H10N2	PIPERAZINE	-6.7819	3.0412E-01	-4.4349E-05	379	1000	--
293	C4H10O	n-BUTANOL	-11.1444	2.8790E-01	-5.6275E-05	391	1000	--
294	C4H10O	sec-BUTANOL	-14.9920	3.1418E-01	-5.5185E-05	373	993	--
295	C4H10O	tert-BUTANOL	-10.0391	2.8178E-01	-7.7623E-05	303	473	--
296	C4H10O	DIETHYL ETHER	-7.9324	3.0235E-01	-7.3858E-05	200	1000	75.6
297	C4H10O	METHYL ISOPROPYL ETHER	-3.0159	2.9496E-01	-6.0300E-05	128	1000	79.6
298	C4H10O	ISOBUTANOL	-11.4120	2.7821E-01	-2.9510E-06	175	1100	71.3
299	C4H10O2	1,3-BUTANEDIOL	-7.5525	2.7198E-01	-4.0704E-05	480	990	--
300	C4H10O2	1,4-BUTANEDIOL	-16.1158	2.6953E-01	-3.0297E-05	500	1000	--
301	C4H10O2	2,3-BUTANEDIOL	-5.4697	2.7493E-01	-4.0547E-05	281	1000	72.9
302	C4H10O2	t-BUTYL HYDROPEROXIDE	-5.7894	2.9373E-01	-4.8826E-05	277	1000	77.4
303	C4H10O2	1,2-DIMETHOXYETHANE	8.6411	2.4096E-01	-1.7203E-05	357	1000	--
304	C4H10O2	2-ETHOXYETHANOL	-6.4825	2.7545E-01	-4.6641E-05	408	1000	--
305	C4H10O3	DIETHYLENE GLYCOL	-3.8630	2.5115E-01	-2.6000E-05	263	1000	68.7
306	C4H10O4S	DIETHYL SULFATE	---	---	---	--	--	--
307	C4H10S	n-BUTYL MERCAPTAN	-13.9936	3.0281E-01	-4.7931E-05	372	1000	--
308	C4H10S	ISOBUTYL MERCAPTAN	-3.4476	2.7288E-01	-2.5320E-05	128	1000	75.7
309	C4H10S	sec-BUTYL MERCAPTAN	-3.6779	2.7489E-01	-2.6200E-05	133	1000	76.0
310	C4H10S	tert-BUTYL MERCAPTAN	-10.8877	3.0565E-01	-4.7129E-05	274	1000	76.1
311	C4H10S	DIETHYL SULFIDE	-13.5781	3.0021E-01	-4.9223E-05	365	1000	--
312	C4H10S2	DIETHYL DISULFIDE	-3.4181	2.6320E-01	-1.6484E-05	172	1000	73.6
313	C4H11N	n-BUTYLAMINE	-12.4389	2.8092E-01	-4.8899E-05	351	991	--
314	C4H11N	ISOBUTYLAMINE	-3.6388	2.6244E-01	-4.5505E-05	189	1000	70.6
315	C4H11N	sec-BUTYLAMINE	-4.8500	2.6464E-01	-4.7781E-05	336	1000	--
316	C4H11N	tert-BUTYLAMINE	-5.7506	2.8198E-01	-5.6921E-05	318	1000	--
317	C4H11N	DIETHYLAMINE	-6.0328	2.8574E-01	-5.6208E-05	329	1000	--
318	C4H11NO	DIMETHYLETHANOLAMINE	-6.3923	2.7958E-01	-3.0439E-05	214	1000	74.3
319	C4H11NO2	DIETHANOLAMINE	---	---	---	--	--	--
320	C4H11NO2	2-AMINOETHOXYETHANOL	-4.7496	2.4628E-01	-3.2317E-05	293	1000	65.8
321	C4H12N2O	N-AMINOETHYL ETHANOLAMINE	-3.9691	2.4086E-01	-2.7574E-05	273	1000	65.4
322	C4H12Si	TETRAMETHYLSILANE	-6.7223	2.9090E-01	-5.0708E-05	174	1000	75.5
323	C4H13N3	DIETHYLENE TRIAMINE	-3.9947	2.3224E-01	-2.8336E-05	234	1000	62.7

* A computer program, containing coefficients for all compounds, is available for a nominal fee. The computer program is in ASCII which can be accessed by other software.

Appendix C

DATA CODE FOR COMPOUNDS

NO	FORMULA	NAME	LIQUID VISCOSITY	GAS VISCOSITY
1	CBrClF2	BROMOCHLORODIFLUOROMETHANE	experimental/estimate	estimate
2	CBrCl3	BROMOTRICHLOROMETHANE	estimate	estimate
3	CBrF3	BROMOTRIFLUOROMETHANE	experimental/estimate	experimental
4	CBr2F2	DIBROMODIFLUOROMETHANE	experimental/estimate	estimate
5	CClF3	CHLOROTRIFLUOROMETHANE	experimental/estimate	experimental
6	CClN	CYANOGEN CHLORIDE	estimate	estimate
7	CCl2F2	DICHLORODIFLUOROMETHANE	experimental/estimate	experimental
8	CCl2O	PHOSGENE	experimental/estimate	experimental
9	CCl3F	TRICHLOROFLUOROMETHANE	experimental/estimate	estimate
10	CCl4	CARBON TETRACHLORIDE	experimental/estimate	experimental
11	CF2O	CARBONYL FLUORIDE	estimate	estimate
12	CF4	CARBON TETRAFLUORIDE	estimate	experimental
13	CHBr3	TRIBROMOMETHANE	experimental/estimate	experimental
14	CHClF2	CHLORODIFLUOROMETHANE	experimental/estimate	experimental
15	CHCl2F	DICHLOROFLUOROMETHANE	experimental/estimate	experimental
16	CHCl3	CHLOROFORM	experimental/estimate	experimental
17	CHF3	TRIFLUOROMETHANE	experimental/estimate	experimental
18	CHN	HYDROGEN CYANIDE	experimental/estimate	experimental
19	CH2BrCl	BROMOCHLOROMETHANE	estimate	estimate
20	CH2Br2	DIBROMOMETHANE	experimental/estimate	estimate
21	CH2Cl2	DICHLOROMETHANE	experimental/estimate	estimate
22	CH2F2	DIFLUOROMETHANE	experimental/estimate	experimental
23	CH2I2	DIIODOMETHANE	experimental/estimate	experimental
24	CH2O	FORMALDEHYDE	estimate	estimate
25	CH2O2	FORMIC ACID	experimental/estimate	estimate
26	CH3Br	METHYL BROMIDE	experimental/estimate	experimental
27	CH3Cl	METHYL CHLORIDE	experimental/estimate	experimental
28	CH3Cl3Si	METHYL TRICHLOROSILANE	experimental/estimate	estimate
29	CH3F	METHYL FLUORIDE	estimate	experimental
30	CH3I	METHYL IODIDE	experimental/estimate	estimate
31	CH3NO	FORMAMIDE	experimental/estimate	---
32	CH3NO2	NITROMETHANE	experimental/estimate	estimate
33	CH4	METHANE	experimental/estimate	experimental
34	CH4Cl2Si	METHYL DICHLOROSILANE	experimental/estimate	estimate
35	CH4O	METHANOL	experimental/estimate	experimental
36	CH4O3S	METHANESULFONIC ACID	---	---
37	CH4S	METHYL MERCAPTAN	estimate	experimental
38	CH5ClSi	METHYL CHLOROSILANE	experimental/estimate	estimate
39	CH5N	METHYLAMINE	experimental/estimate	estimate
40	CH6Si	METHYL SILANE	---	estimate
41	CN4O8	TETRANITROMETHANE	experimental/estimate	estimate
42	CO	CARBON MONOXIDE	experimental/estimate	experimental
43	COS	CARBONYL SULFIDE	estimate	estimate
44	CO2	CARBON DIOXIDE	experimental/estimate	experimental
45	CS2	CARBON DISULFIDE	experimental/estimate	experimental
46	C2BrF3	BROMOTRIFLUOROETHYLENE	estimate	estimate
47	C2Br2F4	1,2-DIBROMOTETRAFLUOROETHANE	experimental/estimate	estimate
48	C2ClF3	CHLOROTRIFLUOROETHYLENE	estimate	estimate
49	C2ClF5	CHLOROPENTAFLUOROETHANE	experimental/estimate	experimental
50	C2Cl2F4	1,2-DICHLOROTETRAFLUOROETHANE	experimental/estimate	experimental
51	C2Cl3F3	1,1,2-TRICHLOROTRIFLUOROETHANE	experimental/estimate	experimental
52	C2Cl4	TETRACHLOROETHYLENE	experimental/estimate	experimental
53	C2Cl4F2	1,1,2,2-TETRACHLORODIFLUOROETHANE	estimate	estimate
54	C2Cl4O	TRICHLOROACETYL CHLORIDE	estimate	estimate
55	C2Cl6	HEXACHLOROETHANE	---	estimate
56	C2F4	TETRAFLUOROETHYLENE	estimate	estimate
57	C2F6	HEXAFLUOROETHANE	estimate	estimate
58	C2HBrClF3	HALOTHANE	estimate	estimate
59	C2HClF2	2-CHLORO-1,1-DIFLUOROETHYLENE	estimate	estimate
60	C2HCl3	TRICHLOROETHYLENE	experimental/estimate	experimental
61	C2HCl3O	DICHLOROACETYL CHLORIDE	estimate	estimate
62	C2HCl3O	TRICHLOROACETALDEHYDE	experimental/estimate	estimate
63	C2HCl5	PENTACHLOROETHANE	estimate	estimate
64	C2HF3O2	TRIFLUOROACETIC ACID	experimental/estimate	estimate
65	C2HF5	PENTAFLUOROETHANE	estimate	estimate
66	C2H2	ACETYLENE	experimental/estimate	experimental
67	C2H2Br4	1,1,2,2-TETRABROMOETHANE	estimate	estimate

NO	FORMULA	NAME	LIQUID VISCOSITY	GAS VISCOSITY
68	C2H2Cl2	1,1-DICHLOROETHYLENE	estimate	estimate
69	C2H2Cl2	cis-1,2-DICHLOROETHYLENE	experimental/estimate	estimate
70	C2H2Cl2	trans-1,2-DICHLOROETHYLENE	experimental/estimate	estimate
71	C2H2Cl2O	CHLOROACETYL CHLORIDE	estimate	estimate
72	C2H2Cl2O	DICHLOROACETALDEHYDE	estimate	estimate
73	C2H2Cl2O2	DICHLOROACETIC ACID	experimental/estimate	estimate
74	C2H2Cl3F	1,1,1-TRICHLOROFLUOROETHANE	estimate	estimate
75	C2H2Cl4	1,1,1,2-TETRACHLOROETHANE	estimate	estimate
76	C2H2Cl4	1,1,2,2-TETRACHLOROETHANE	experimental/estimate	estimate
77	C2H2F2	1,1-DIFLUOROETHYLENE	estimate	estimate
78	C2H2F4	1,1,1,2-TETRAFLUOROETHANE	estimate	estimate
79	C2H2O	KETENE	estimate	estimate
80	C2H2O4	OXALIC ACID	---	estimate
81	C2H3Br	VINYL BROMIDE	estimate	estimate
82	C2H3Cl	VINYL CHLORIDE	experimental/estimate	estimate
83	C2H3ClF2	1-CHLORO-1,1-DIFLUOROETHANE	experimental/estimate	estimate
84	C2H3ClO	ACETYL CHLORIDE	experimental/estimate	estimate
85	C2H3ClO	CHLOROACETALDEHYDE	estimate	estimate
86	C2H3ClO2	CHLOROACETIC ACID	experimental/estimate	estimate
87	C2H3ClO2	METHYL CHLOROFORMATE	estimate	estimate
88	C2H3Cl3	1,1,1-TRICHLOROETHANE	experimental/estimate	estimate
89	C2H3Cl3	1,1,2-TRICHLOROETHANE	estimate	estimate
90	C2H3F	VINYL FLUORIDE	estimate	estimate
91	C2H3F3	1,1,1-TRIFLUOROETHANE	estimate	estimate
92	C2H3N	ACETONITRILE	experimental/estimate	estimate
93	C2H3NO	METHYL ISOCYANATE	---	estimate
94	C2H4	ETHYLENE	experimental/estimate	experimental
95	C2H4Br2	1,1-DIBROMOETHANE	estimate	estimate
96	C2H4Br2	1,2-DIBROMOETHANE	experimental/estimate	estimate
97	C2H4Cl2	1,1-DICHLOROETHANE	experimental/estimate	experimental
98	C2H4Cl2	1,2-DICHLOROETHANE	experimental/estimate	experimental
99	C2H4Cl2O	BIS(CHLOROMETHYL)ETHER	estimate	estimate
100	C2H4F2	1,1-DIFLUOROETHANE	experimental/estimate	estimate
101	C2H4F2	1,2-DIFLUOROETHANE	estimate	estimate
102	C2H4O	ACETALDEHYDE	experimental/estimate	estimate
103	C2H4O	ETHYLENE OXIDE	experimental/estimate	experimental
104	C2H4O2	ACETIC ACID	experimental/estimate	estimate
105	C2H4O2	METHYL FORMATE	experimental/estimate	estimate
106	C2H5Br	BROMOETHANE	experimental/estimate	estimate
107	C2H5Cl	ETHYL CHLORIDE	experimental/estimate	experimental
108	C2H5ClO	2-CHLOROETHANOL	experimental/estimate	estimate
109	C2H5F	ETHYL FLUORIDE	estimate	estimate
110	C2H5I	ETHYL IODIDE	experimental/estimate	estimate
111	C2H5N	ETHYLENEIMINE	experimental/estimate	estimate
112	C2H5NO	ACETAMIDE	estimate	estimate
113	C2H5NO	N-METHYLFORMAMIDE	experimental/estimate	estimate
114	C2H5NO2	NITROETHANE	experimental/estimate	estimate
115	C2H6	ETHANE	experimental/estimate	experimental
116	C2H6AlCl	DIMETHYLALUMINUM CHLORIDE	experimental	estimate
117	C2H6O	DIMETHYL ETHER	estimate	experimental
118	C2H6O	ETHANOL	experimental/estimate	experimental
119	C2H6OS	DIMETHYL SULFOXIDE	experimental/estimate	estimate
120	C2H6O2	ETHYLENE GLYCOL	experimental/estimate	estimate
121	C2H6O4S	DIMETHYL SULFATE	---	---
122	C2H6S	DIMETHYL SULFIDE	experimental/estimate	estimate
123	C2H6S	ETHYL MERCAPTAN	estimate	estimate
124	C2H6S2	DIMETHYL DISULFIDE	estimate	estimate
125	C2H7N	DIMETHYLAMINE	experimental/estimate	experimental
126	C2H7N	ETHYLAMINE	estimate	estimate
127	C2H7NO	MONOETHANOLAMINE	experimental/estimate	estimate
128	C2H8N2	ETHYLENEDIAMINE	experimental/estimate	estimate
129	C2H8Si	DIMETHYL SILANE	---	estimate
130	C2N2	CYANOGEN	estimate	experimental
131	C3F6	HEXAFLUOROPROPYLENE	estimate	estimate
132	C3F6O	HEXAFLUOROACETONE	estimate	estimate
133	C3F8	OCTAFLUOROPROPANE	estimate	estimate
134	C3H2N2	MALONONITRILE	estimate	estimate
135	C3H3Cl	PROPARGYL CHLORIDE	estimate	estimate
136	C3H3N	ACRYLONITRILE	experimental/estimate	estimate
137	C3H3NO	OXAZOLE	---	estimate
138	C3H4	METHYLACETYLENE	experimental/estimate	experimental
139	C3H4	PROPADIENE	experimental/estimate	experimental
140	C3H4Cl2	2,3-DICHLOROPROPENE	estimate	estimate
141	C3H4O	ACROLEIN	experimental/estimate	estimate
142	C3H4O	PROPARGYL ALCOHOL	estimate	estimate

NO	FORMULA	NAME	LIQUID VISCOSITY	GAS VISCOSITY
143	C3H4O2	ACRYLIC ACID	estimate	estimate
144	C3H4O2	beta-PROPIOLACTONE	estimate	estimate
145	C3H4O2	VINYL FORMATE	estimate	estimate
146	C3H4O3	ETHYLENE CARBONATE	experimental/estimate	estimate
147	C3H4O3	PYRUVIC ACID	estimate	estimate
148	C3H5Cl	2-CHLOROPROPENE	estimate	estimate
149	C3H5Cl	3-CHLOROPROPENE	experimental/estimate	estimate
150	C3H5ClO	alpha-EPICHLOROHYDRIN	experimental/estimate	estimate
151	C3H5ClO2	METHYL CHLOROACETATE	estimate	estimate
152	C3H5ClO2	ETHYL CHLOROFORMATE	estimate	estimate
153	C3H5Cl3	1,2,3-TRICHLOROPROPANE	estimate	estimate
154	C3H5N	PROPIONITRILE	experimental/estimate	estimate
155	C3H5NO	ACRYLAMIDE	estimate	estimate
156	C3H5NO	HYDRACRYLONITRILE	estimate	---
157	C3H5NO	LACTONITRILE	estimate	estimate
158	C3H5N3O9	NITROGLYCERINE	experimental/estimate	estimate
159	C3H6	CYCLOPROPANE	estimate	experimental
160	C3H6	PROPYLENE	experimental/estimate	experimental
161	C3H6Cl2	1,1-DICHLOROPROPANE	estimate	estimate
162	C3H6Cl2	1,2-DICHLOROPROPANE	estimate	estimate
163	C3H6Cl2	1,3-DICHLOROPROPANE	estimate	estimate
164	C3H6O	ACETONE	experimental/estimate	experimental
165	C3H6O	ALLYL ALCOHOL	experimental/estimate	estimate
166	C3H6O	METHYL VINYL ETHER	estimate	estimate
167	C3H6O	n-PROPIONALDEHYDE	experimental/estimate	estimate
168	C3H6O	1,2-PROPYLENE OXIDE	experimental/estimate	estimate
169	C3H6O	1,3-PROPYLENE OXIDE	estimate	estimate
170	C3H6O2	ETHYL FORMATE	experimental/estimate	estimate
171	C3H6O2	METHYL ACETATE	experimental/estimate	experimental
172	C3H6O2	PROPIONIC ACID	experimental/estimate	estimate
173	C3H6O2S	3-MERCAPTOPROPIONIC ACID	estimate	estimate
174	C3H6O3	LACTIC ACID	estimate	estimate
175	C3H6O3	METHOXYACETIC ACID	estimate	estimate
176	C3H6O3	TRIOXANE	experimental	estimate
177	C3H7Br	1-BROMOPROPANE	experimental/estimate	estimate
178	C3H7Br	2-BROMOPROPANE	experimental/estimate	estimate
179	C3H7Cl	ISOPROPYL CHLORIDE	experimental/estimate	estimate
180	C3H7Cl	n-PROPYL CHLORIDE	experimental/estimate	estimate
181	C3H7I	ISOPROPYL IODIDE	experimental/estimate	estimate
182	C3H7I	n-PROPYL IODIDE	experimental/estimate	estimate
183	C3H7N	ALLYLAMINE	estimate	estimate
184	C3H7N	PROPYLENEIMINE	estimate	estimate
185	C3H7NO	N,N-DIMETHYLFORMAMIDE	experimental/estimate	estimate
186	C3H7NO	N-METHYLACETAMIDE	experimental/estimate	estimate
187	C3H7NO2	1-NITROPROPANE	experimental/estimate	estimate
188	C3H7NO2	2-NITROPROPANE	estimate	estimate
189	C3H8	PROPANE	experimental/estimate	experimental
190	C3H8O	ISOPROPANOL	experimental/estimate	experimental
191	C3H8O	METHYL ETHYL ETHER	estimate	estimate
192	C3H8O	n-PROPANOL	experimental/estimate	experimental
193	C3H8O2	2-METHOXYETHANOL	estimate	estimate
194	C3H8O2	METHYLAL	estimate	estimate
195	C3H8O2	1,2-PROPYLENE GLYCOL	experimental/estimate	estimate
196	C3H8O2	1,3-PROPYLENE GLYCOL	estimate	estimate
197	C3H8O3	GLYCEROL	experimental/estimate	estimate
198	C3H8S	n-PROPYLMERCAPTAN	estimate	estimate
199	C3H8S	ISOPROPYL MERCAPTAN	estimate	estimate
200	C3H9N	n-PROPYLAMINE	estimate	estimate
201	C3H9N	ISOPROPYLAMINE	experimental/estimate	estimate
202	C3H9N	TRIMETHYLAMINE	experimental/estimate	estimate
203	C3H9NO	1-AMINO-2-PROPANOL	experimental/estimate	estimate
204	C3H9NO	3-AMINO-1-PROPANOL	estimate	estimate
205	C3H9NO	METHYLETHANOLAMINE	estimate	estimate
206	C3H9O4P	TRIMETHYL PHOSPHATE	experimental/estimate	---
207	C3H10N2	1,2-PROPANEDIAMINE	experimental/estimate	estimate
208	C3H10Si	TRIMETHYL SILANE	---	estimate
209	C4Cl4S	TETRACHLOROTHIOPHENE	experimental	estimate
210	C4Cl6	HEXACHLORO-1,3-BUTADIENE	estimate	estimate
211	C4F8	OCTAFLUORO-2-BUTENE	estimate	estimate
212	C4F8	OCTAFLUOROCYCLOBUTANE	experimental/estimate	estimate
213	C4F10	DECAFLUOROBUTANE	estimate	estimate
214	C4H2O3	MALEIC ANHYDRIDE	experimental/estimate	estimate
215	C4H4	VINYLACETYLENE	---	---
216	C4H4N2	SUCCINONITRILE	estimate	estimate
217	C4H4O	FURAN	experimental/estimate	estimate

NO	FORMULA	NAME	LIQUID VISCOSITY	GAS VISCOSITY
218	C4H4O2	DIKETENE	experimental	estimate
219	C4H8O3	SUCCINIC ANHYDRIDE	estimate	estimate
220	C4H4O4	FUMARIC ACID	estimate	estimate
221	C4H4O4	MALEIC ACID	estimate	estimate
222	C4H4S	THIOPHENE	experimental/estimate	estimate
223	C4H5Cl	CHLOROPRENE	estimate	estimate
224	C4H5N	trans-CROTONITRILE	estimate	estimate
225	C4H5N	cis-CROTONITRILE	estimate	estimate
226	C4H5N	METHACRYLONITRILE	estimate	estimate
227	C4H5N	PYRROLE	experimental/estimate	estimate
228	C4H5N	VINYLACETONITRILE	estimate	estimate
229	C4H5NO2	METHYL CYANOACETATE	experimental/estimate	estimate
230	C4H6	1,2-BUTADIENE	experimental/estimate	estimate
231	C4H6	1,3-BUTADIENE	experimental/estimate	experimental
232	C4H6	DIMETHYLACETYLENE	---	estimate
233	C4H6	ETHYLACETYLENE	experimental/estimate	experimental
234	C4H6Cl2	1,3-DICHLORO-trans-2-BUTENE	estimate	estimate
235	C4H6Cl2	1,4-DICHLORO-cis-2-BUTENE	estimate	estimate
236	C4H6Cl2	1,4-DICHLORO-trans-2-BUTENE	estimate	estimate
237	C4H6Cl2	3,4-DICHLORO-1-BUTENE	estimate	estimate
238	C4H6O	trans-CROTONALDEHYDE	estimate	estimate
239	C4H6O	2,5-DIHYDROFURAN	estimate	estimate
240	C4H6O	DIVINYL ETHER	estimate	estimate
241	C4H6O	METHACROLEIN	estimate	estimate
242	C4H6O2	2-BUTYNE-1,4-DIOL	---	estimate
243	C4H6O2	gamma-BUTYROLACTONE	experimental	estimate
244	C4H6O2	cis-CROTONIC ACID	estimate	estimate
245	C4H6O2	trans-CROTONIC ACID	estimate	estimate
246	C4H6O2	METHACRYLIC ACID	estimate	estimate
247	C4H6O2	METHYL ACRYLATE	experimental/estimate	estimate
248	C4H6O2	VINYL ACETATE	experimental/estimate	estimate
249	C4H6O3	ACETIC ANHYDRIDE	experimental/estimate	estimate
250	C4H6O4	SUCCINIC ACID	estimate	estimate
251	C4H6O5	DIGLYCOLIC ACID	estimate	estimate
252	C4H6O5	MALIC ACID	estimate	estimate
253	C4H6O6	TARTARIC ACID	---	estimate
254	C4H7N	n-BUTYRONITRILE	estimate	estimate
255	C4H7N	ISOBUTYRONITRILE	estimate	estimate
256	C4H7NO	ACETONE CYANOHYDRIN	---	estimate
257	C4H7NO	2-METHACRYLAMIDE	estimate	estimate
258	C4H7NO	3-METHOXYPROPIONITRILE	estimate	estimate
259	C4H7NO	2-PYRROLIDONE	experimental/estimate	estimate
260	C4H8	1-BUTENE	experimental/estimate	experimental
261	C4H8	cis-2-BUTENE	experimental/estimate	experimental
262	C4H8	trans-2-BUTENE	experimental/estimate	experimental
263	C4H8	CYCLOBUTANE	estimate	estimate
264	C4H8	ISOBUTENE	experimental/estimate	experimental
265	C4H8Cl2	1,4-DICHLOROBUTANE	estimate	estimate
266	C4H8O	n-BUTYRALDEHYDE	experimental/estimate	experimental
267	C4H8O	ISOBUTYRALDEHYDE	estimate	estimate
268	C4H8O	1,2-EPOXYBUTANE	experimental/estimate	estimate
269	C4H8O	METHYL ETHYL KETONE	experimental/estimate	experimental
270	C4H8O	ETHYL VINYL ETHER	estimate	estimate
271	C4H8O	TETRAHYDROFURAN	experimental/estimate	estimate
272	C4H8O2	cis-2-BUTENE-1,4-DIOL	experimental/estimate	estimate
273	C4H8O2	trans-2-BUTENE-1,4-DIOL	---	estimate
274	C4H8O2	ISOBUTYRIC ACID	experimental/estimate	estimate
275	C4H8O2	n-BUTYRIC ACID	experimental/estimate	estimate
276	C4H8O2	1,4-DIOXANE	experimental/estimate	estimate
277	C4H8O2	ETHYL ACETATE	experimental/estimate	estimate
278	C4H8O2	METHYL PROPIONATE	experimental/estimate	estimate
279	C4H8O2	n-PROPYL FORMATE	experimental/estimate	estimate
280	C4H8O2S	SULFOLANE	experimental/estimate	estimate
281	C4H8S	TETRAHYDROTHIOPHENE	estimate	estimate
282	C4H9Br	1-BROMOBUTANE	experimental/estimate	estimate
283	C4H9Br	2-BROMOBUTANE	experimental/estimate	estimate
284	C4H9Cl	n-BUTYL CHLORIDE	experimental/estimate	estimate
285	C4H9Cl	sec-BUTYL CHLORIDE	experimental/estimate	estimate
286	C4H9Cl	tert-BUTYL CHLORIDE	experimental/estimate	estimate
287	C4H9N	PYRROLIDINE	experimental/estimate	estimate
288	C4H9NO	N,N-DIMETHYLACETAMIDE	experimental/estimate	estimate
289	C4H9NO	MORPHOLINE	experimental/estimate	estimate
290	C4H10	n-BUTANE	experimental/estimate	experimental
291	C4H10	ISOBUTANE	experimental/estimate	experimental
292	C4H10N2	PIPERAZINE	estimate	estimate

```
NO  FORMULA    NAME                      LIQUID VISCOSITY        GAS VISCOSITY
--- ---------  ------------------------  --------------------    -------------
293 C4H10O     n-BUTANOL                 experimental/estimate   estimate
294 C4H10O     sec-BUTANOL               experimental/estimate   estimate
295 C4H10O     tert-BUTANOL              experimental/estimate   experimental
296 C4H10O     DIETHYL ETHER             experimental/estimate   experimental
297 C4H10O     METHYL ISOPROPYL ETHER    estimate                estimate
298 C4H10O     ISOBUTANOL                experimental/estimate   estimate
299 C4H10O2    1,3-BUTANEDIOL            experimental/estimate   estimate
300 C4H10O2    1,4-BUTANEDIOL            experimental/estimate   estimate
301 C4H10O2    2,3-BUTANEDIOL            estimate                estimate
302 C4H10O2    t-BUTYL HYDROPEROXIDE          ---                estimate
303 C4H10O2    1,2-DIMETHOXYETHANE       experimental/estimate   estimate
304 C4H10O2    2-ETHOXYETHANOL           estimate                estimate
305 C4H10O3    DIETHYLENE GLYCOL         experimental/estimate   estimate
306 C4H10O4S   DIETHYL SULFATE                ---                     ---
307 C4H10S     n-BUTYL MERCAPTAN         estimate                estimate
308 C4H10S     ISOBUTYL MERCAPTAN        estimate                estimate
309 C4H10S     sec-BUTYL MERCAPTAN       estimate                estimate
310 C4H10S     tert-BUTYL MERCAPTAN      estimate                estimate
311 C4H10S     DIETHYL SULFIDE           experimental/estimate   estimate
312 C4H10S2    DIETHYL DISULFIDE         estimate                estimate
313 C4H11N     n-BUTYLAMINE              experimental/estimate   estimate
314 C4H11N     ISOBUTYLAMINE             experimental/estimate   estimate
315 C4H11N     sec-BUTYLAMINE            estimate                estimate
316 C4H11N     tert-BUTYLAMINE           experimental/estimate   estimate
317 C4H11N     DIETHYLAMINE              experimental/estimate   estimate
318 C4H11NO    DIMETHYLETHANOLAMINE      experimental            estimate
319 C4H11NO2   DIETHANOLAMINE            experimental/estimate        ---
320 C4H11NO2   2-AMINOETHOXYETHANOL      estimate                estimate
321 C4H12N2O   N-AMINOETHYL ETHANOLAMINE estimate                estimate
322 C4H12Si    TETRAMETHYLSILANE              ---                estimate
323 C4H13N3    DIETHYLENE TRIAMINE       estimate                estimate
```

Appendix D

COMPOUND LIST BY FORMULA

Appendix E

COMPOUND LIST BY NAME

355

Printed and bound by CPI Group (UK) Ltd, Croydon, CR0 4YY

03/10/2024

01040335-0019